The Planning and Scheduling of Production Systems

The Planning and Scheduling of Production Systems

Methodologies and applications

Edited by

A. Artiba

Professor of Industrial Management and Computer Science
FUCAM University
Mons
Belgium

and

S.E. Elmaghraby

University Professor of Operations Research and Industrial Engineering
North Carolina State University
USA

CHAPMAN & HALL

London · Weinheim · New York · Tokyo · Melbourne · Madras

Published by Chapman & Hall, 2–6 Boundary Row, London SE1 8HN, UK

Chapman & Hall, 2–6 Boundary Row, London SE1 8HN, UK

Chapman & Hall GmbH, Pappelallee 3, 69469 Weinheim, Germany

Chapman & Hall USA, 115 Fifth Avenue, New York NY 10003, USA

Chapman & Hall Japan, ITP-Japan, Kyowa Building, 3F, 2-2-1 Hirakawacho, Chiyoda-ku, Tokyo 102, Japan

Chapman & Hall Australia, 102 Dodds Street, South Melbourne, Victoria 3205, Australia

Chapman & Hall India, R. Seshadri, 32 Second Main Road, CIT East, Madras 600 035, India

First edition 1997

Typeset in 10/12 pt Palatino by Cambrian Typesetters, Frimley, Surrey

Printed in Great Britain by T.J. Press (Padstow) Ltd, Padstow, Cornwall

ISBN 0 412 61020 5

A catalogue record for this book is available from the British Library

Library of Congress Catalog Card Number: 94–85862

∞ Printed on permanent acid-free text paper, manufactured in accordance with ANSI/NISO Z39.48-1992 and ANSI/NISO Z39.48-1984 (Permanence of Paper).

Contents

Contributors

F. Alexandre, CRIN (CNRS URA 262) and INRIA Lorraine, Batiment LORIA, BP 239, 54506 Vandoeuvre Cedex, France

A. Artiba, FUCAM, 151 Chaussee de Binche, B7000 Mons, Belgium

D. Bahler, Dept of Computer Science, North Carolina State University, Raleigh, NC 27695–8206, USA

S.K. Banerjee, Dept of Design Manufacture and Engineering Management, University of Strathclyde, Glasgow, G1 1XJ

J. Bowen, Dept of Computer Science, National University of Ireland, Cork, Ireland

C. Cardeira, CRIN (CNRS URA 262) and NIRIA Lorraine, Batiment LORIA, BP 239, 54506 Vandoeuvre Cedex, France

F. Charpillet, CRIN (CNRS URA 262) and NIRIA Lorraine, Batiment LORIA, BP 239, 54506 Vandoeuvre, France

S.E. Elmaghraby, Graduate Program in Operations Research, Box 7913, North Carolina State University, Raleigh, NC 27695–7913, USA

R.E. Karnoub, Graduate Program in Operations Research and Dept of Industrial Engineering, Box 7913, North Carolina State University, Raleigh, NC 27695-7913, USA

Z. Mammeri, CRIN (CNRS URA 262) and INRIA Lorraine, Batiment LORIA, BP 239, 54506 Vandoeuvre Cedex, France

P. Massotte, IBM Academy of Technology, LG12P Laboratory, Ecole des Mines, 6 Avenue de Clavieres, 30319 Ales, France

W.G. Nulty, Vice President, CAPS Logistics, Inc., 2700 Cumberland Parkway, Atlanta, Georgia 30339–3321, USA

M.-C. Portmann, CRIN (CNRS URA 262) Ecole des Mines de Nancy, Parc de Saurupt, 54042 Nancy Cedex, France

A.A.B. Pritsker, President, Pritsker Corporation, 8910 Purdue Road, Suite 500, Indianapolis, IN 46268–1170, USA

H.D. Ratliff, Regent's and UPS Professor, School of Industrial and Systems Engineering, Georgia Inst. of Technology, Atlanta, Georgia 30332–0360, USA

J. Slomp, University of Groningen, Faculty of Management and Organization, PO Box 800, 9700 AV Groningen, The Netherlands

K. Snyder, Senior Consultant, Pritsker Corporation, 8910 Purdue Road, Suite 500, Indianapolis, IN 46268–1170, USA

Preface

If one accepts the premise that there is no wealth without production, whether at the individual or national level, one is immediately led to the conclusion that the study of productive systems lies at the forefront of subjects that should be intensively, as well as rationally and extensively, studied to achieve the desired 'sustainable growth' of society, where the latter is defined as growth in the quality of life that does not waste the available resources in the long run.

Since the end of World War II there has been a remarkable evolution in thinking about production, abetted to a large measure by the nascent field of **informatics**: the computer technology and the edifices that have been built around it, such as information gathering and dissemination worldwide through communication networks, software products, peripheral interfaces, etc. Additionally, the very thought processes that guide and motivate studies in production have undergone fundamental changes which verge on being revolutionary, thanks to developments in **operations research** and **cybernetics**. A vast amount of theory, and an equally vast amount of 'canned' computer software (*logiciel* in French terminology) that did not exist even ten years ago, are now abundantly available, with the promise (or threat) of more to come!

This book arose from our desire to provide a select collection of articles written by those expert in their own fields which can help the practitioner 'see the wood for the trees' and sift out what is relevant to each particular domain of activities. The fact is, the problems arising in production planning, scheduling and control are as diverse as the techniques and tools proposed for solving them. Some problems can be coped with in practice by simple heuristics, or 'rules of thumb'; they satisfy, and everyone is happy. Others, due to competitive pressures and the need to 'squeeze the last possible drop' from the manufacturing system, may require more sophisticated optimization or logical modeling techniques, with extensive use of the computer, either in the execution or simulation modes. Still others need to be tackled with the appropriate combination of different tools and techniques. Which is the more appropriate approach? And has it been used before? How, and in what context? What does the future hold?

We hope that this book responds to these questions by, first, explaining the fundamentals, strengths and weaknesses of the more recent methodologies – each essay in its own domain – for the planning and scheduling of operations, and, second, giving examples of real-life applications of these approaches. Guidance to more in-depth reading in production/manufacturing systems is also provided via up-to-date bibliographies at the end of each chapter.

The book should also be of interest, as supplementary reading, to students at the senior/first-year graduate level in the field of **operations management**, in departments of industrial engineering or in colleges of business or management. Many of the chapters are reports of real-life applications, replete with the constraints and multiple objectives that one finds in such processes.

After stating what this book **is** about, it may be helpful to also mention what it **is not** about. This is not a textbook. While it discusses methodology, it does not pretend to demonstrate the validity of each proposition through logical argument of the theorem-proof type. It does not claim to be comprehensive: we readily admit that there are numerous facets of production/manufacturing processes that have not been treated here. We did not set out to compile an encyclopedia on planning and scheduling of production systems. Finally, it is not a collection of **prescriptive** remedies telling you, the reader, what you should, or should not, think or do. In fact, you may find the views expressed in one chapter diametrically opposed to the views in another. We leave it to you to decide, since we do not believe in the 'one best method' myth.

The requirements background of the reader varies from chapter to chapter, depending necessarily, on subject matter. All chapters are accessible to a college senior student from a 'what is it all about?' viewpoint. Some chapters require additional concentration and perseverance, and a few demand additional familiarity with mathematical notation and logical deduction. To the serious reader, the book should be instructive as well as informative.

We are indebted to all the contributors for their patience with what may have appeared sometimes as demanding editors, and for their professionalism and devotion to providing chapters that distill each author's years of experience into some 20 pages of print! We also thank Mark Hammond, our editor at Chapman & Hall, for his continued support throughout this undertaking. Finally, we wish to thank the three anonymous reviewers whose comments were most helpful in shaping the final version of the manuscript.

Salah E. Elmaghraby
Raleigh, NC, USA

Abdelhakim Artiba
Mons, Belgium

December 1996

CHAPTER 1

Introduction

A. Artiba and S.E. Elmaghraby

1.1 BACKGROUND

The title of this book is 'The Planning and Scheduling of Production Systems' and we feel obliged to preface its contents with a few comments on **production planning, scheduling and control** (PPS&C) in general, before delving into the discussion of the chapters that constitute this volume.

1.1.1 Production planning and control (PP&C)

Competitive firms are operating today in global and worldwide markets. Productivity enhancement is thus a major objective for these manufacturing enterprises. Manufacturers are experiencing fluctuating market demand for their products, with ever shorter lead times and smaller but more frequent order quantities, accompanied by more frequent changes in product specifications – even during manufacture (Artiba and Tahon, 1992). Product life cycles are shorter, and forecasting of demand is becoming increasingly difficult. All this requires agility and flexibility on the part of the factory, which runs counter to its built-in inertia and the gravitational force to 'conduct business as usual', combined with the limited ability of management to reconsider decisions almost 'in real time'.

To achieve the rationalization of manufacturing management with the desired attributes of agility, flexibility and continuous productivity enhancement, many methods, tools and techniques have emerged over the past two decades which deserve attention. Traditional **material requirement planning** (MRP I) methods (Orlicky, 1970, 1975), are recognized to be efficient for supporting the stock management and purchasing functions, but inadequate for manufacturing shop-floor management. This push-based approach supports much flexibility relative to fluctuations in market demand and product options, but it fails to impose any order on the shop floor or to take into consideration the limitations of shop capacities. Consequently, some production stages may starve because of lack of parts, while in-process inventory of other parts abounds! The shortcomings of MRP I gave rise to **manufacturing resources planning** (MRP II), whose potential seemed unlimited. Now that the euphoria is over, however, manufacturing enterprises are once again dissatisfied; it appears that MRP (I or II) is a mixture of assumptions which seem to contradict each other (Wortmann and Wijngaard, 1992):

- the assumption that the market can be forecasted contradicts the assumption that demand should be managed;
- the assumption that leadtime is known in advance contradicts the assumption that leadtime is controlled by priorities;
- the assumption that MRP I is a tool for decision support contradicts the assumption that its plans will be executed;
- the assumption that material management and master production scheduling are organizational functions contradicts the assumption that MRP is neutral with respects to organizational structure.

On the other hand, **just-in-time** (JIT) and **total quality control** (TQC) are complementary philosophies which aim at the gradual elimination of intermediate stock buffers between workstations, wasteful processes and production disruptions. JIT and TQC frequently 'pull' materials through a manufacturing plant. In practice, this means that when the stock level of a finished product drops below a predefined threshold, the last workstation receives authorization to produce a batch of this product. The JIT method focuses on the management of the flow of materials with fixed production rates, limited stock buffers in the workshop, and with pulling demand mechanism in the manufacturing plant. The application of this method can lead to good results in industries where the value-added labor portion is low (typically less than 10%), and the materials portion is high. Furthermore, JIT requires the demand to be stable, the product to be standard, and the workshop to be organized according to materials flow, which can impose machine duplication and poor usage of expensive resources. The result is a reduction of work in process, and hence of the leadtime (Bourgeois *et al.*, 1993).

The assumption of stability of demand and product design required by the JIT approach is in contradiction with the inherent variability in today's market behavior. Production planners are therefore looking for alternative support methodologies in coordinating their activities. The literature on production planning and control is diverse and prolific. In addition to material management and scheduling problems, the field of PP&C covers (and overlaps with) the areas of 'operations management', 'materials handling', 'finished goods and in-process inventory control' and 'maintenance management'. Tools and techniques vary from single-machine sequencing algorithms to large-scale computer-integrated manufacturing systems.

Several comparative studies of specific techniques have been published. Krajweski *et al.* (1987) have reported a comparison of MRP and Kanban. Gelders and Van Wassenhove (1985) have presented a critical comparison of OPT (optimized production technology), MRP and JIT in capacity-constrained situations while Shivran *et al.* (1987) have traced the evolution of production control approaches from these techniques. This evolution describes a transitional trend from strictly limited scope quantitative models to a systems perspective. These studies highlight the importance of selecting the right technique for the right problem because there is no global best technique.

It has been reported that there is a considerable gap between theory and practice in production scheduling (Buxey, 1989). Barancsi *et al.* (1983), discussing inventory modeling, identified five causes of this gap:

1. The organizational structure of companies does not allow for the creation of comprehensive models.
2. Decision makers are skeptical of theoretical models and are averse to their application.
3. Systems designers are not aware of available models and when and how they are applicable.
4. Current computer support is oriented toward database activities and not decision making.
5. A great proportion of models have been created in the light of the mathematical aspect of the problem without considering the possibility of application.

The planning and control of manufacturing systems involve organizing and managing the process of converting raw materials into a predesigned finished product. Burbidge (1978) defines production control as the function of management which plans, directs and controls the materials supply and processing activities of an enterprise. Where planning is the process of deciding what to do in the future, directing comprises the operation of issuing orders, and control can be described as the constraining of events to follow plans.

The framework usually used in PP&C consists of three levels of decision making: strategic, tactical and operational. These levels are best reflected in the **hierarchical production planning** (HPP) approach of Hax and Meal (1975), which approaches the global problem of production planning by first solving an 'aggregate planning problem' of **types** of products, which is subsequently disaggregated into a mid-level planning problem of **families** of products, which, in turn, is disaggregated into a low-level planning problem of individual **items**. The main problem with the HPP approach is that it cannot run alone; it must be implemented in conjunction with other decision and information systems to provide it with the large amount of data required. Meal *et al.* (1987) have proposed integrating HPP with MRP to overcome the problems in both. They recommend the use of MRP to generate the 'master schedule' and the HPP as the capacity planning module. As stated above, the actual situation is that there is an increase in product variety (which leads to a decreasing of lot sizes), a decrease in order quantities and leadtimes. Therefore, finite capacity short-term scheduling tools are advisable (if not necessary).

1.1.2 Scheduling

Scheduling deals with determining the time-sequencing of jobs or orders, and with the allocation of required resources (personnel, machines, tools, etc.) to accomplish the related set of operations.

The production environment is dynamic and unforeseen events can happen at any time. These disturbances can have an impact on the current schedule; reactions to them result in rescheduling (practitioners/planners say there is no scheduling, rather a perpetual rescheduling problem).

Due to its combinatorial nature, scheduling is a 'difficult' task in the technical sense: most real-life scheduling problems are NP-hard, meaning that it is not likely that an algorithm can be developed which can solve these problems (where 'solve' means achieving the optimum, on a digital computer of current vintage) in 'reasonable' time. Here, 'reasonable' is defined in terms of being 'bounded by a polynomial in the length of the input'. Indeed, one expects that computing effort increases as the 'size' of the problem increases; the question is, how fast? The general consensus is that if the increase can be bounded by a polynomial in the size of the problem then the problem is 'solvable'; otherwise, computing power, no matter how vast, will be overwhelmed sooner or later. Complete enumeration of possible schedules is usually out of the question, as can be easily verified. Methods for coping with such complexity may be classified into two broad categories: **optimization** and **logical modeling** approaches (Roy and Meikle, 1995). Optimization techniques depend on the objectives of scheduling being explicitly expressed as a single mathematical function. Commonly used techniques are mathematical programming and implicit enumeration (which uses branch-and-bound algorithms). More recent approaches investigate the use of 'controlled random search', which we have termed 'compu-search techniques' because of their reliance on the computer, to achieve 'good', albeit not necessarily optimal, solutions.

Optimization techniques, whether mathematical or enumerative, suffer from one fundamental problem: the difficulty of expressing the business goals of scheduling (minimize costs, meet due dates, minimize makespan, minimize in-process inventory, etc.) in the form of a single mathematical function. Logical modeling, on the other hand, represents the activities of a manufacturing system as a set of logical constraints or rules (e.g. precedence constraints, resource requirements for operations, capacity restrictions). The objectives of scheduling, however, are not explicitly expressed but, instead, heuristic procedures/rules are incorporated in the model in order to derive 'good' acceptable schedules. Discrete event simulation is usually used in one of two modes: system design and operating mode. In the system design mode the simulation is used to test how a proposed system will function. In the operating mode the simulation is used to test how an existing system will function under a proposed control directive. The simulation model is then used to estimate the operating performance of a proposed schedule.

The performance of such techniques in meeting business goals mainly depends on the heuristics used. The efficiency of the production planning and control system is heavily influenced by the capabilities of the information and communication systems in terms of the quality of information.

1.2 OVERVIEW

The aim of this book is to give a state-of-the-art report on modern approaches to problems of planning, scheduling and control of production systems, and in particular manufacturing systems. (We use the word 'production' in its wide sense, to mean the activities that lead to the enhancement of the utility of an idea, a function, or an object. Hence it encompasses manufacturing, transportation, distribution and storage.) It presents both the theoretical basis and practical aspects of the different methods and techniques used, and illustrate these by means of instructive examples as industrial applications.

Apart from the introductory remarks of this chapter, the contents may be divided into three main groups: a chapter on logistics, five chapters that address practical problems of aggregate production planning and control, and four chapters that deal with modern methodologies of short-term scheduling.

1.2.1 Logistics

'Logistics' refers to all planning and coordination activities required to deliver the desired product (however that 'product' is defined). It is central to the theme of this book, and no discussion of the design and operation of production systems is complete without thorough understanding of the logistics function and what it entails.

This is treated by Ratliff and Nulty in Chapter 2. The presentation is at a conceptual level, though the concepts are exemplified whenever needed. The authors define 'logistics object modeling' as the framework in which the logistical functions are addressed, and guide the reader through the design and implementation of an object-oriented logistics model. This chapter should be considered when structuring your logistical operations, the pitfalls that you should avoid, and the methodologies available to you in performing that function.

1.2.2 Aggregate production planning and control

We have argued that the complexity of the production systems under consideration precludes their optimization. Consequently, we must lower our sights and the arguments should revolve around how to achieve a 'better' operational scheme. Several strategies are suggested and you have to decide as to which is the most appropriate for your operations.

The first, eloquently championed by Banerjee in Chapter 3, is a strategy based on a more-or-less *laissez faire* attitude: specify the boundaries of behavior required from individual components of the system and then let go! Do not try to 'micro-manage' by specifying schedules, replete with start times and finish times of each job in the enterprise. The rationale is based on the premise that the people closest to the operational problems are better qualified to resolve them than others far removed from the 'shop floor'. The resulting paradigm requires continuous exchange of timely (and accurate) information among the various components of the system.

The second, championed by Massotte in Chapter 4, is a strategy that takes the opposite point of view: it advocates centralized planning and control, with data and information flowing from the shop floor to 'the planner' and decisions flowing back from 'the planner' to the shop floor. Its rationale is equally simple: myopic decisions are never conducive to good overall performance, let alone global optimum, except under conditions of stationarity, and then they are only asymptotically so. But real life is not stationary, and we cannot proceed asymptotically hoping for the good result 'in the end'.

The third is a strategy that lies inbetween these two extremes. It is based on a hierarchical paradigm of decisions, with the functional rather than administrative. It is presented in Chapters 5 (by Artiba) and 6 (by Elmaghraby and Karnoub):

- Chapter 5 describes a production control system that was designed for, and implemented at, a pharmaceutical manufacturing enterprise. The issues raised may be taken as generic to all process industry, which encompasses the chemical, food and petroleum industries. The approach is knowledge-based as well as rule-based, in the sense of appealing to a decision support system that is the infrastructure of the control mechanism.

- Chapter 6 treats a textile manufacturing plant as an example of the so-called 'hybrid flowshops' – that is, manufacturing plants in which material flow is unidirectional from start to finish, but at each stage of processing there are several facilities with varying capabilities and economics. Production managers in many parts manufacturing plants as well as in industries such as metal machining and fabrication, electronics and machinery can easily identify with the issues raised in this application.

The rise of FMSs (flexible manufacturing systems) in many manufacturing plants necessitated that we devote a chapter to them; they are dealt with in Chapter 7, by Slomp. This chapter distills the author's experience with FMSs and addresses issues such as: what are their performance characteristics? How do you design one, and what elements should you consider to ensure success of your design? How do you control them?

Studying these five chapters should give a well-rounded view of the authors' experiences in coping with a variety of PP&C contexts and problems in the real world. We have refrained from advocating a particular point of view, or a particular paradigm, as the 'best', for the good reason that we believe that such a universally 'best' approach does not exist: different contexts demand different approaches. You will have to decide on which is the most appropriate for your enterprise.

1.2.3 Methodologies of short-term scheduling

Recent years have witnessed the evolution of a number of methodologies distinctly different from the 'run of the mill' methodologies of operations research that had been fashionable for the previous 30 or 40 years. These new methodologies are characterized by two important features: first, they are almost model free and, second, their very existence is crucially dependent on the existence of the computer. A modern book on production planning and scheduling would be incomplete if it did not attempt to bring these new technologies within the reach of practitioners.

Chapter 8, by Bahler and Bowen, discusses 'constraint logic' and its application to scheduling problems. The objective in such an approach is to find a feasible solution, where feasibility is defined in terms of satisfying a set of constraints, which may later be improved upon by any of the methods of operations research already available. Its greatest value lies in rendering concurrent engineering an achievable goal. Anyone who has tried to coordinate the functions of product design, material and parts acquisition, manufacturing, distribution and marketing knows very well how difficult it is to accomplish the task. But if manufacturing is to be agile, customized and economical, there is no escape from insisting on such total coordination of these functions. Constraint logic plays a central role in this game.

Chapters 9, by Portmann, and 10, by Alexandre *et al.*, should be taken together. First, Chapter 9 gives an updated review of the scheduling function and the various mathematical models that have been offered in operations research for the solution of many scheduling/sequencing problems. Then, still within Chapter 9, the discussion passes on to the different compu-search methodologies that have been recently proposed for the solution of different scheduling problems. This discussion is continued in Chapter 10. Basically, these two chapters try to answer the following (seemingly simple) questions: (1) how do I schedule? and (2) what is meant by [. . .], where the dots are filled with any one of the recent approaches proposed for the solution of scheduling problems: namely, simulated annealing, Tabu search, genetic algorithms and neural nets; and how do I go about applying it?

These two chapters should provide you with an introduction to the nascent **compu-search** methodologies so that you may decide their applicability, or otherwise, to your problems and, if you desire, seek more information from the references cited at the end of each chapter. The plethora of examples cited should facilitate your transition from a passing knowledge of the terms to a deeper understanding of what is involved in using any of these methodologies.

Finally, Chapter 11, by Pritsker and Snyder, 'puts it all together' by discussing a mechanism for visualizing the progress of work in a facility, acting upon such visualization, and witnessing the results of your decisions. The premise of this discussion is simply the following: there is nothing better than seeing what is happening to act correctly.

1.3 CONCLUSION

The content of this book is focused on the production planning and scheduling of manufacturing systems. Several different methodologies, tools and techniques to aid decision-making at different levels are presented and illustrated with pedagogical examples. Advantages and limitations of each approach are highlighted.

For effective decision-making, a decision support system needs to combine appropriate methods and tools that would, collectively, provide the capabilities needed for making decisions at different levels and horizons.

Manufacturing planning and control systems design defines the way of distributing and coordinating the different decision functions. The performance of the production planning and control system is also dependent on the availability and reliability of the information system.

REFERENCES

Artiba, A. and Tahon, C. (1992) Production planning knowledge-based system for pharmaceutical manufacturing lines, *European Journal of Operational Research*, **61**(1–2), 18–29.

Barancsi, E., Ritzman, L.P. and Gupta, J.N.D. (1983) Determining lot sizes and resource requirements: A review, *Operations Research*, **33**(3), 329–45.

Bourgeois, S., Artiba, A. and Tahon, C. (1993) Decision support system for dynamic job-shop scheduling, *Journal of Decision Systems*, **2**(3–4), 277–92.

Burbidge, J.L. (1978) *The Principles of Production Control*, 4th edn, McDonald & Evans, UK.

Buxey, G. (1989) Production scheduling: Practice and theory, *European Journal of Operational Research*, **39**, 17–31.

Gelders, L.F. and Van Wassenhove, L.N. (1985) Capacity planning in MRP, JIT, and OPT: A critique, *Engineering Costs and Production Economics*, **9**, 201–9.

Hax, A.C. and Meal, H.C. (1975) Hierarchical integration of production planning and scheduling, in M.A. Geisler (ed.), *Logistics*, North-Holland, Amsterdam, pp. 53–69.

Krajweski, L.J., King, B.E., Ritzman, L.P. and Wong, D.S. (1987) Kanban, MRP and shaping the manufacturing environment, *Management Science*, **33**(1), 39–57.

Meal, H.C., Wachter, M.H. and Whybark, D.D. (1987) Materials requirements planning and hierarchical production planning, *International Journal of Production Research*, **25**(7), 947–56.

Orlicky, J.P. (1970) *Requirements Planning Systems*, Proceedings of the 13th APICS International Conference, Cincinnati, Ohio.

Orlicky, J.P. (1975) *Material Requirements Planning*, McGraw-Hill, New York.

Roy, R. and Meikle, S.E. (1995) The role of discrete event simulation techniques in finite capacity scheduling, *Journal of the Operational Research Society*, **46**, 1310–21.

Shivran, J., Joyce, R. and Browne, J. (1987) Production and inventory management techniques – a systems perspective, in A. Kusiak (ed.) *Modern Production Management Systems*, North-Holland, Amsterdam.

Wortmann, J.C. and Wijngaard, J. (1992) Beyond MRP. Guest editorial, *Production Planning and Control*, **3**(3), 225–6.

CHAPTER 2

Logistics composite modeling

H.D. Ratliff and W.G. Nulty

2.1 INTRODUCTION

2.1.1 Logistics

What is a supply chain?

A **supply chain** is the collection of all components and functions associated with the creation and ultimate delivery of a product or service. Figure 2.1 illustrates an example product supply chain.

What is logistics?

Logistics is the collection of activities associated with acquiring, moving, storing and delivering supply chain commodities (i.e. products in all stages of manufacture, services and information) [1]. Logistics encompasses the business functions of transportation, distribution, warehousing, material handling and inventory management, and interfaces closely with manufacturing and marketing.

Logistics supply chains, (also called **logistics systems** or **logistics networks**) arise in numerous business segments and government functions, including: manufacturing firms, retailing firms, food producers and distributors, the military, transportation carriers (such as trucking and railroad companies), service companies, postal delivery, utilities, petroleum pipelines and public transportation, among others.

2.1.2 Significance of logistics

Cost of logistics

Logistics is a key business function for many reasons, including the high cost of operating a supply chain. Estimated total logistics costs incurred

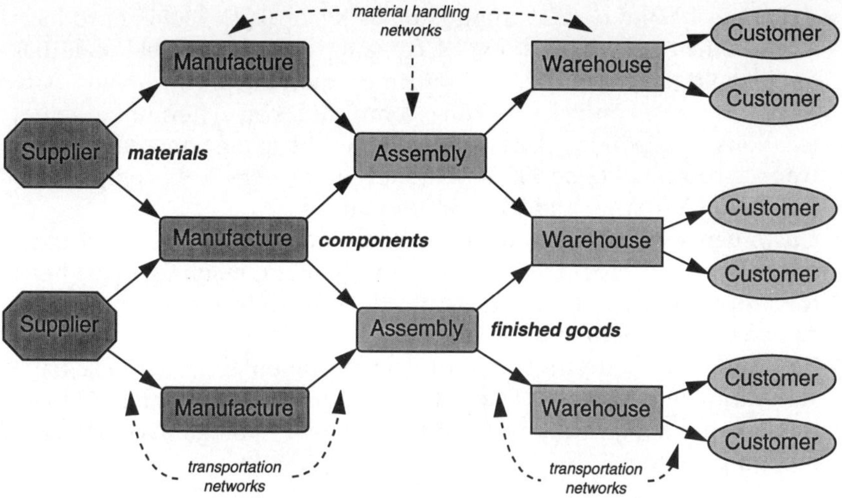

Figure 2.1 Example logistics supply chain.

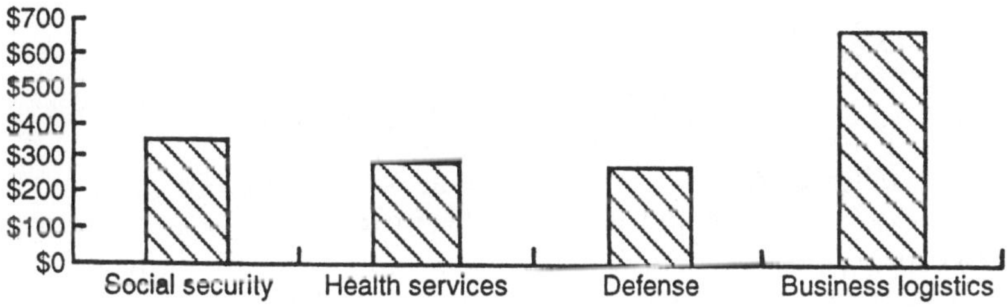

Figure 2.2 Significance of logistics costs (1993, in billions).

by United States businesses in 1993 was 670 billion dollars [2], or roughly 11% of the US Gross Domestic Product (GDP). This cost is higher than the annual US government expenditures in social security, health and human services, and defense (Figure 2.2) [3].

Beyond costs, business logistics is increasing in importance due to the following:

- **Deregulation**. In the US, transportation (including rail, trucking and air modes) has shifted from a highly regulated to an increasingly free market industry. The result is more choices and complexity regarding logistics services and costs, and more opportunities to improve business operations.
- **Global markets**. The business marketplace is increasingly global in scope, with world trade projected to increase from four trillion US

dollars in 1993 to over 16 trillion dollars by 2010 [4]. Moving products from point of origin to point of consumption on a global scale has obvious logistical challenges. Further, manufacturing wages vary widely among countries, adding complexities in determining smart locations to produce goods. For example, the average manufacturing wage is projected to be $25.40 in the US in the year 2010, compared to $45.80 in Germany, and $4.00 in Mexico [5].

- **Customer service**. Deregulation, global markets, and other factors create a more competitive business environment, resulting in the need for supply chains that can deliver products quickly and accurately and can adapt to rapid market changes.
- **Environment**. Current and future environmental regulations have significant implications on logistics, and may fundamentally impact the locations of facilities including plants, storage facilities and recycling centers.

2.2 LOGISTICS DECISIONS

2.2.1 Types of logistics questions

What are key logistics decisions?

Figure 2.3 illustrates common questions arising in the design and operation of a logistics supply chain. While this is only a sampling of logistics questions, the questions lead to a wide variety of difficult issues

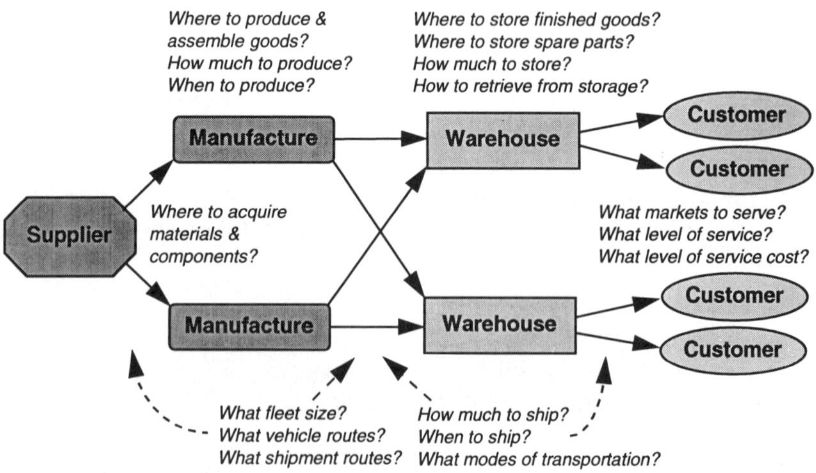

Figure 2.3 Logistics questions.

involving the design and operation of logistics systems. Most of these logistics decisions embody five fundamental characteristics:

1. **Multiple** business functions are impacted.
2. There are **tradeoffs** among conflicting objectives.
3. Logistics system impacts are difficult to evaluate **precisely**.
4. There are business issues **unique** to each logistics system.
5. **Quantitative** analysis is essential for intelligent decisions.

Figure 2.3 illustrates specific logistics decisions – at a higher level, an organization must select appropriate logistics policies or **strategies** to support the company's financial, service, or other goals. Logistics strategies provide a framework for the type and scope of specific logistics decisions. Often, choosing the right strategy is more significant (from a financial or customer service standpoint) than optimizing specific lower-level decisions.

As an example, a common distribution strategy is to ship all products to a customer from a single distribution center (DC). Another common strategy is to ship to a customer from multiple DCs. Choosing the best DC(s) to serve the customer is a specific logistics decision in either strategy, but the allowable choices are shaped by the respective strategies. (Section 2.4 discusses contemporary logistics strategies in more detail.)

Section 2.2.2 describes a simple case study designed to illustrate a particular set of decisions and a sample analysis. Subsequent sections generalize and organize ideas in this case study analysis into a logistics modeling framework.

2.2.2 Case study illustration

Introduction

Sheridan Technologies, Inc. is an industrial products company operating three plants in the United States, located in Huntsville, Alabama; Fort Wayne, Indiana; and Tucson, Arizona. The plants are dedicated to product groups A, B and C, respectively.

The plants ship finished products in truckload (TL) quantities to five DCs, located in Allentown, Pennsylvania; Atlanta, Georgia; Columbus, Ohio; Richardson, Texas; and Covina, California. The company groups customers into three-digit ZIP code territories, with each ZIP3 assigned to a single DC. The company ships via less-than-truckload (LTL) common carriers out of the DCs, typically weekly for each customer.

Figure 2.4 illustrates the company's current supply chain infrastructure (plants are triangles, DCs are circles, and customer territories are squares). Note the customer territories are scaled to relative average order quantities. Figure 2.4 also illustrates the company's current

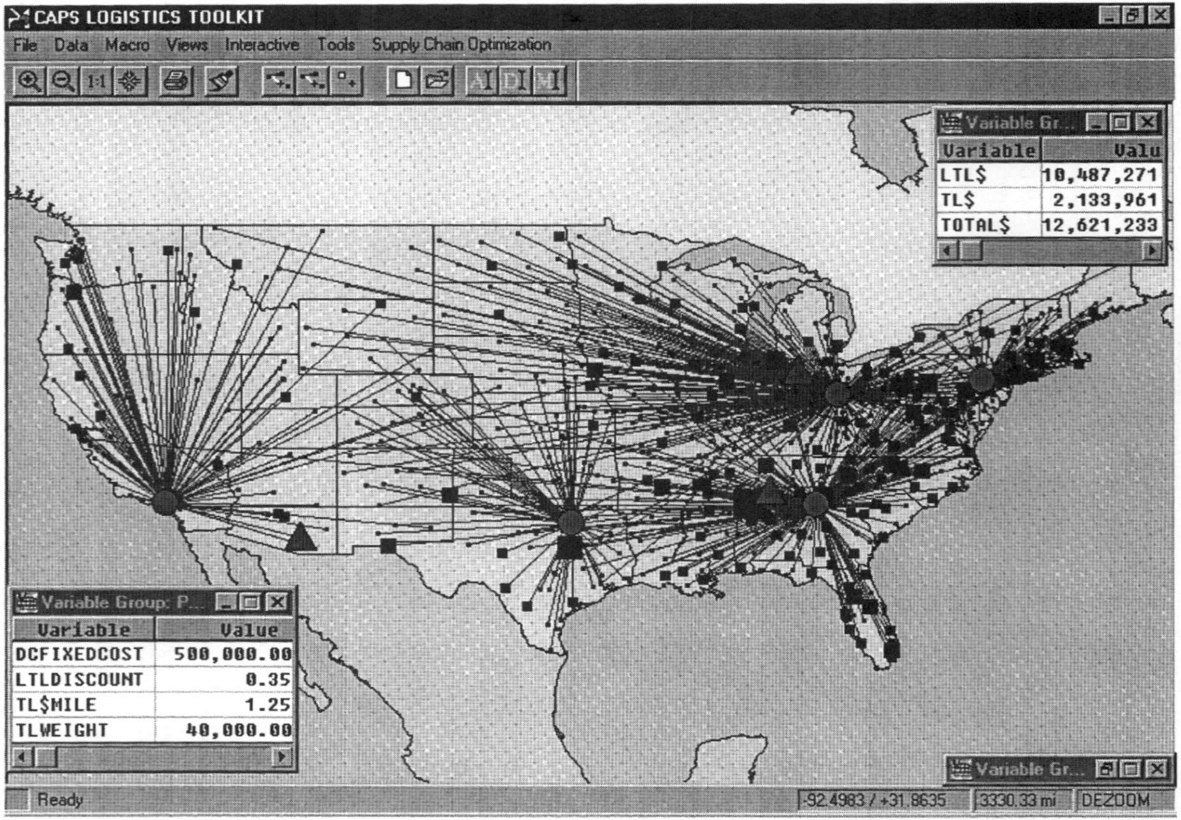

Figure 2.4 Supply chain infrastructure and current market-to-DC assignments.

assignment of market territories to DCs. The current sourcing assignments have developed historically over several years, and have been influenced by various factors including workload balance, company growth, politics and historical partnerships.

The new Vice President of Logistics at Sheridan Technologies has initiated a study of the company's logistics supply chain operations, and formed a project team to analyze the following:

1. Given the company's single sourcing distribution strategy (supplying all products shipped to a customer from a single DC), are the customer territories being supplied from the **right** DCs?
2. Should the company consider changing to a **split-sourcing** distribution strategy? Which customers should be served from which DCs under this strategy?
3. Under the company's current single sourcing distribution strategy, what is the optimal **number** and **location** of DCs that minimizes logistics costs?

Developing a model of the logistics system

The project team decides to develop a computer-based decision support model of the company's logistics supply chain, so potential changes to the system can be quickly generated and evaluated (both interactively and using automated algorithms). Graphics are needed to better understand the supply chain structure and tradeoffs of possible alternatives and to specify alternatives interactively.

Before generating and analyzing any changes to the current system, the project team first wants to create a simplified model representation and ensure the model accurately represents the actual logistics system. A simplified model is desirable to better understand the significant elements and costs of the supply chain and to allow alternatives to be rapidly generated and easily interpreted.

One year's historical shipping information will be analyzed to capture any monthly or quarterly seasonal variations in customer ordering patterns. The company's mainframe computer holds over 100 000 freight bills paid to trucking companies last year, so the team decides to simplify the analysis by calculating the average order quantity and order frequency by each three-digit ZIP region from the historical data.

For each average order quantity the corresponding outbound LTL cost is determined using LTL freight rating tables. The team notes that there may be some error introduced by calculating costs in this manner (as the LTL rates are not linear but are discounted for higher volumes), but the error should be small as the company's just-in-time policy requires a fairly steady flow of products.

As TL shipments inbound to the distribution centers are actually composed of orders from many different customer regions, the associated inbound TL costs must be fairly allocated over individual customer territories and products. The project team uses the average order quantity by product family to estimate a customer's portion of a TL shipment.

Using average order quantities, the estimated annual LTL and TL costs are about $10.5 million and about $2.1 million, respectively. The team decides to ignore storage and handling costs as they are roughly comparable among the DCs. The estimated transportation costs are very close to the actual company TL and LTL expenses for the past year. The team also checks several customer territories and compares the estimated LTL costs to the actual LTL freight costs to that customer. In all cases the estimates are within a few per cent, so the team believes the cost estimation method based on average quantity shipping costs and allocated TL costs are a reasonable model of true transportation costs.

Generating and evaluating alternatives

Armed with a reasonable model of the company's logistics supply chain, the project team sets out to analyze and improve the transportation configuration. It is not clear if the current assignment of customer territories to distribution centers is smart – many of the current assignments in Figure 2.4 do not look very intuitive, but the team knows LTL transportation rates are influenced by factors other than just shipping distance, such as the trucking company's own transportation infrastructure.

For examples, trucking rates are disproportionately more expensive shipping to Florida, because Florida is a consuming state and trucks must often leave the state empty. The team generates graphics of LTL rate contours to better understand the relationship of current DCs and customer territories – Figure 2.5 illustrates the rate contours for 1000–2000 pound shipments originating from the company's Richardson, Texas DC.

Next, the best assignment of markets of DCs is evaluated, given the company's current single sourcing strategy. The DCs are uncapacitated, so the best assignment for each market is simply the DC delivering the average market shipment at minimal total transportation cost. The team calculates the inbound TL costs to each DC and weights the TL costs to each market depending on individual product volume.

Figure 2.6 illustrates the assignment of markets to DCs minimizing total transportation costs. The total annual LTL and TL costs for this solution are roughly $10 million and $2.1 million respectively, a saving of roughly $500 000 annually. The team notes the influence of the LTL rate structure and inbound TL costs on market assignments – obviously the DC nearest a market is not always the best.

The best split-sourcing solution is calculated in the same manner, with

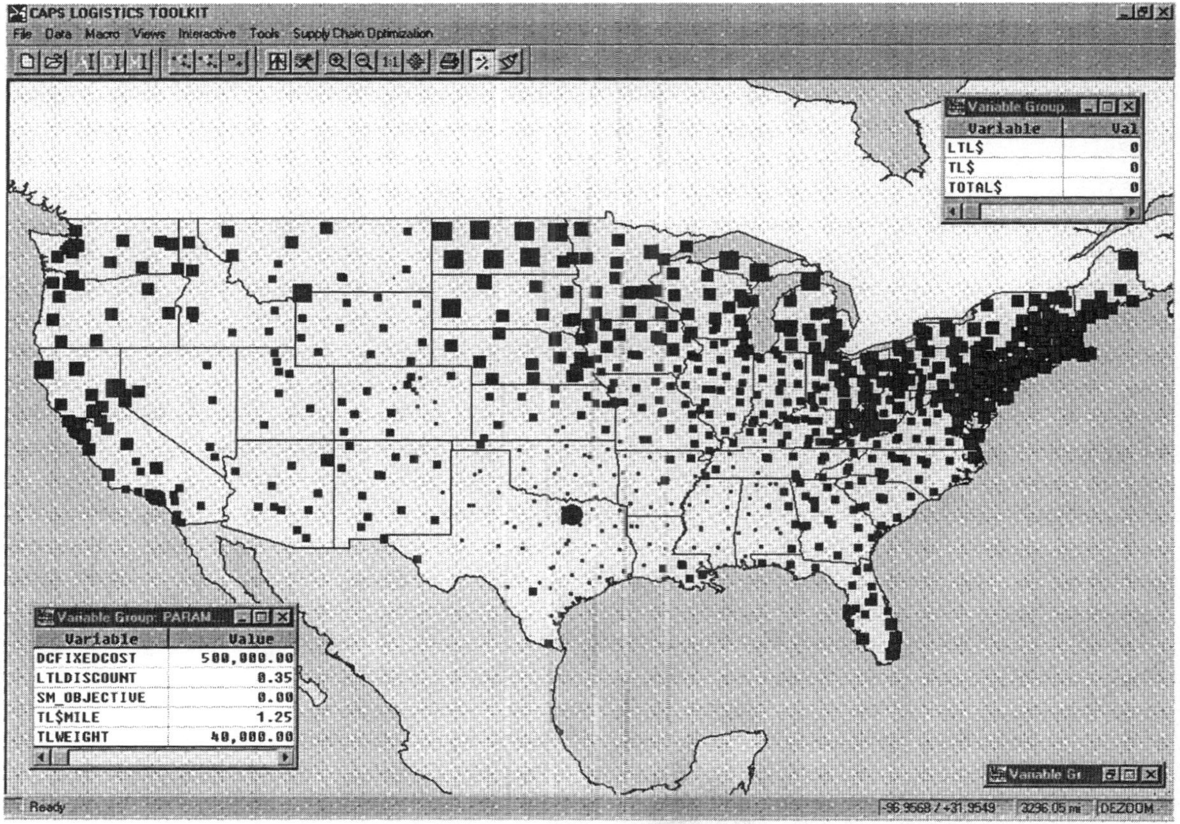

Figure 2.5 LTL rate contours (from ZIP 750, 1000 lbs+).

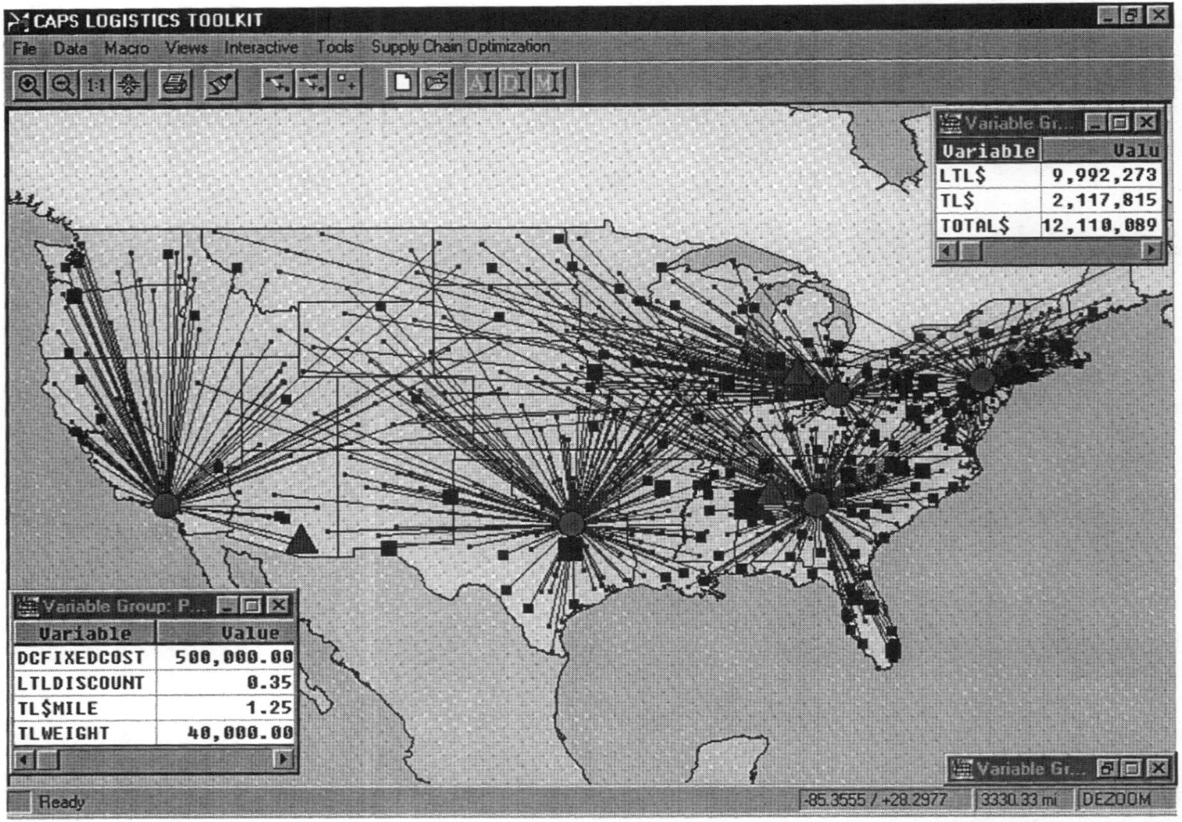

Figure 2.6 Best single sourcing assignment.

total annual LTL and TL costs roughly \$11.2 million and \$2.1 million, respectively. Thus the split-sourcing solution increases costs by roughly \$700 000 annually over the current configuration. The project team rationalizes that single sourcing reduces costs because shipping all products together in larger shipment volumes is less expensive (though individual products may be sourced from a more expensive DC).

The project team next decides to investigate the effect of consolidating existing distribution centers. Since there are only five DCs, the solutions with each DC closed by itself can be easily enumerated. Table 2.1 illustrates the total inbound TL and outbound LTL costs associated with closing each existing DC one by one.

Closing Columbus increases transportation costs by the least amount, resulting in the allocation illustrated in Figure 2.7. If the cost to operate this facility is greater than \$400 000 annually, then closing this DC reduces total costs. The team judges intuitively this is the least important DC, as it is close to Atlanta and Allentown and these DCs are needed for the heavy Southeast and Northeast regions.

The project team next decides to investigate a supply chain configuration not restricted to the current distribution centers. By visual inspection of customer geographical proximity and average order volumes, the team selects 25 DC locations to be analyzed as candidate sites. Each DC is estimated to cost \$200 000 annually to operate, independent of the actual shipment volume handled by the DC.

The team formulates this problem as a mixed-integer mathematical optimization model, with open/close integer variables representing opening/closing candidate DCs. Figure 2.8 illustrates the candidate DCs and the optimal selection of DCs to open and associated market assignments.

Note the existing Covina facility is selected, but Charlotte is selected to handle the Southeast and Northeast, and Denver and Minneapolis are opened to handle the Midwest and parts of the Northwest and Southwest. The total transportation cost of this solution is roughly \$10 million annually, a savings of over \$2 million annually compared to either the current supply chain configuration or the best single sourcing solution. The project team notes the severe overlap of DC-customer allocations due to the disproportionate structure of LTL rates.

Table 2.1 Total annual transportation costs (millions) – one DC closed

Best	*Allentown*	*Atlanta*	*Columbus*	*Richardson*	*Covina*
\$12.1m	\$12.9m	\$13.4m	\$12.5m	\$12.75m	\$12.9m

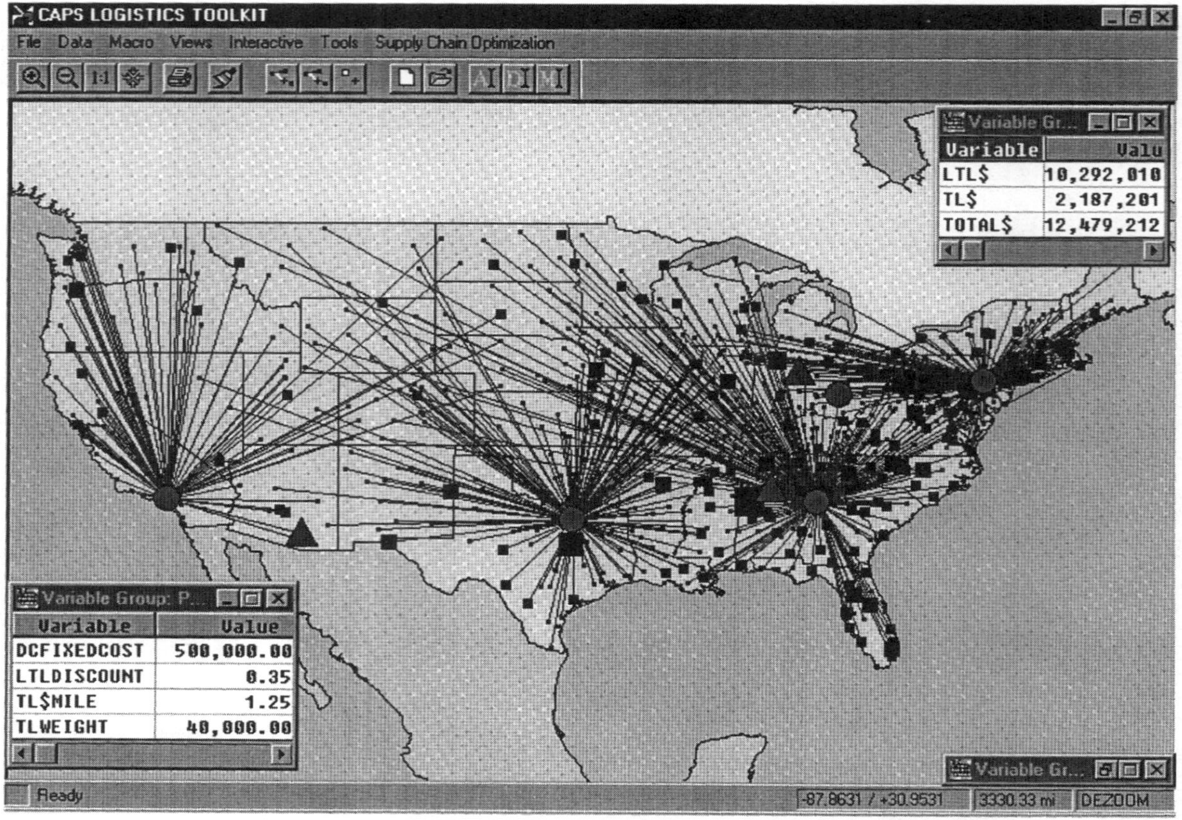

Figure 2.7 Customer allocation with Columbus DC closed.

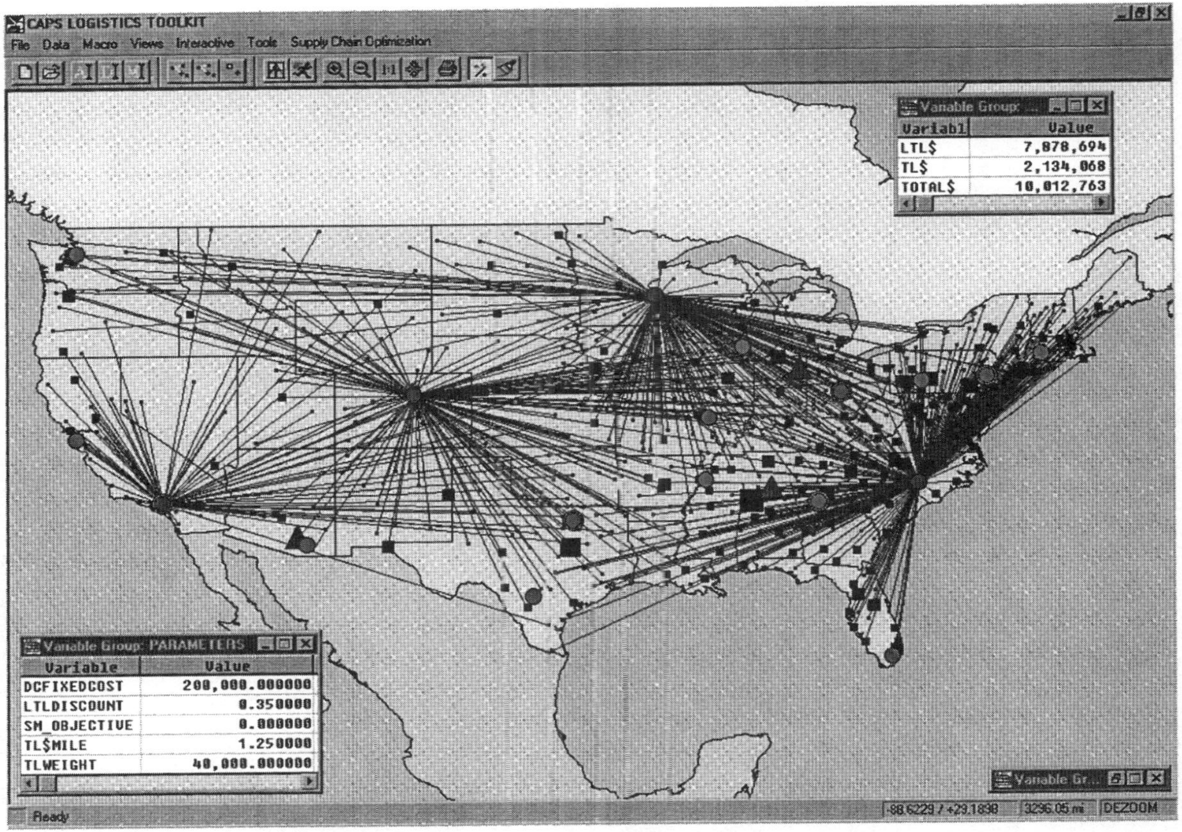

Figure 2.8 Optimal selection/allocation of candidate distribution centers.

2.3 OVERVIEW: LOGISTICS COMPOSITE MODELING

2.3.1 Elements of composite modeling

From the scope and complexity of the logistics questions there is no single best approach, best representation, best model, or best algorithm for optimizing logistics decisions. Each logistics supply chain has some unique characteristics which will always frustrate and complicate the job of the logistics decision maker.

However, there are a growing array of decision support concepts and tools from operations research, geographic information systems, database management and graphical user interfaces that (when properly brought together under the decision maker's control) immensely improve the quality and timeliness of logistics decisions. Figure 2.9 illustrates concepts and tools useful in logistics analysis.

Logistics analysts are naturally biased to their particular modeling expertise: for example, a logistics practitioner may focus on benchmarking models; an operations research analyst may focus on mathematical optimization models; and a computer scientist may focus on object-oriented data models. However, each of these elements is important and should be included in a **composite** model. The next section organizes these elements into a framework for logistics analysis.

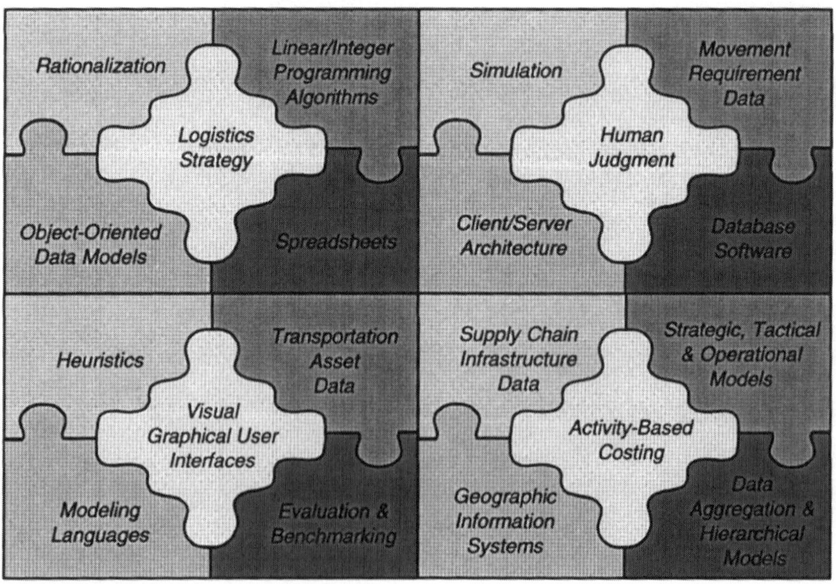

Figure 2.9 Elements of composite modeling.

2.3.2 The composite modeling process

Given the complex questions and tradeoffs involved in a logistics supply chain, the only practical way to determine how to improve logistics operations is to generate and evaluate logical alternatives. The structured approach presented here brings together a variety of tools and the logistics decision maker into a **logistics composite model** (LCM) for optimizing logistics decision-making. Figure 2.10 illustrates the major elements of the LCM analysis process.

These modeling concepts and tools of LCM are described in detail in the following sections.

2.4 LOGISTICS STRATEGIES

2.4.1 Introduction

Logistics strategies includes the business goals, requirements, allowable decisions, tactics and vision for designing and operating a logistics system. Although some logistics strategies impact decisions throughout the supply chain, for clarity the application areas of strategies can be generally organized as illustrated in Figure 2.11.

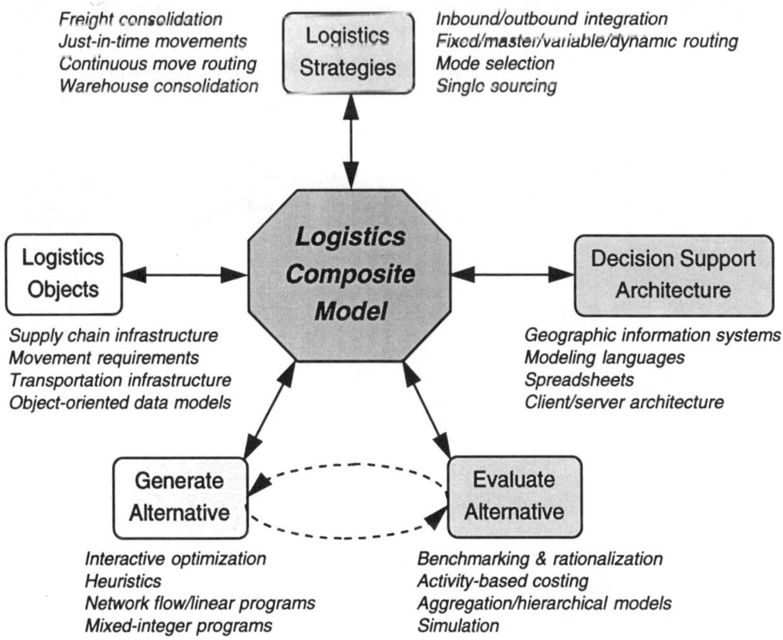

Figure 2.10 Composite modeling process.

- **Supply chain planning** includes the location, sizing and configuration of plants and distribution centers, the configuration of shipping lanes and sourcing assignments, and customer profitability and service issues.
- **Shipment planning** is the routing and scheduling of shipments through the supply chain, including freight consolidation and transportation mode selection.
- **Transportation systems planning** includes the location, sizing and configuration of the transportation infrastructure, including fleet sizing and network alignment.
- **Vehicle routing and scheduling** includes the routing and scheduling of drivers, vehicles, trailers, etc. Other applications include dynamic dispatching, customer zone alignment, and frequency of delivery questions.
- **Warehousing** includes the layout design and storage/picking operations of distribution centers.

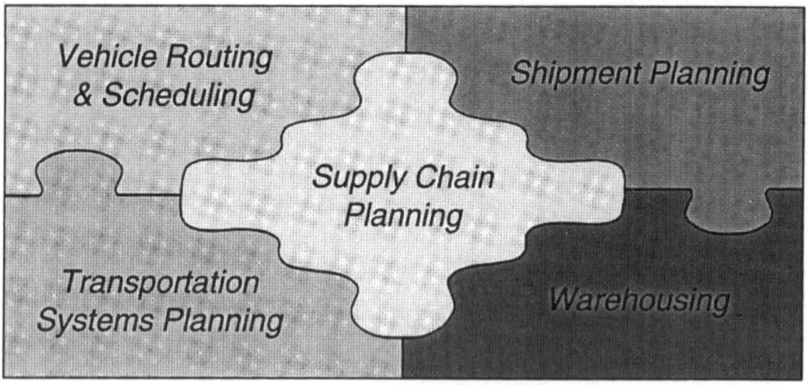

Figure 2.11 Application areas of logistics modeling.

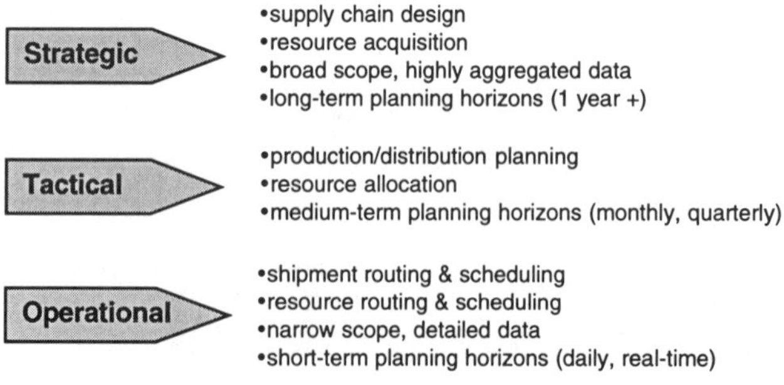

Figure 2.12 Strategic, tactical and operational models.

2.4.2 Strategic, tactical and operational model views

Analyzing the various logistics strategies requires the appropriate modeling views of a logistics supply chain. **Strategic**, **tactical** and **operational** models are three fundamental classes of modeling views, with general properties shown in Figure 2.12.

The logistics application areas in Figure 2.11 can be organized into modeling views as shown in Table 2.2. Examples of strategies for these application areas are illustrated in the next section.

2.4.3 Contemporary logistics strategies

Just-in-time logistics

Historically, products have been 'pushed' through a supply chain based on forecasts of future customer demand. This strategy allows scale economies in the purchasing of raw materials, manufacturing batch runs, and transportation shipments. However, costly inventories build up to protect errors in forecasts, and the logistics system is slow-moving and inflexible to rapid market changes.

If we knew precisely where, when and how much material is needed at each stage of a logistics supply chain, goods could be moved through the supply chain **just-in-time** (JIT) for use by the next process, without a need to build up inventories. Thus product replenishments are 'pulled' all the way through the supply chain from the point of sale. To control the precise movements of products, computerized integration and tracking of supply chain operations is necessary.

JIT is a shift in thinking from inventory **levels** to inventory **velocity** or 'turns'. For a specified time period, the turn rate for a product is calculated by dividing total throughput by the average inventory level. Note the turn rate is only one performance indicator of a logistics supply chain, and by itself is not a very good measure. Often, higher inventory turn rates also mean higher transportation and service costs – Figure 2.13 illustrates this tradeoff.

JIT logistics impacts all five application areas illustrated in Figure 2.11, particularly shipment planning and supply chain design. Shipment planning is fundamentally affected as smaller and more frequent shipments impact transportation mode selection and freight consolidation opportunities. The design of a supply chain is also impacted as there is less emphasis on product storage.

Freight consolidation

Freight consolidation strategies are one of the key concepts in shipment planning. Shipments in the logistics system can be routed and

Table 2.2 Logistics application areas by modeling views

	Supply chain planning	*Transportation planning*	*Shipment planning*	*Vehicle routing*	*Warehousing*
Strategic	Site location Capacity sizing Sourcing	Site location Fleet sizing	Outsourcing Bid analysis Fleet sizing	Fleet sizing	Warehouse layout Material handling design
Tactical	Production planning Sourcing	Routing strategy Network alignment	Consolidation strategy Mode strategy	Routing strategy Zone alignment	Storage allocation Order picking strategies
Operational	MRP, DRP, ERP	Load matching	Shipment dispatching	Vehicle dispatching	Order picking

scheduled independently of each other or can be combined to try and achieve transportation economies-of-scale. There are many ways to consolidate freight, including:

1. **Vehicle routing**. Individual shipments can be combined to share a transportation asset making pickup or delivery stops at different facilities. This type of consolidation is called **multi-stop vehicle routing** (Figure 2.14).
2. **Pooling**. Individual shipments can be brought to a central location or **pooled**, creating large shipments suitable for economy-of-scale transportation modes such as truckload or rail carload (Figure 2.15).
3. **Scheduling**. Sometimes shipment schedules can be adjusted forward or backward in time so they can be combined with other shipments.

Integration of inbound and distribution logistics

Historically the purchasing and scheduling of supplier-to-plant **inbound** shipments have been treated independently of the distribution of goods

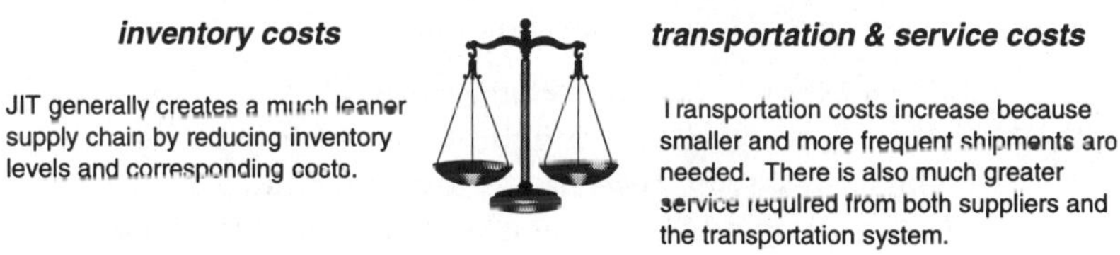

Figure 2.13 Tradeoffs of just-in-time statistics.

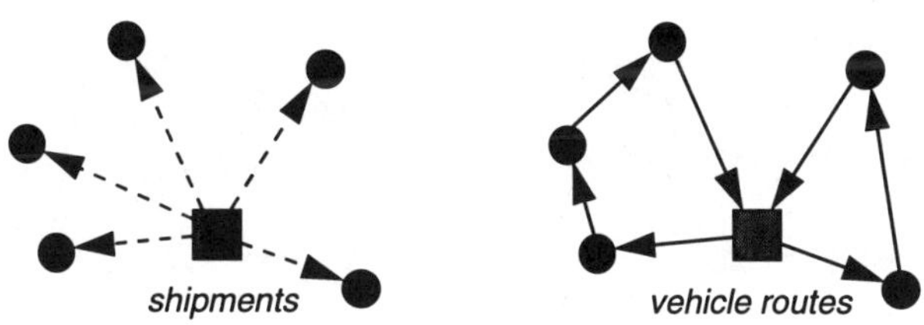

Figure 2.14 Freight consolidation (vehicle routing).

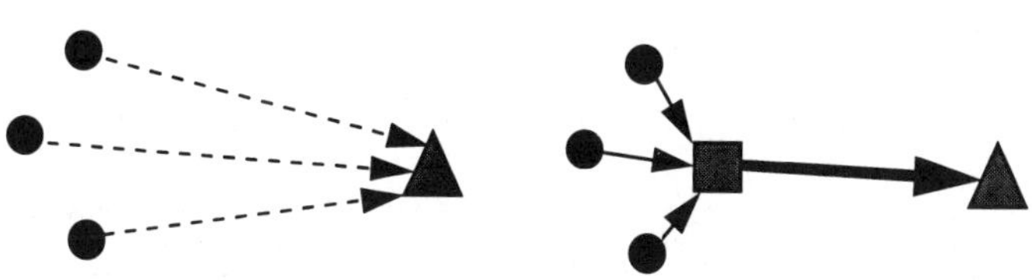

Figure 2.15 Freight consolidation (pooling).

coming out of the plant. Coordinating inbound and outbound shipments and resources requires more control of the logistics system, but can increase the utilization of resources.

This strategy particularly impacts shipment planning and vehicle routing and scheduling. For example, Figure 2.16 illustrates separate delivery and pickup routes (left), and integrated delivery/pickup routes (right).

Fixed/master routes and variable/dynamic routes

Fixed and master routes are regular vehicle route sequences and schedules developed using average demand forecasts. Fixed routes are regular run each period without considering actual customer demand, while master routes are adjusted slightly based on actual demand. In contrast, variable or irregular routes are tailored to actual customer demand information. The extreme case of variable routes is dynamic routes, which are adjusted dynamically as the routes are run. Figure 2.17 illustrates the tradeoffs of these strategies.

Distribution center consolidation vs decentralization

Fundamental decisions in supply chain design include the number, location, sizing, and product configuration of distribution centers. Figure 2.18 illustrates two basic strategies: **consolidated** distribution (fewer but larger distribution centers) and **decentralized** distribution (more but smaller distribution centers).

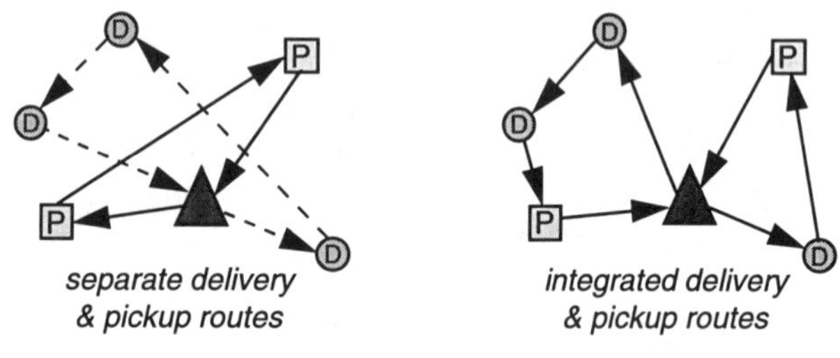

Figure 2.16 Integrating inbound and distribution vehicle routes.

Figure 2.17 Fixed/master vs variable/dynamic route tradeoffs.

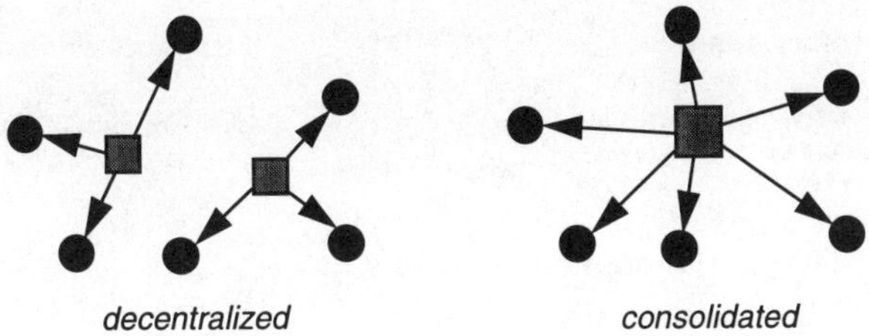

Figure 2.18 Decentralized and consolidated distribution.

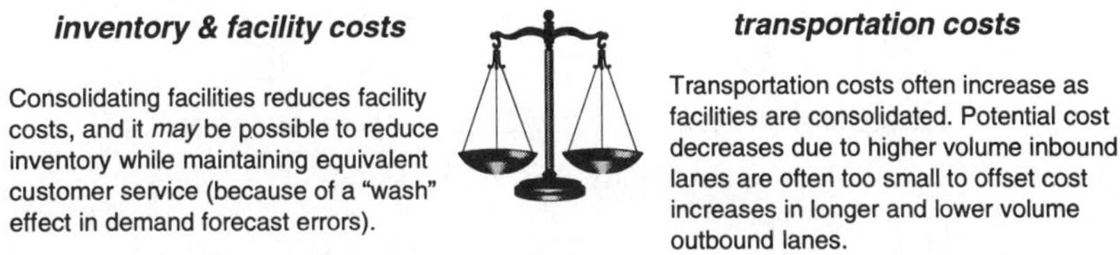

Figure 2.19 Distribution center consolidation tradeoffs.

Figure 2.19 illustrates the decision tradeoffs involved in the consolidation of distribution centers. Note also that an increased number of distribution centers allows closer positioning of inventory to customers, reducing delivery time to customers.

Private fleet vs for-hire fleet

Some manufacturing and retailing firms choose to own and operate their own transportation fleet. This gives more control over transportation costs and service, but forces the firm into operations secondary to the business. Private fleets are becoming less attractive as competition from transportation deregulation has resulted in better service and lower costs from transportation providers. Additionally, deregulation allows organizations to negotiate discounts for longer-term **dedicated** services from transportation providers.

Transportation mode selection

Mode selection is another fundamental concept in shipment planning. Common transportation modes include overnight package, parcel, less-than-truckload (LTL), truckload (TL), and rail carload (CL), for example. Figure 2.20 illustrates the fundamental tradeoffs with choosing the appropriate transportation mode for shipments.

Transportation modes and inventory
Slower transportation modes may also raise **safety stock** inventories needed to protect uncertainties in supply and demand. A slower

Figure 2.20 Transportation mode tradeoffs.

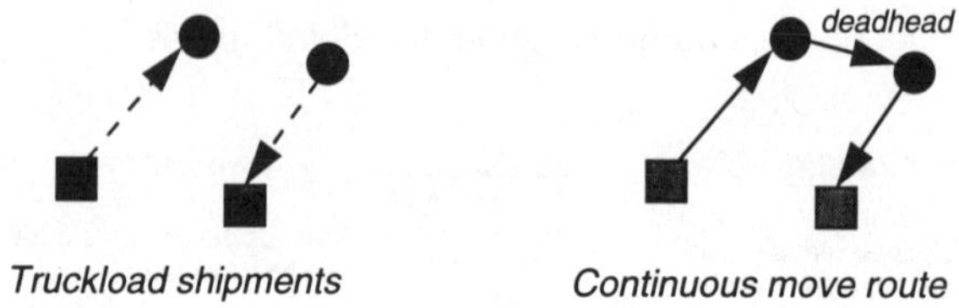

Figure 2.21 Continuous move route.

transportation mode increases the **order lead time** (the time between placing an order and actually receiving the shipment), so more safety stock may be needed to protect against the lack of knowledge about demand during the lead time.

Continuous move routing

Some trucking companies offer discounts for **continuous move** routes, where drivers and tractors are kept highly utilized by coordinating the dropoff of an inbound trailer with the pickup of an outbound trailer. Figure 2.21 illustrates combining two truckload shipments into a continuous movement route. This route reduces costs if the continuous move discount is more than the deadhead cost to travel to the origin of the second shipment.

Single sourcing

Single sourcing refers to satisfying all product demand at a location from one supplier – in contrast, **split sourcing** refers to multiple suppliers satisfying the same demand location. The most common form of single sourcing is between distribution centers and customers or markets, with each customer assigned a single distribution center. Figure 2.22 illustrates examples of split sourcing.

Single sourcing simplifies the logistics supply chain which can reduce management and operational costs. Single sourcing also creates larger volume shipments along lanes, which may reduce transportation costs. However, single sourcing requires each supplier to stock all products – split sourcing allows each product to be shipped via the cheapest shipping route to a customer. Split sourcing can also reduce costs if the supply points are capacitated, as the least cost allocation of supply may require split shipments.

2.5 LOGISTICS OBJECTS

2.5.1 Introduction

What is object-oriented modeling?

To develop a computerized model of a logistics supply chain, a strategy for representing logistics information and supply chain operations is needed. **Object-oriented modeling** is one approach which focuses on the natural elements or building blocks of a logistics supply chain. In this approach the **data** and **operations** of a logistics entity are combined to form a **logistics object**. Figure 2.23 illustrates an example plant object.

There are three basic families of objects in a logistics system:

1. **Supply chain infrastructure**. This family includes physical sites such as suppliers, plants, distribution centers, and customers. This family also includes connections and territory groupings among sites, such as shipping lanes, facility-to-facility assignments, and customer zones.
2. **Movement requirements**. This family is all shipment information

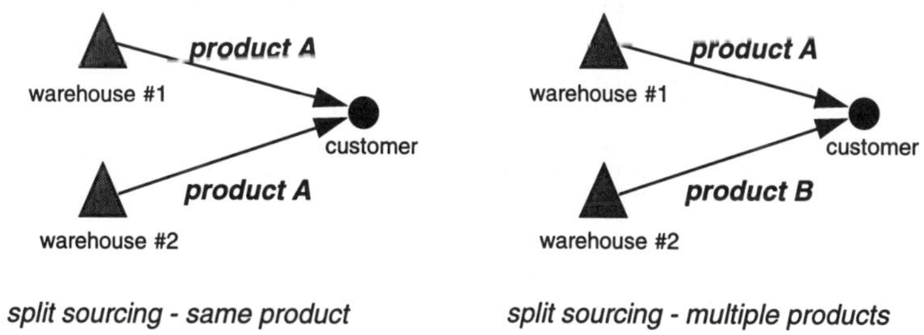

Figure 2.22 Split sourcing examples.

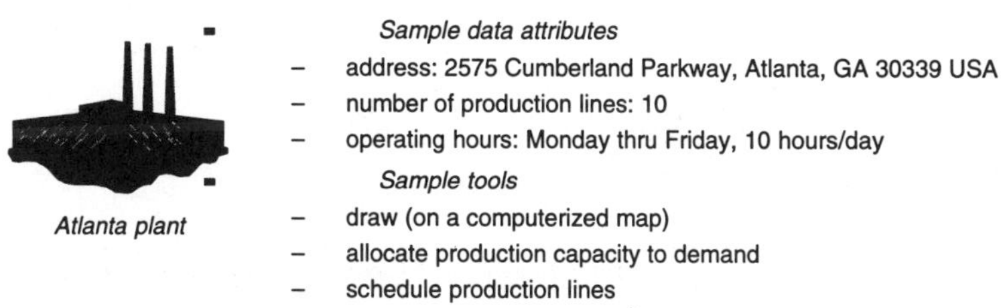

Figure 2.23 Example of a logistics object.

including what commodities to move, when commodities are needed, special instructions or requirements for movement, etc.

3. **Transportation network**. Transportation network objects include the physical components of the transportation infrastructure (road and rail networks, ports, depots, pool points, intermodal exchange locations, etc.), assets that are either owned or available for hire (drivers, trucks, trailers, containers, planes, ships, rail cars, intermodal containers, pallets, etc.), and asset locations and transportation capabilities (such as maximum loads or transit speeds).

Class hierarchies of logistics objects

Objects in these three families can be **hierarchically** organized into **classes** to provide an intuitive representation of a logistics system. Figure 2.24 illustrates one example hierarchical organization of logistics objects, representing the logistics system of a manufacturing firm (note there are many different and valid ways to describe a logistics system using object hierarchies).

Other properties of hierarchical models

A hierarchical model provides a systematic way to adjust the **resolution** of a model. By navigating a hierarchical model we can back up and look at the 'big picture' in a logistics system, or focus in on detailed components. A hierarchical representation also provides a systematic way to **extend** a model – a new object class (such as European manufacturing plants) can be quickly created by first **inheriting** data and operations from an existing object class.

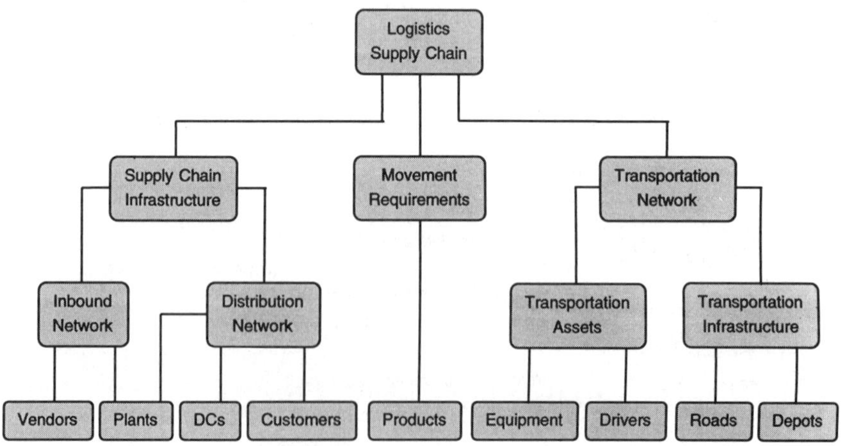

Figure 2.24 Example hierarchy of logistics objects.

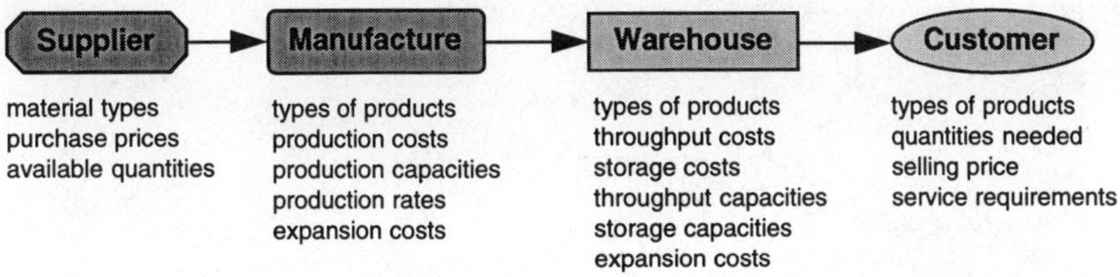

Figure 2.25 Facility data needed or useful in logistics models.

Other benefits of object-oriented modeling

There are many other benefits of an object-oriented modeling approach. Combining data and operations (or **tools**) together into objects allows the tools to protect or **encapsulate** what can be done with data. Further, the binding of data and tools makes it clear how data can be manipulated. Object-oriented modeling is also particularly suitable as a basis for computerized decision support systems. Modularity makes objects easier to maintain and extend, and the hierarchies of object classes encourage reusable programming code.

2.5.2 Developing a supply chain model

Facilities

Supply chain infrastructure
Facilities are the foundation of the supply chain infrastructure and include suppliers, plants, distribution centers, and customers. Figure 2.25 illustrates facility data that is needed or useful in logistics modeling.

What is geocoding?
Geocoding is the process of determining the geographic (longitude and latitude) coordinates or geocodes of a facility, given a description of the facility (Figure 2.26). The description of a location could be a street address, city name or postal code.

Why are geocodes needed?
Geocodes are needed for many types of quantitative analysis tools where nearness among facilities is important, such as the routing of vehicles or the location of new facilities. Geocodes also allow the logistics supply chain to be visually represented using map-based graphical user interfaces (discussed in section 2.7.3).

A large number of commercial databases are available to support geocoding, including databases of postal codes and metropolitan streets – an extensive reference of available databases is the GIS World Sourcebook [5].

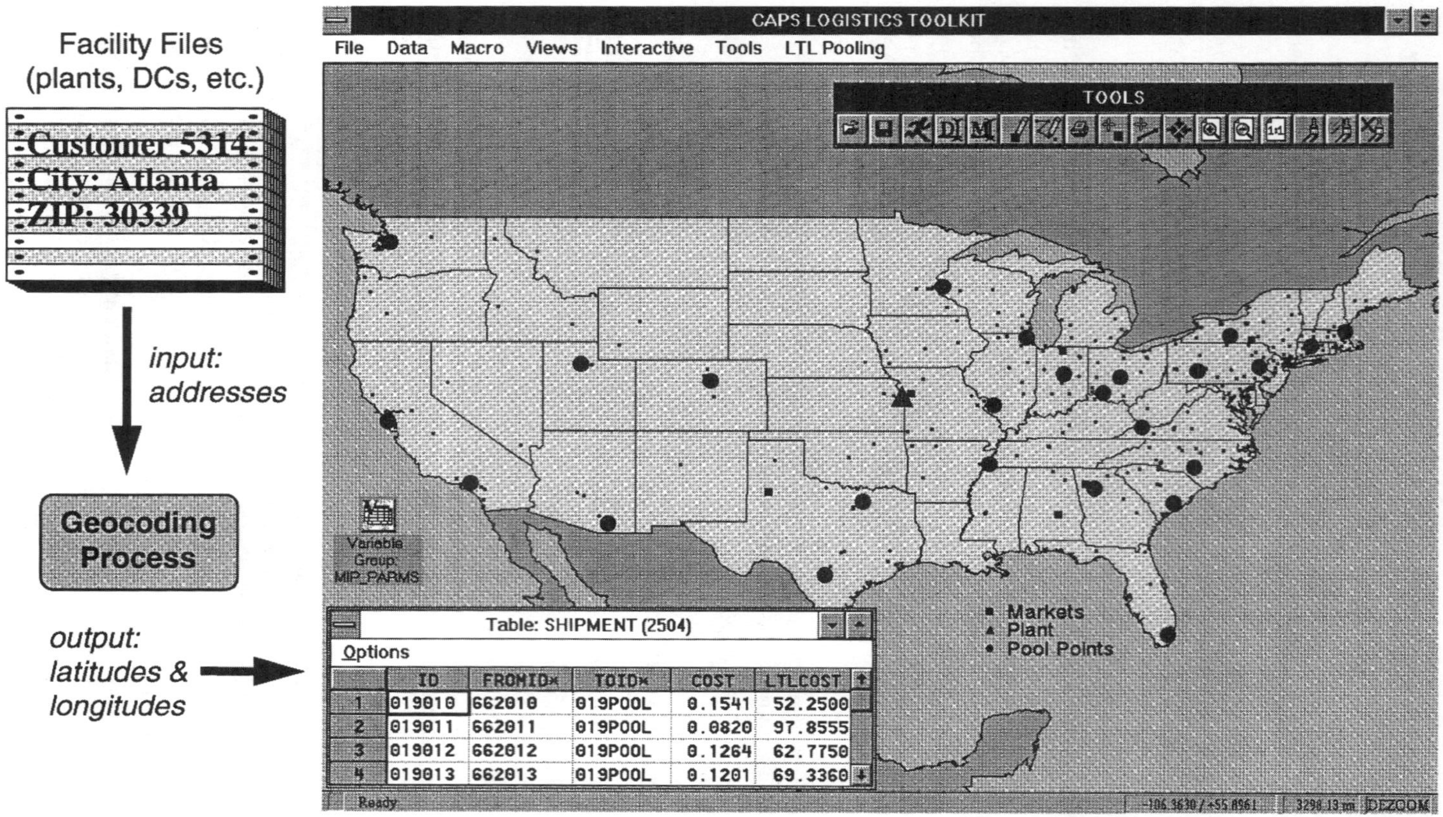

Figure 2.26 Geocoding process.

Facility zones

Zones define the geographical territories of facilities, such as sales regions, customer territories, or distribution center areas. Zones can be predetermined (such as marketing territories) or can be created automatically by rules or algorithms. Figure 2.27 illustrates example zones defined for a set of customers – individual customer data such as demand is aggregated to determine total demand by zone.

Why define facility zones?

Facility zones play an important role in simplifying a logistics model, as aggregate regions can represent the demand of hundreds or even thousands of individual customers. Effective zones usually define logical geographical clusters of facilities, adjusted to balance some attribute (such as total zone demand or transportation workload, for example).

Shipping lanes

Shipping lanes are the product movement connections between supply chain facilities. Figure 2.28 illustrates shipping lane data relevant in logistics modeling.

Shipping lane **generation** defines the shipping lanes between facilities

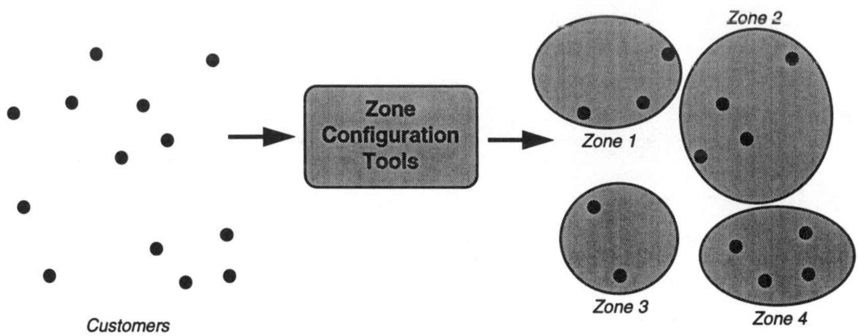

Figure 2.27 Generating facility zones.

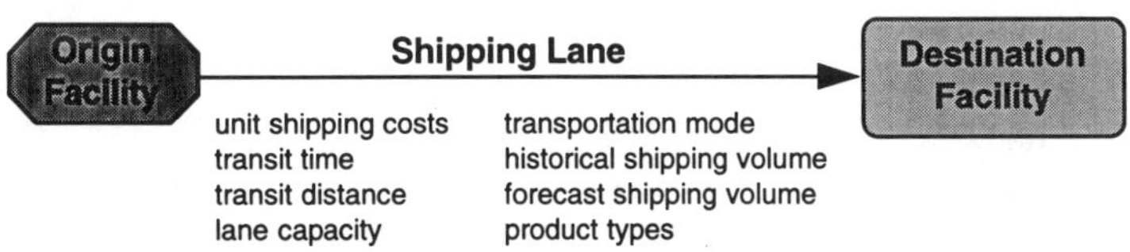

Figure 2.28 Shipping lane data.

that are **candidates** for product flow. Lane generation tools help to pare down the large number of possible transportation lane combinations to a practical or logical set. These candidate lanes are inputs to analytical tools deciding actual product flows and schedules.

Shipping lanes can either be predetermined or generated by rules or algorithms (such as all lanes of distance less than 500 miles, or by product compatibility issues). Figure 2.29 illustrates example lanes generated between facilities.

Representing movement requirements

Movement requirements indicate product demand or shipments, including when and how many products or components are needed at specific facilities or zones. One way to indicate movement requirements is to specify the **aggregate** supply and demand for products at facilities or zones (illustrated in Figure 2.30). Aggregate supply-and-demand models are appropriate where the origins and/or destinations of movements are unknown. Aggregate models are also useful for answering strategic design questions.

Another fundamental way to represent movement requirements is to describe **explicit** shipments, including origins, destinations, products, volumes, due dates and pickup dates. Origin–destination shipment

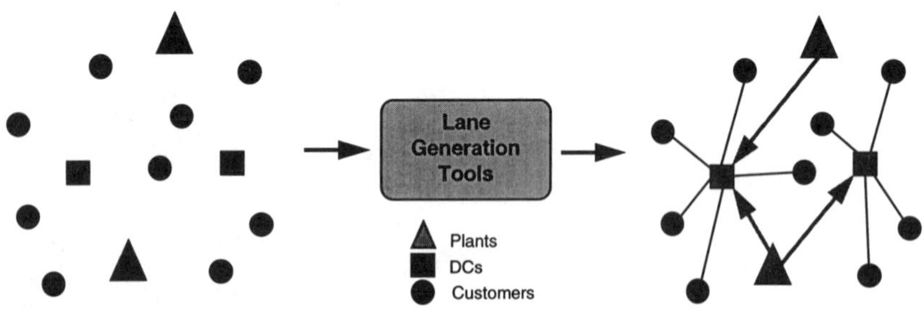

Figure 2.29 Generating candidate shipping lanes between facilities.

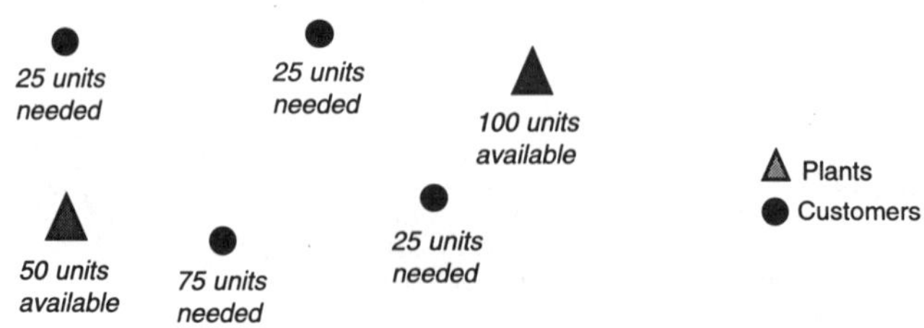

Figure 2.30 Supply–demand movement requirements.

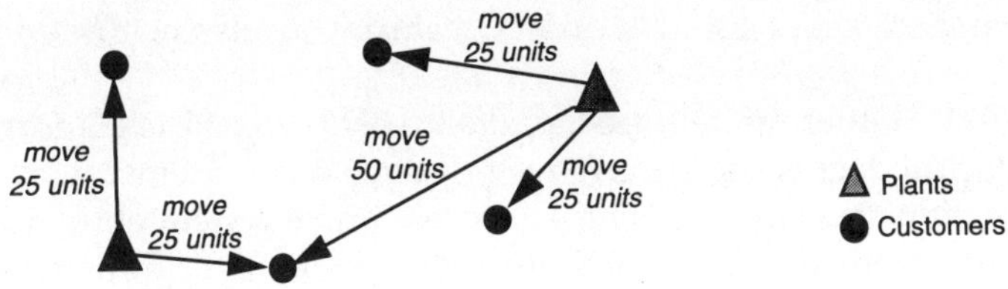

Figure 2.31 Origin–destination movement requirements.

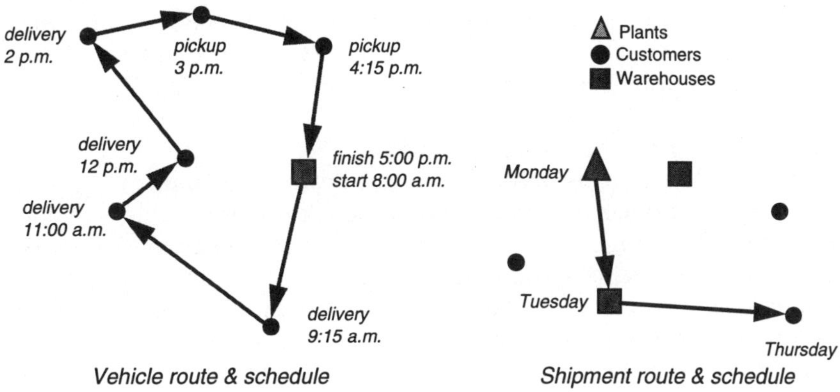

Figure 2.32 Example paths, routes and schedules.

models are useful for answering routing and scheduling questions involving explicit shipments and transportation assets. Figure 2.31 illustrates origin–destination movement requirements.

Representing the flow of shipments and assets

Logistics objects that are **moving** include shipments and transportation assets. **Paths** and **routes** are used to represent the movement of goods and transportation equipment among facilities. **Schedules** describe timing information associated with the movements. Figure 2.32 illustrates example paths, routes and schedules. Note that transportation assets include trucks, drivers and trailers, each with possibly distinct routings and schedules.

Other data issues

Sources of data
Data for logistics objects can be generated in three basic ways:

1. **Current information**. Some logistics models are based on current logistics information. For example, vehicle dispatching models need

information about today's orders, vehicles available, driver status, etc.

2. **Forecasts**. Other models are based on forecasts of future information – historical data is used to predict future customer demand, available production capacity, etc. The estimates can be generated in a variety of ways, from using sophisticated forecasting algorithms to simply rolling up a year's worth of historical data to give an annual view of a supply chain.
3. **Historical information**. Still other models use actual historical data to calibrate model accuracy – model outputs can be compared to what actually happened to ensure the model is a valid representation of the logistics supply chain.

Time-related data: the modeling horizon

A key modeling issue is defining the time span or **horizon** of a logistics model. Some models are **single-period** models – there is only one time period, so data in these models does not change over time. A popular single-period model is a one-year view of a supply chain, with relevant facility data including the total production capacity or demand forecast for the entire year. Single-period models are useful for analyzing solutions to strategic design models.

In contrast, some models are **multi-period** models, with data potentially changing from one time period to the next. For example, the customer demand for a soft drink products increases during the summer months. A popular multi-period model is a one-year view of a supply chain by month, with relevant facility data including the production capacity or demand forecast for each month, for example. Multi-period models are useful for analyzing solutions to resource scheduling models.

2.6 EVALUATING ALTERNATIVES

2.6.1 Introduction

Evaluating Alternatives is 'playing out' or **simulating** the operation of a logistics supply chain using a model and analyzing the attractiveness of the supply chain configuration. Cost and service performance measures, resource utilizations and bottlenecks, and other statistics of the logistics system are calculated in this phase of LCM.

Evaluating Alternatives is composed of the **Evaluate**, **Benchmark** and **Rationalize** steps, each geared towards answering particular analysis questions. These three steps combined with the **Generate Alternatives** step (discussed later) form an iterative analysis cycle, illustrated in Figure 2.33. The analysis process is naturally iterative because evaluating one alternative often suggests new alternatives to investigate.

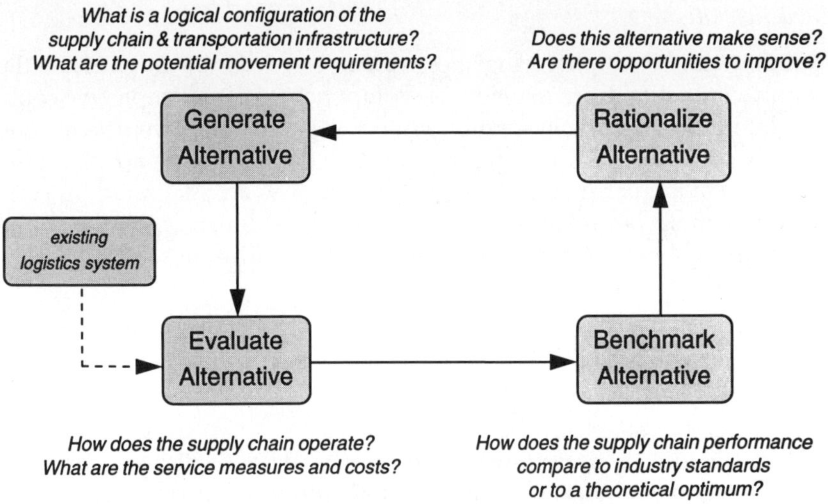

Figure 2.33 Analysis of alternatives cycle.

2.6.2 Model simplification

Logistics supply chains can be very large systems composed of hundreds of facilities moving tens of thousands of products or more. These systems are far too large and complex to work with all at once in the **Generate**, **Evaluate**, **Benchmark** and **Rationalize** steps. The only practical way to analyze and improve a logistics system is to simplify the logistics decisions into smaller interrelated and manageable components. Two model simplification techniques are **aggregation** and **partitioning**.

Model aggregation

Aggregation is collecting or 'rolling up' related data up to a simpler or more approximate representation. Examples of logistics supply chain aggregation are the following:

- Grouping individual products or stock-keeping units (SKUs into **product families**, representing groups of similar product items.
- Adding up the individual product demand for customers to get the total demand by **customer zone**.
- Adding up the manufacturing capabilities of individual production lines and assembly stations into a **total production capacity** for a manufacturing plant.
- Representing large numbers of individual truck trailers by a few basic **trailer types**, such as refrigerated, 48 foot, etc.

Model partitioning

Another way to simplify a logistics system is to decouple or **partition** the supply chain into more manageable components. For example, we could divide the distribution system into regions, and develop vehicle routing models separately within each region. Of course, a key part of supply chain modeling is treating the logistics system as an integrated process, so care must be taken to provide enough 'linkage' between the components to capture the relevant decisions and issues.

Model accuracy

In the ideal logistics model:

- All data is available and correct.
- There is no error in forecasts of future data (including customer demand, availability of supply, availability of resources, etc.).
- The model exactly captures all of the relevant issues in the logistics supply chain.

Unfortunately, in most situations some data is missing or incorrect, the forecasts of future data are wrong, and some supply chain characteristics are too fuzzy to capture precisely in a model. Thus most logistics models are at best approximate representations of the actual logistics system.

How do we know if an approximate and simplified model is an accurate representation of the logistics supply chain? This is the heart of the Evaluate Alternatives step. This modeling step 'plays out' a given logistics system configuration, so more detailed data can be used. The result is that baseline statistics can be calculated and used to gauge the precision of more simplified models (Figure 2.34).

For example, in aggregate models it is common to use average or approximate values for costs and demand quantities. Given the movement of actual shipments, we can evaluate the true shipping costs

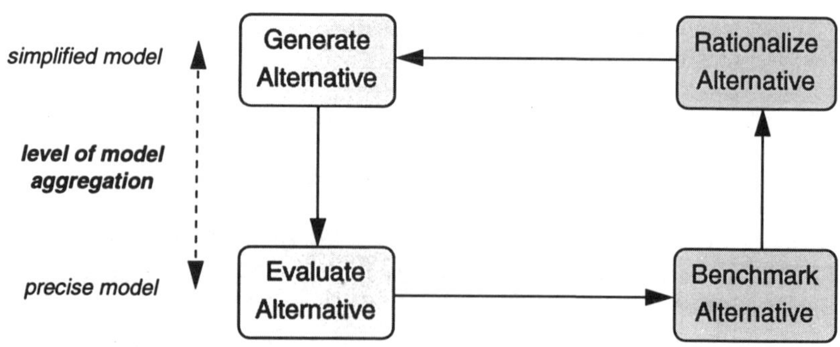

Figure 2.34 Simple and precise modeling levels.

and compare with the approximate costs. We can then modify and improve how we estimate the approximate costs and demands based on what actually happened. Thus the evaluation step measures the accuracy of a simplified model.

Simplified models are particularly useful in the Generate step, as solution generation tools can examine a greater number and variety of decision alternatives using more aggregate models. It is important to note that the output of the Generate phase is the input to the Evaluate phase. For example, the optimal solution generated by a mathematical optimization model is not necessarily the 'answer' but rather must be played out and evaluated to judge the solution's true attractiveness. More aggregate model views are also useful in the Rationalize step, as these models are easier to understand and manipulate.

In general, the right level of model simplification balances accuracy (so that judgments based on the model are correct judgments about the actual logistics system) with practicality (illustrated in Figure 2.35).

Simulation

Simulation is a general term for a class of tools and models that play out a given logistics system. While these tools are descriptive only (and do not prescribe smart alternatives), simulation tools can handle a large amount of detail, and can effectively represent the **probabilistic** elements of a logistics system. Thus these tools are effective for evaluating the actual behavior of a logistics system and calibrating the accuracy of more approximate models.

2.6.3 Costing

Landed customer costs

The process of determining the product cost delivered to the customer forces the specification of cost models for the various components of the entire logistics supply chain. The cost to deliver a unit of product to the customer is called the **landed customer cost**.

Figure 2.36 illustrates the various types of costs incurred as a product moves through a logistics supply chain. The result is the cost to get a product to two different customers is almost always different.

Figure 2.35 Model accuracy tradeoffs.

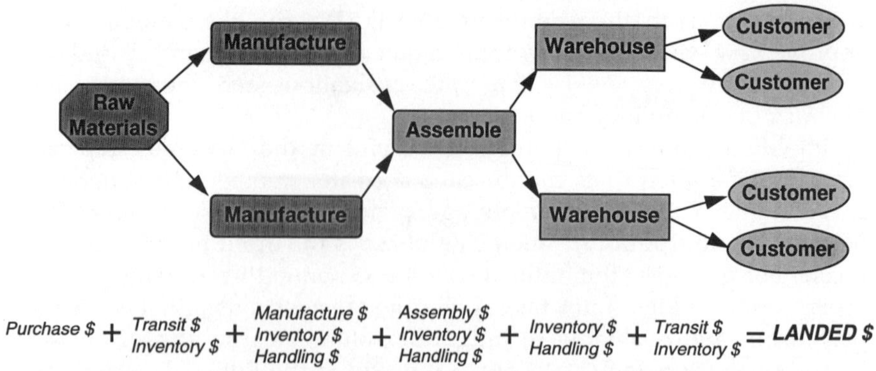

Figure 2.36 Cost elements in a logistics supply chain.

Activity-based costing

Some cost components are easy to determine for specific products and customers, but other costs are shared among products or customers and must be fairly allocated. **Activity-based costing** is one allocation method that attempts to allocate resource costs accurately by focusing on the activities performed by the resources. Costs are then allocated based on the activity **levels** needed by individual products or customers.

For example, we could allocate the cost of a vehicle route over a set of customers by first identifying specific route activities, such as driving, loading, and unloading. Some activities are tied to individual customers (such as unloading), and thus the corresponding costs are easy to allocate. Other activities (such as driving) are jointly influenced by customers and must be allocated using some estimate of an individual customer's contribution to the activity.

2.6.4 Benchmarking and rationalization

Benchmarking is comparing the performance of a logistics supply chain to organizational or industry standards or to some theoretical 'ideal'. If data is available, it may be possible to compare the supply chain to so-called 'best-practice' standards or corporate supply chains that are recognized as industry leaders in logistics operations.

Benchmarking metrics generally fall into two basic groups: **costs** and **service measures**. Sometimes costs and service measures can be measured directly, but frequently **surrogate** indicators must be used to estimate performance (particularly for service). For example, the inventory turnover rate, the total cycle time of a product in a supply chain, and the movement accuracy (timeliness of actual shipment movements compared to predicted movements) are commonly used as estimates of the level of customer service provided by the supply chain.

In general, the more surrogate the metric, the more carefully it should be treated when evaluating supply chain performance.

The Rationalize step is the interpretation of the Evaluation and Benchmark results, and the justification of the logistics supply chain configuration. Tools to use in this step include cost reports, service metrics, and the utilization of resources. Model aggregation is important in this step, as it is important to see the 'big picture' of the logistics supply chain and focus on the key opportunities for improvement. Rationalization relies heavily on strong graphical user interfaces that can illuminate resource bottlenecks, high cost elements, service problems, etc.

2.7 GENERATING ALTERNATIVES

2.7.1 Introduction

Generating alternatives includes any change to the logistics strategy, supply chain infrastructure, transportation infrastructure, movement requirements, or the relevant operating parameters. There are four fundamental ways to generate logistics supply chain alternatives:

1. **Existing system**. If the logistics supply chain already exists, then the first alternative analyzed is the current system, and the analysis proceeds directly to the Evaluating alternatives phase.
2. **Specified**. The alternative to investigate could be given, such as a strategic plan that the organization's management would like to evaluate.
3. **Automatic**. The alternative could be generated automatically, using computer algorithms based on mathematical optimization, heuristics, rules, etc.
4. **Interactive**. The alternative could be generated interactively, in an exploratory or 'what-if' style.

2.7.2 Automatic generation of alternatives

Mathematical optimization

Analytical models

One of the most important steps of LCM is developing an **analytical** or mathematical model of the logistics supply chain. An analytical representation is natural because of the many quantifiable elements in logistics (such as shipping costs, storage costs, transit times, inventory levels, production capacities, and demand forecasts). **Mathematical optimization** is a powerful class of quantitative models, tools, and algorithms that can be used to generate automatically and examine vast numbers of decision alternatives and pinpoint smart alternatives.

Elements of optimization models

A mathematical optimization model consists of the following three components:

- **Objective**. Usually we wish to maximize or minimize some quantifiable goal. For example, common logistics objectives include maximizing profitability, minimizing landed costs, maximizing on-time shipments or minimizing the number of trucks needed.
- **Decision variables**. Decision variables represent choices in a logistics supply chain. For example, common logistics decision variables include where to locate facilities, how to route freight, and when to send shipments.
- **Constraints**. Constraints represent restrictions or requirements of the logistics supply chain. For example, common logistics constraints are storage space in a warehouse, available manufacturing capacity at a plant, the number of trucks available, and the shipment delivery time required by a customer.

'Easy' optimization models

Some mathematical optimization models are 'easy' in the sense that there are algorithms available that can consistently find the optimal solution in a predictable amount of time. The most useful models in this class are **linear programming** (LP) models. In an LP model the objective and all constraints are linear equations, and all decision variables are 'continuous' (i.e. fractions are okay). Very large linear programs with tens of thousands of decision variables or more can be optimized quickly using efficient computer algorithms.

A special class of linear programs are **network** linear programs which have many natural applications in modeling supply chain networks. Minimum cost network flows, shortest paths and matching tools belong to this class, and have applications in resource allocation, production scheduling, and supply chain design.

'Hard' optimization models

Some mathematical optimization models are 'hard' in the sense that there are algorithms available that can consistently find the optimal solution in a reasonable amount of time, if the problem size is sufficiently small. Thus these are 'limited size solvable' models. For these models we can optimize small problems but either cannot optimize large problems or cannot solve them with consistency.

Many of the most important logistics models fall into the 'hard' class. This includes most models of vehicle routing and scheduling, facility location and sizing, shipment routing and scheduling, freight consolidation and transportation mode selection. These problems can be represented as **mixed-integer programming models**, a class of models with some of the decision variables restricted to integer values. For example, the number of drivers and trucks assigned to drive a certain distribution lane could be 0, 1, 2, etc., but could never be 2.7; a manufacturing plant can either be constructed or not constructed, but not partially built.

Mixed-integer models are often difficult to optimize, as there may be an exponential number of possible decision alternatives. For example, the number of possible combinations of opening or closing n distribution centers is 2^n. There is no algorithm available which can guarantee finding the optimal alternative without the possibility of examining many of these alternatives.

A further complication is that the efforts required to solve a mixed-integer program is often dependent on the specific problem data, and a very slight change to a model may transform a solvable problem to an unsolvable problem. Thus mixed-integer programming models are often better suited for planning when there is sufficient time to use alternative approaches if the solution effort becomes too great.

Heuristics

Heuristics are another important class of methods for automatically generating supply chain alternatives and decisions. A heuristic is simply any intelligent approach that attempts to find good or plausible solutions. The heuristic may be based on mathematical optimization, rules, or any other method that can generate alternatives.

The word 'heuristic' sometimes implies a 'seat-of-the-pants' solution approach, with little or no intelligence or sophistication used to make decisions. This is unfortunate, as analytical heuristics can be as technically sophisticated as mathematical optimization approaches. Many heuristics are actually based on mathematical optimization methods and algorithms such as using practical rules to formulate a mathematical optimization model. A powerful heuristical approach is to modify a mixed-integer program by temporarily treating the integer variables as continuous variables, creating an approximate but much more solvable logistics model. The solution to this problem is then used as a basis for constructing a solution to the integer program.

2.7.3 Interactive generation of alternatives

One of the most powerful techniques for generating and analyzing alternatives is **visual logistics modeling**. Visual logistics modeling allows logistics analysts to specify decision alternatives via a combination of mathematical optimization, heuristics, and graphical user interfaces (Figure 2.37). The visual interface shows computerized maps, supply chain infrastructures, transportation infrastructures, flow requirements, schedules, etc. Visual logistics modeling is also ideally suited for understanding a supply chain, as graphical solution representations can often best portray resource limitations, service or cost problems, structural problems with the supply chain, inefficient vehicle routes, and other improvement opportunities.

Digital geographic data is an important part of visual logistics modeling, and is also used directly in computations for many types of logistics models. Examples include computing transportation distances, routes and schedules over a highway network, or determining the closest distribution center for a set of customers. Another common application is the use of geographic **zones** as a part of the modeling process (e.g. assign all customers in this area to a particular distribution center).

2.7.4 Comparison of solution generation approaches

Mathematical optimization, heuristics, and visual logistics modeling are all tools that can generate logistics alternatives. Which method is best?

No single solution generation approach is appropriate for all logistics modeling situations, and each method has certain complementary benefits. In LCM we first try to represent logistics decisions using a mathematical optimization model because of the power of quantitative models to consider large numbers of alternatives and pinpoint optimal solutions. Note that this means we may wish to use simplified models in order to pose a quantitative model that can be solved in a reasonable amount of time.

We can rely on the mathematical optimization solution if we are confident that our model is a precise representation of the logistics supply chain. But what if:

- the quantitative model is only an approximate representation of the actual logistics supply chain?
- the data is estimated and likely contains errors, or there is operational variability in the supply chain which cannot be predicted?
- there are objectives, decisions, or constraints which are not naturally quantifiable, and require human judgment?
- the model is a 'limited size solvable' optimization model?

These are all common aspects of logistics modeling, and require the 'composite' approach of LCM to bring together various complementary tools. If the quantitative model is a high-level approximation of the real logistics system, then it is critical that the output of a mathematical

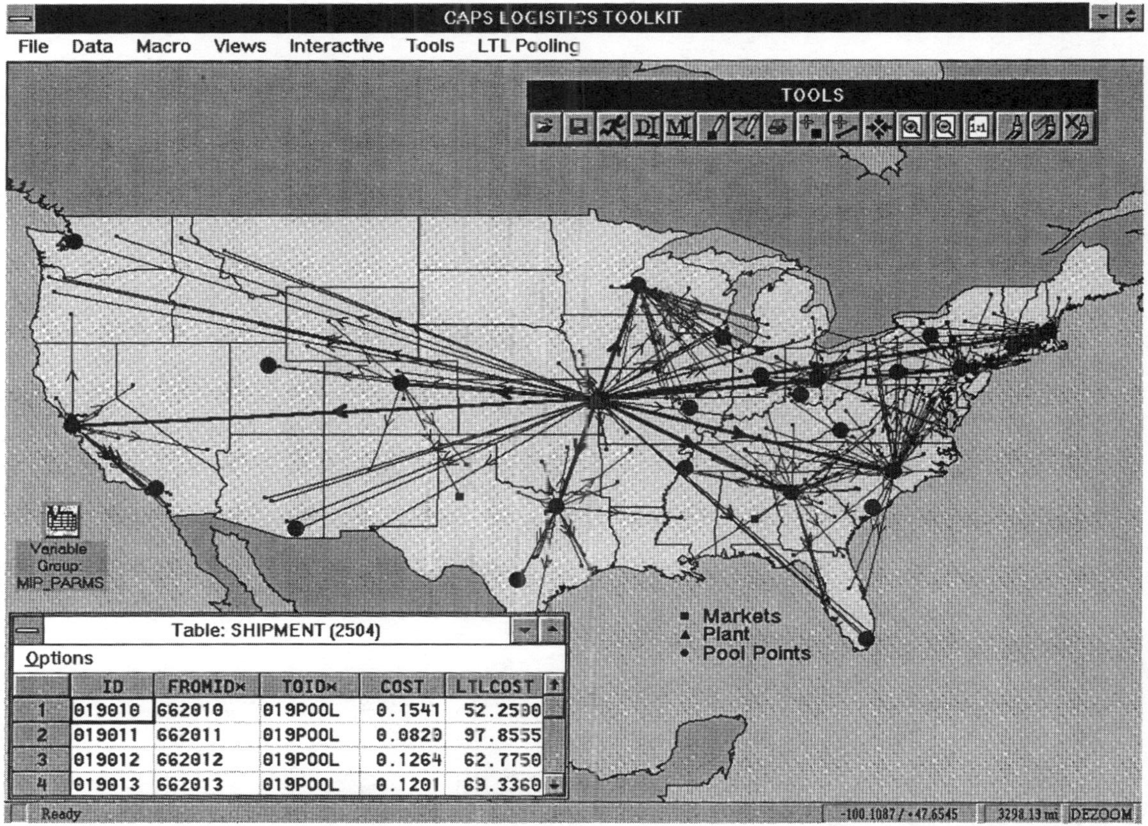

Figure 2.37 Visual logistics modeling.

optimization model is treated as the **input** to the Evaluate Alternatives step rather than the final solution.

Heuristics are a key part of LCM, as heuristics may be able to best handle non-quantitative business issues or rules, imperfect data, and limits on solution time and computing capacities. Generating alternatives using visual logistics modeling software is an excellent way to take advantage of human judgment and control of the decision-making process, increasing the understanding and control of mathematical optimization models.

2.8 DECISION SUPPORT ARCHITECTURE

2.8.1 Introduction: Evolution of decision support tools

Logistics decision support tools have advanced steadily since the development of operations research, and very rapidly in the last ten years. Major milestones are illustrated in Figure 2.38. The next sections summarize classes of these tools particularly useful in logistics modeling.

2.8.2 Computing architecture

Logistics modeling languages

The scope and complexity of logistics systems and models necessitate the use of computer-based decision support systems. Two fundamental classes of software applications are 'custom-built' systems developed for unique situations, and 'off-the-shelf' systems developed for more

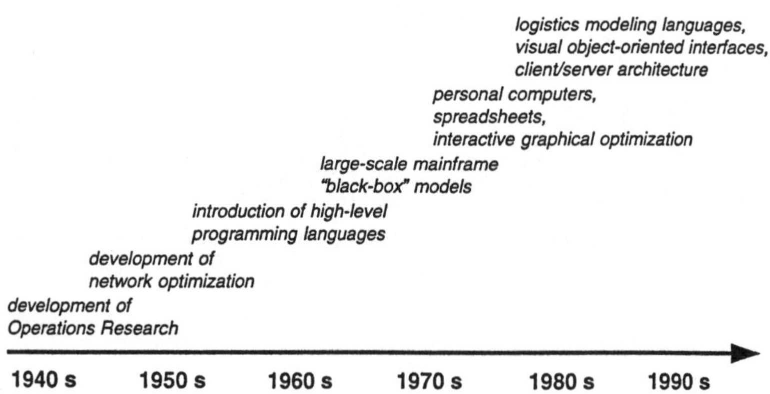

Figure 2.38 Evolution of logistics analysis tools.

general use. Each type of system has inherent problems for use in logistics decision support. Custom systems can be tailored to the unique needs of a business, but generally take a long time to build, are expensive, and are difficult to change as the business needs change. Off-the-shelf systems are less expensive and quicker to implement, but often do not fit the unique logistics issues of an organization.

Combining custom and off-the-shelf benefits

One successful approach for developing flexible and tailored software quickly and cost-effectively features high-level, reusable tools and data objects that can be configured and 'programmed' by business analysts and end users. This type of programmable software system features a **high productivity** language. Spreadsheet and database software applications are two excellent examples of high productivity software languages.

Spreadsheet and database systems are useful in logistics modeling, but a richer architecture is needed as a foundation for LCM. Figure 2.39 illustrates a hierarchical architecture for logistics decision support based on a logistics modeling language. At the lowest level, a low-level programming language such as C++ is used to develop the logistics modeling language, comprised of logistics data objects such as those described in section 2.5, analytical tools and a macro control language.

Next, a 'platform' or set of macro libraries is assembled, providing an almost off-the-shelf software application to a well-defined set of logistics issues (such as vehicle route dispatching, fleet sizing, supply chain facility location, and production planning). Note that a platform can be easily modified or extended to form a unique application (such as an organization's private fleet routing scenario), as platforms are written entirely in the modeling macro language.

Finally, scenario alternatives to a specific application can be represented as distinct projects. Note each layer of the decision support

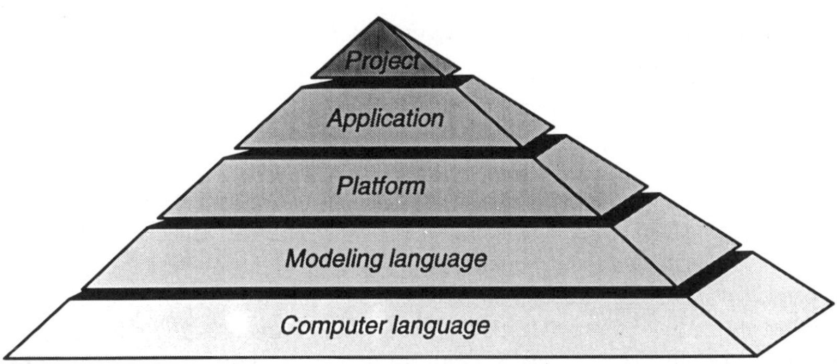

Figure 2.39 Layered architecture for logistics decision support.

architecture becomes more focused towards a specific set of logistics issues, culminating in a well-defined problem and set of logistics decisions.

The advantages of the layered architecture approach are many: a custom software solution is possible; systems are developed quickly; platforms and applications can be changed and enhanced; and software quality is high as reusable tools and data objects comprise the foundation. The first commercial logistics decision support system based on a layered architecture supporting LCM is the CAPS LOGISTICS TOOLKIT® [6], introduced in 1989.

Visual logistics modeling

Contemporary software applications are characterized by highly visual and object-oriented user interfaces, providing a natural conceptual representation of a logistics problem. Data representations are raised to natural and intuitive representations (icons for ports, planes, etc.), and commands are issued through direct manipulation of these visual objects. Figure 2.40 illustrates this type of user interface.

A visual and object-oriented user interface includes galleries or libraries of logistics objects, strategies and model templates. The interface manages the various model representations useful in logistics modeling, including map-based geography, time-based scheduling charts, algebra-based mathematical optimization formulations, row-and-column-based spreadsheets, and table-and-record-based database views.

Modeling-by-example

The user can sketch out a conceptual picture of a logistics problem using this type of user interface in a 'modeling-by-example' style. Lower-level and more procedural details such as model formulation and generation, data connections, data validation, etc. are handled automatically. Object-oriented approaches in particular are more productive because we can efficiently specify objectives, costs and constraints for entire classes of logistics elements.

Client/server computing architecture

Computing architecture has evolved from mainframe computers, to personal computers, to networks of desktop client computers linked by servers to form a **client/server** architecture. This type of computing architecture is ideally suited to LCM. Desktop computers are suitable for highly interactive personal productivity tools such as a logistics decision support system based on visual logistics modeling concepts. Host database servers allow large logistics databases to be shared throughout

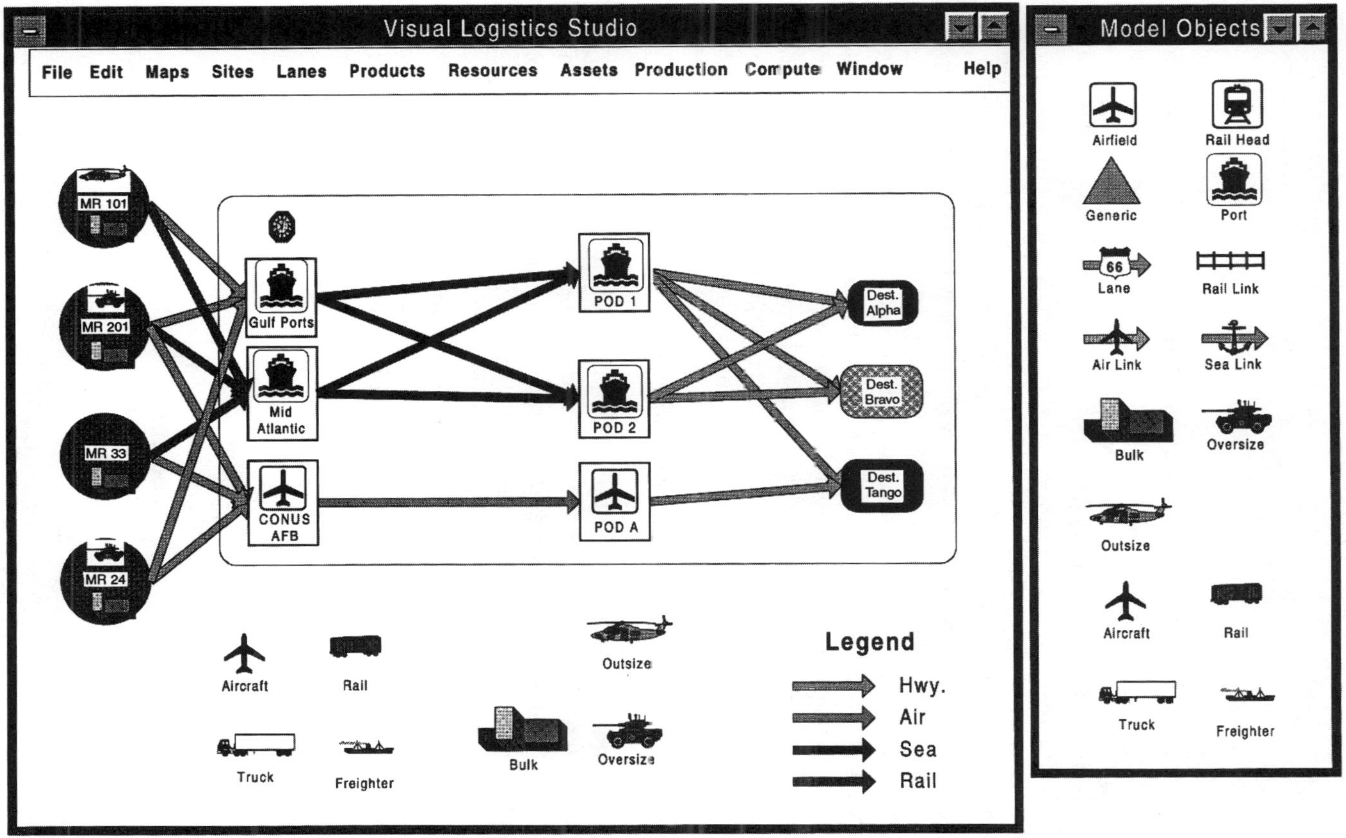

Figure 2.40 Visual logistics modeling interface.

the various business functions forming an enterprise's logistics supply chain.

2.9 SUMMARY

LCM is suitable for logistics business decisions spanning a range from **planning** to **operations**. Operations refers to the actual management and execution of a logistics supply chain. Planning includes all of the analysis and design studies undertaken prior to system operation. Planning also includes an understanding of how the logistics system is expected to operate after system implementation and/or modification, although not in as much detail as required during actual operation.

More time is available in planning, so many decision tradeoffs and alternatives can be evaluated and there is greater opportunity for user interaction with models. This is important, because changing a logistics supply chain is complicated, disruptive, time-consuming and expensive. In contrast, supply chain operations require immediate decisions, so little time is available for generating and testing alternatives. Hence automation is more important here, but pre-planning is still extremely important in order to control and limit the scope of operational decisions. Note planning and analysis activities should continue even after a logistics system becomes operational, in a continuous review and improvement style (Figure 2.41).

Another opportunity to apply LCM is in the integration of supply chains across entire enterprises. Similar to the integration of inbound and distribution logistics within an enterprise, enterprise supply chains such as vendor systems, manufacturing systems, and customer systems can be viewed as interacting processes. Further, logistics customers and service providers are increasingly sharing information about future needs and capabilities, allowing service providers time to anticipate and plan for efficient resource utilization, which creates logistics savings that can be passed back to the logistics customer. LCM provides a common

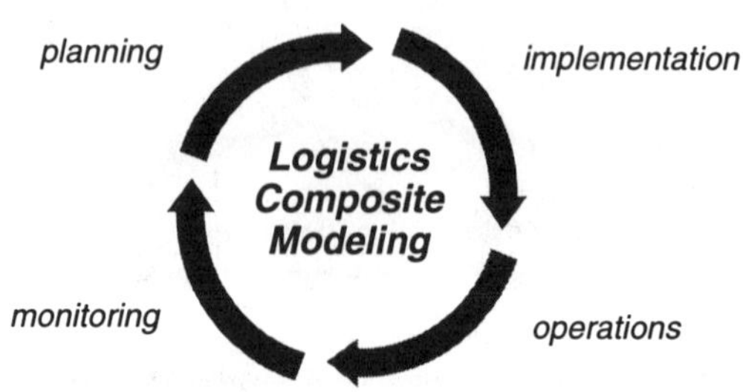

Figure 2.41 Continuous improvement cycle for planning and operations.

foundation of logistics objects and analytical processes, allowing closer integration of data and decision support models across enterprises.

NOTES

1. The United States Council of Logistics Management, Oakbrook, Illinois, defines logistics: 'Logistics is the process of planning, implementing, and controlling the efficient, effective flow and storage of raw materials, in-process inventory, finished goods, services, and related information from point of origin to [point of consumption (including inbound, outbound, internal, and external movements)] for the purpose of conforming to customer requirements.'
2. *Fortune*, November 28, 1994.
3. *Business Week*, January, 1994.
4. Ibid.
5. *The GIS World Sourcebook*, published by GIS World Inc., Fort Collins, Colorado, or *http://www.gisworld.com*.
6. CAPS Logistics, Inc., Atlanta, Georgia, USA, or *http://www.caps.com*.

FURTHER READING

Ballou, R.H. (1987) *Basic Business Logistics*, Prentice-Hall, Englewood Cliffs, NJ.

Foley, J.D. *et al.* (1995) *Computer Graphics: Principles and Practice*, Addison-Wesley, Reading, MA.

Francis, R.L., McGinnis, L.F. and White, J.A. (1992) *Facility Layout and Location*, 2nd edn, Prentice-Hall, Englewood Cliffs, NJ.

Golden, B.L. and Assad, A.A. (eds) (1988) *Vehicle Routing: Methods and Studies*, North-Holland, Amsterdam, Netherlands.

Nemhauser, G.L. and Wolsey, L.A. (1988) *Integer and Combinatorial Optimization*, John Wiley & Sons, New York.

Ratliff, H.D. and Nulty, W.G. (in preparation) *Introduction to Logistics Modeling*.

Rumbaugh, J. *et al.* (1991) *Object-Oriented Modeling and Design*, Prentice-Hall, Englewood Cliffs, NJ.

CHAPTER 3

Methodology for integrated manufacturing planning and control systems design

S.K. Banerjee

ABSTRACT

Manufacturing planning and control systems (MPC) design or redesign in the current dynamic industrial environment is a complex process and must be based on a true understanding of the business needs of a particular enterprise and the physical facilities and resources available to satisfy those needs. Interdomain relationships between various elements, both within and outside the system, must be considered in this context; a necessary requirement for any integrated systems design. Many past efforts in the field seem to lack this approach and as a result changes in the business needs are difficult, if not impossible, to incorporate in the current MPC systems without a major redesign. This chapter outlines the major design issues in this context and proposes a methodology to develop an integrated system's model for implementation. The methodology starts with an analysis of the business needs and considers the coordination and communications requirements between various disparate units and domains of an organization. To aid the design process, the methodology uses many well-known tools and methods in its various stages; their functions and usefulness are discussed. The chapter gives an example to show how the methodology can be used to develop a MPC system in real life.

3.1 INTRODUCTION

Manufacturing industries are experiencing a market place dominated by supply of products; gone are the days of abundance of demand in a relatively stable and predictable environment. Market now calls for rapid changes in product and its technology content, asks for higher quality and reliability, and requires flexibility in order quantity and delivery time. Time-based competition is therefore the new reality and thus time to market has become the current management challenge.

The transition from mass production era to marketing era necessitated a fundamental review of the management procedure and practices that are concerned with organizing, planning and controlling activities of an enterprise. The conclusion of this review (Ansoff and McDonnell, 1990) carried out at a macro level was that to deal with the ever-increasing environmental choices and uncertainties, a more responsive and creative management is required: stable, reactive and even anticipative procedures such as long-range and aggregate planning methods on their own are not able to deal with the current market needs. This conclusion applies both at the business and detail operational levels. It is no longer advisable to specify and/or plan all activities of a system on the anticipation of the future that is based on the extrapolation of historical data. It is widely accepted in the business community that these environmental changes have become more complex and two aspects of these changes, namely the frequency of changes and the rate of diffusion of change, have impacted on the business. The frequency of change refers to the exponential growth since the 1960s in the number of products and services and the new technologies. Diffusion of change, on the other hand, refers to the speed with which new products and services invade the market. When these two aspects are combined with the changing customer demand pattern it is difficult for management to anticipate change, including 'surprises', sufficiently in advance to plan a timely response.

To assist managers with such requirements, manufacturing systems must be designed with the necessary level of flexibility in order to increase the speed of implementation of the required responses. This revolves around the **variability handling capability** of a system. Variability handling capability of a system can be provided by incorporating the appropriate level of resource flexibility in the system by design; it can also be provided through the development and incorporation of rules, procedures and guidelines during systems operation which can reduce the consequences of change during manufacture at the point of impact (Duimering *et al.*, 1992).

The above conclusions imply that the acceleration of change and the resulting environmental turbulence call for a more dynamic approach to both the design and operation of an organization. To avoid long reaction time in business, the relevant tasks for a specific organization have to be clearly understood, defined and managed within the required time frame. The term 'manage' in this context is concerned with two things: first, it is about the frequent monitoring and controlling of a system's activities and its elements against predetermined goals and objectives, based on the true understanding of its business needs; and second, on a periodic basis it is concerned with auditing and evaluating the capability of the physical system and its organization for competitiveness.

The main aims of the chapter are to examine the effects of these issues in manufacturing systems design in general and manufacturing planning and control (MPC) systems in particular. On the basis of industrial research and applications, the objectives are to describe briefly a methodology which can assist design and redesign of systems and to show an exemplar of an MPC system design in this context.

3.2 BACKGROUND

In attempting to become more dynamic in their operations and trying to improve overall approach to efficiency, many organizations have adopted decentralized structures and evolved into disparate and autonomous production departments or units. However, in trying to optimize functional efficiency of a unit one may lose sight of the strategies and goals of the organization as a whole. Coordination of these units is therefore a primary concern.

Advances in technology, systems, methods and tools in the last two or three decades have provided the managers and systems designers with a variety of concepts and solutions in this regard, and terminologies such as FMS (flexible manufacturing systems), CIM (computer-integrated manufacturing), SMED (Single Minute Exchange of Dies – a Japanese concept for reduction of setup time (Robinson, 1990)), GT (group technology – a manufacturing facility organization concept), JIT (just-in-time; a systems design concept), MRPII (manufacturing resource planning – a manufacturing planning and control method; some may consider this as a concept) and *Kanban* (a manufacturing control tool – part of the JIT concept) have become well known within the user community. In many cases, however, applications of these concepts and tools for design of manufacturing systems were found to be lacking in the true understanding of the strengths and weakness of each method and the interrelationships that exist among them in a complex environment. It is true that application of many stand-alone concepts or tools on their own can bring about considerable benefits to a company when used for the first time, but the real competitive strength can only come about when these concepts and technologies are implemented in an integrated manner that achieves a true synergy for the enterprise. To accomplish the required synergy the design process needs to consider interrelationships between relevant domains and their tasks in the system and not rely on specialisms and expertise of a single unit or a function to satisfy the uncertain environmental needs of the company as a whole. Moreover a procedure is needed to evaluate the designed system's effectiveness periodically so that necessary adjustments can be made or considered.

The above assertions are based on a recent industrial survey analysis (FORCAST, 1991) of 400 manufacturing organizations in Europe; this survey was carried out as an investigative part of a European project called FORCAST (Flexible Operational Real-time Control And Scheduling Tools – a Eureka project). The key points that emerge from this survey analysis are:

1. Businesses are often concerned with manufacture of more than one product family and the physical facilities are often organized into cells and/or flow lines (69%); however the overall schedule adherence is found to be no more than 35% at best.
2. When an individual product group's relationships with its market, demand pattern and processing facilities are considered, only about 32% shows a good manufacturing strategy match (Hill, 1993). It was found that in almost all of these cases there is no formal attempt to characterize an enterprise in this regard. Often manufacturing is found to be subjected to a corporate control system that adequately addresses the needs of an organization's principal product, leaving the other product groups at a disadvantage and less competitive than their rivals.
3. The responsiveness and flexibility of the system were identified as business success factors by all organizations but there was very little evidence (15%) of a formal mechanism to ascertain the variability handling capability of a system on a periodic basis.
4. Most companies (70%) are unable to adhere to generated schedules even though the survey portrays a significant investment in MPC. An analysis of this investment shows that the heaviest investment had been made in the area of long- to medium-range planning tools with approximately 85% of sites having sales order processing systems installed. Investment in short-range planning tools was much less, with only about 30% having some form of procedures in place. Evidence suggests that most schedule repairs (77%) are carried out at the shop floor level by the supervisors, based on their own perception or awareness of scheduling problem.
5. Non-availability of material (32%), customer change (18%), and workstation breakdowns (12%) were identified by the users as three principal causes of schedule disturbances; the shop floor supervisors had very few, if any, tools to deal with these situations in the business context. Procedures such as substitution of orders for orders, 'stealing' of components from other orders and overtime to reduce backlogs were found to be widely used at the local level.
6. Even though there is a general recognition (60%) that improvements at the lower-level control systems are needed, only 6% of the sample

thought that a shop floor control system would contribute significantly to the operational goals of the company. The future investment preference in the field supports this finding; approximately 70% appears to be aimed at more of the higher-level systems.

7. Organizations that adhere to schedules well (approximately 30%) in the short term make a significantly greater use of order sequencing and release, shop floor data collection and unplanned deviation analysis.

A further analysis was undertaken using a much more detailed questionnaire that included all aspects of the organization such as the physical system and its business characteristics at six selective sites. The main aim of this analysis was to identify the issues (FORCAST, 1992) that are relevant in this context and that ought to be considered by the systems designers. The results of this study and the available knowledge in the field gathered from literature and in-house research are summarized in section 3.3.

3.3 MANUFACTURING PLANNING AND CONTROL (MPC) SYSTEMS DESIGN ISSUES

Experience has shown that while millions have been spent on manufacturing systems in this regard no real solution to the need of greater responsiveness and flexibility has been found (FORCAST, 1991). Indeed after a period of time has elapsed since the installation of a system, many large and small-to-medium size companies have had great difficulties in generating, implementing and adhering to schedules. Changes in business needs may precipitate such a situation and therefore the companies are looking for better methods or techniques that address these issues but preferably without the complete abandonment of existing resources and technology.

It is widely accepted that all manufacturing systems are systems of organized complexity, not by design but of necessity, as many elements within the system have to be grouped into convenient functioning that suits the requirements of a particular company best. These elements interact with each other as well as within each to develop the system's states. These systems' states, whenever possible, are planned and monitored, and control actions are taken where and when deemed necessary to achieve the desired outcome.

It is probably correct to say that even within the same organization manufacturing activities are viewed differently by different people depending on their world view and perception. Therefore a common understanding is necessary, based on the organization's ethos, culture and business needs. This common understanding can be used for both problem identification and solution. Since a large part of manufacturing systems in general and MPC systems in particular comprise human activity, it is imperative that this common understanding is made known or communicated to all concerned, ranging from shop floor users to management, as they should all play their respective roles to achieve the stated goals or the desired outcomes. The outcome is measured by the 'value' that someone else, either within or outside the organization, puts on the throughput of the system. This throughput's value can be measured by the utility it possesses with cost, efficiency, profitability, price, delivery, service, reliability, perception of inherent quality and so on. Well-known terms such as performance measure and benchmarking are commonly used in this context; but it is important to realize that these are not fixed measures and their relative values depend on the dynamics of the environment.

A prerequisite for such systems development is therefore to understand the organization's objectives clearly so that a particular architecture of manufacturing can be established. This known architecture can be monitored and audited periodically against the prevailing 'measures' that are determined in the main by the environment. Establishing such an architecture is no mean task, as it involves true understanding of many manufacturing dimensions such as the company's products, markets, investment, cost, infrastructure, competition, etc. Technology solutions and computer-aided systems and methods are not panaceas in this respect; isolated and incremental applications of these may have been, for some organizations, instrumental in their uncompetitive positions in the market place. Many companies in the past have knowingly overextended themselves in a single dimension only to realize that, at best, their overall competitive position has not improved. Evidence of this state of affairs also can be found in many current solutions such as TQM (Dar-El, 1993) and Business Process Re-engineering (Hall *et al.*, 1993).

The way in which a business decides to organize itself to meet customer needs should not be based upon a single dimension; this may create a situation in which important manufacturing and business perspectives may be lost or at best given inadequate attention. Unfortunately, all too often one sees such examples both in academic research (Carrie and McIntosh, 1992) and industrial systems design (Higgins and Browne, 1990). Informed insights are needed to understand the complexity of business and strategic directions that are required in order to succeed. The manufacturing architecture of a company can provide such insights, as it can be used to visualize relationships between functions and processes; it is also useful for identifying conflicts within a business. This architecture can also assist with the resolutions of possible conflicts and differences found in the system.

There is a vast number of manufacturing architectures, configured to meet a wide range of market or customer needs, which are almost impossible or even dangerous to standardize (Will *et al.*, 1992). Attempts made in the past to classify these, such as MTO (make to order – products are manufactured against a firm customer order), MTS (make to stock – products are made for stock), CTO (customized to order – products are manufactured according to a particular customer specification) and so on (Ingham, 1971, Miles and Snow, 1978) have been useful in systems design. However, in today's environment, more often than not industries are having to cope with a divergence of demand patterns in the market place and thus having to provide for a combination of these in their systems; this reason alone makes it impossible to develop a standard architecture to be used in all occasions. What is therefore required is a set of references and perhaps partial architectures or models of systems from which a particular architecture may be derived.

Many of the recent efforts in this respect unfortunately appear to concentrate on developing such a reference architecture (Kochhar, 1992) from a single domain perspective and appear to assume that there is very little or no design impact or implication from interdependent areas within the system. This process may in most cases result in confusion and impairs the effectiveness of the system in the long run. For example, the MPC systems design must take due cognizance of the physical systems architecture; the support systems design such as information and organization should follow this process. It is necessary that these interrelated domain design processes follow a logical pattern and that adequate mapping and/or matching between them is carried out, perhaps iteratively, to ensure their compatibility in real world situations. The process of designing such a system should therefore be based on a clear understanding of the interdependencies that exist among functional groups or domains within the organization or enterprise. This should result in a specific domain architecture that exhibits appropriate interconnectivity, coordination and communication with other business practices, functions and processes that fulfil a predetermined set of domain objectives.

Enterprise modelling and integration research groups recognize these issues (Gorranson, 1992) and advocate and suggest design methods that encompass more than one dimension of an enterprise. However, suggested models and methods are at various stages of development and use and are not easily understood or usable by the practitioners in industry at present (Williams, 1994).

3.4 TOWARDS AN INTEGRATED METHODOLOGY

On the basis of the above a systems design methodology should therefore critically examine the operating environment of the company as a whole. This involves two primary tasks: an analysis of externally imposed requirements on the business by the chosen market place and an internal appraisal of what is available or exists to fulfil these requirements. By addressing these strategic issues a requirement's specification for a particular company can be established. Such a business specification can then be used as a primary input to the MPC system's design. This specification can also be used to audit the existing physical systems design and/or structure of the company. In some circumstances it can trigger a redesign of the physical system; in that event the MPC systems design should await such outcome before going further. The relationship between the business specification and the consequent imposed technical and functional requirements on both the physical and the MPC systems design is shown schematically in Fig. 3.1. An interdependency exists between the MPC system, the manufacturing process and the manufacturing task (Kieffer and Thiel, 1992) and needs to be carefully considered before design choices can be made. Moreover, these three areas must be seen as under constant threat of change due to new market requirements, new strategic goals in business and implementation of new processing technology: any of these changes can mandate a change in the MPC systems design.

It follows, therefore, that any MPC system's design methodology should incorporate:

- a means for matching the MPC systems design with the ongoing needs of a company's market, the relevant manufacturing tasks and the manufacturing process architecture or layout, and
- a means for the realization of an integration path between these interacting domains; this is required to ensure that the effect of any changes can be assessed and, if necessary, the relevant system or its elements can be amended.

Furthermore, as was indicated above, the requirements are not static and the adjustment of the system is an ongoing process; systems modularity that will facilitate dynamic reconfigurability is a desirable feature. The methodology should therefore provide the means for configuring the MPC system in such a way that it will be possible to change, adjust and/or restructure, with relative ease, a component(s) of the system for the required behaviour without interfering with the entire system or its functionality.

In addition, because of the magnitude of the investment in MPC systems and the time required to implement any change, the methodology should provide a transition route, as shown in Fig. 3.1, by recognizing the differences between the existing and the desired systems. This proposition supports the argument that in most real life applications implementation of a design solution for an integrated MPC system will be an evolution of the existing system. This requirement also indicates that the methodology should have:

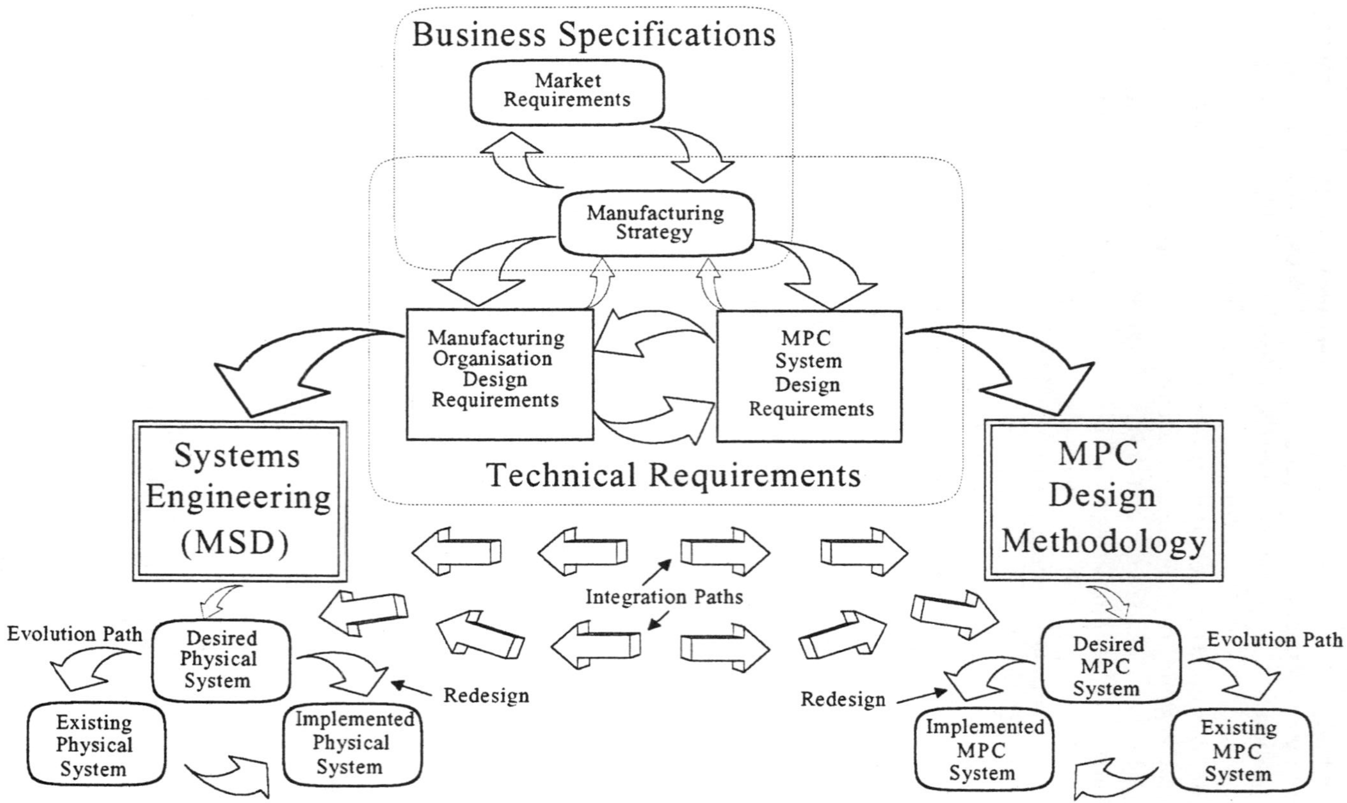

Figure 3.1 MPC design methodology.

- a means for constructing a conceptual model of a desired system, and
- a process for the selection of modelling tools and techniques to facilitate the representation of both the existing system and the desired system.

Finally, what emerges from the above is that it would be highly desirable if the methodology could have some provision for expressing systems requirements in a formal specification language, which can be used as a means for selecting appropriate decision support tools for a particular system and its unique business needs.

3.5 METHODOLOGY DEVELOPMENT

The feasibility of such an approach was investigated and its viability was confirmed in a recent collaborative project (FORCAST, 1992). The conclusions confirmed that dynamic niche application tools are commercially viable but a formalized method of implementation is desirable for linking the enabling technology modules within a design framework. Based on the required features outlined above, a methodology has been developed and has been tested in several industrial situations and found to have the necessary robustness for real life application. It consists of several stages as described below. Figure 3.2 shows the flow diagram of the methodology.

3.5.1 Diagnostic stage

The starting point of the methodology is the business specification or the business analysis process. This fundamental analysis process reviews the current and intended customer and market segments targeted by the enterprise, their present and projected needs with regard to the company's products and services, assessment of competition, etc. This stage also reviews the manufacturing task statement in terms of volume, delivery flexibility, critical product cost and quality, and other manufacturing dimensions in terms of consistency with the adopted marketing strategy. An extensive questionnaire, tested and refined, has been developed to perform this analysis and has been used to survey several enterprises. The output of this stage gives a unique identity to the enterprise and provides a firm basis for the definition of technical requirements for the manufacturing processing organization and MPC systems design. In many respects this stage could be compared with a typical QFD (quality function deployment) (Jacobs and Kethers, 1994) model of a company, the main goal of which is to ensure that ambiguities and misinterpretations do not become a part of the requirements that are produced. In this phase it is essential that the level

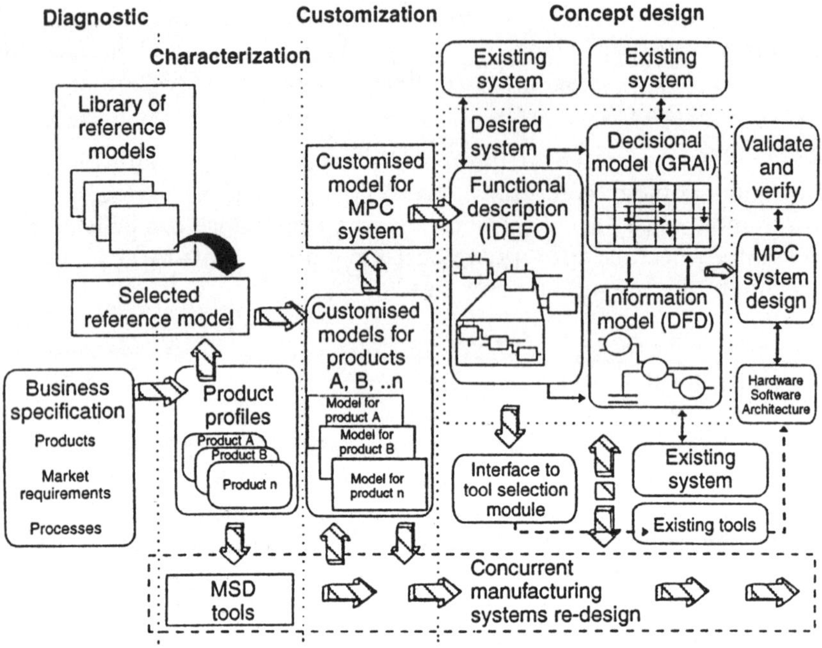

Figure 3.2 Methodology flow chart.

of communication between the requirements' developers and the company management is high.

3.5.2 Characterization stage

Following the business analysis, it is considered essential to pursue a much more detailed characterization of the company in the form of product profiles. The objective of this stage is to produce a minimum set of variables, in the main quantifiable, which will serve as an input to the selection of features and activities required by the MPC system for a particular company.

The idea here is to identify, from the information, natural groupings of products based on similar markets, product and/or process characteristics. Natural groupings form a basis for both manufacturing processing and MPC systems design and therefore can assist with the task of auditing the physical system before proceeding with the MPC systems design. In extreme cases this may trigger a redesign of the physical system and in that case MPC systems design must await such outcome to ensure overall systems conformance. Within the MPC domain boundary, planning and control decisions can be categorized by distinct aspects or phases. It can be represented by a closed hierarchical structure with varying scope of control and time horizon (Figs. 3.3 and 3.4). The framework shown in Fig. 3.3 can be considered as a generic representation of an MPC system in the sense that although each component should be accomplished by any company, it does not capture the emphasis or importance each may have in a particular company. Figure 3.4 shows the time relationship between various levels of planning. In a hierarchical planning system each level has a fixed planning period (*P*) which may be divided into a number of review periods (*T*). The time relationship between different level is important as shown in the figure. The planning horizon (*H*) of a lower level is equal to one fixed planning period of the immediate higher level; review periods within a planning period are used to monitor the system's status against a predetermined plan of that level so that appropriate control actions can be taken, if deemed necessary.

It was realized that the creation of conceptual models for the identified product profiles could be assisted and accelerated considerably by employing the concept of reference models or, if available, a library of relevant partial models. Reference and partial models can be used as templates and may assist the designers of systems in the identification of tasks and activities that are appropriate and relevant for the particular system.

3.5.3 Reference models for MPC

The notion of reference models can be found in many scientific domains; they serve as a catalyst in the requirements definition phase. Reference models are general, type of trade or type of industry oriented models, which conform to the requirements of many potential users. Reference models can be adapted to the specific situation by examining the variability of the particular case – a process that could be termed as customization. A reference model that describes a complex system, such as MPC, can be thought of as a configuration of components; each component executes its own, globally defined, distinct tasks but interacts to realize the functionality of the system as a whole.

CIM–OSA (Computer Integrated Manufacture–Open Systems Architecture), developed by the AMICE corporation as part of ESPRIT (CIM–OSA, 1991), advocates such a process but so far has not quite come up with simple and easily understood general tools or utilities for the community of users. It was therefore decided to decompose each component within the hierarchy (Fig. 3.3) into a set of primary activities and groups of related tasks. This nested structure of activities allows the customization of the reference model for each product profile. An example of this activity decomposition, for the component of phase one 'demand management', is shown in Fig. 3.5. Figure 3.5 represents the long-, medium- and short-range planning activities of Fig. 3.3 as creation

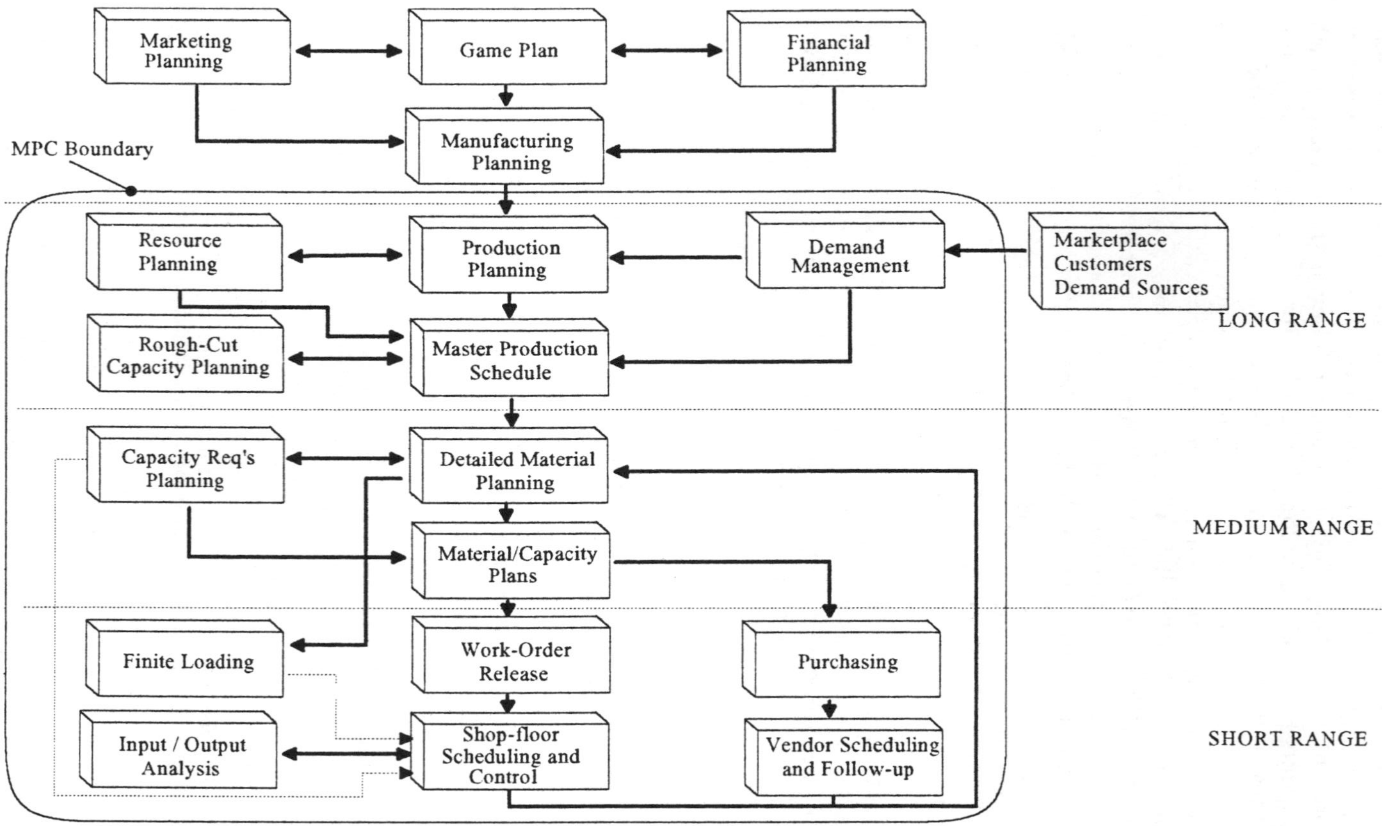

Figure 3.3 Generic representation of an MPC system.

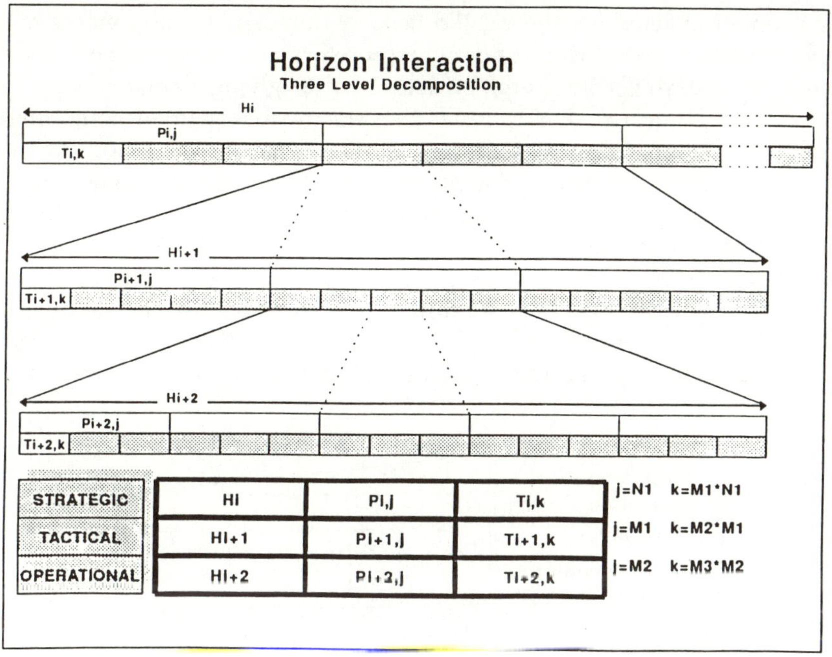

Figure 3.4 Hierarchical relationships between MPC system phases.

of overall manufacturing (phase 1), material and capacity (phase 2) and execution (phase 3) plans. Any of the five planning components of phase 1 can be decomposed into a set of further activities as shown in Fig. 3.5. Reference models for different industry sectors and/or different types of companies can be developed by using this method. Once available designers can identify specific activities, from an appropriate reference model, that are relevant for a particular company's operations; this is part of the customization process.

3.5.4 Customization

The output of this stage is an optimized MPC model. To create this, for each product profile a reference model is used to develop a customized MPC model according to the requirements as identified by the characterization stage and within the rules and constraints that result from the variability analysis. The result of this process could be a number of conceptual MPC systems configurations for a single enterprise, each directly related to the requirements of a specific product and its market profile. However, for feasible implementation, an optimized version or solution is needed and can be obtained by combining these configurations into a single aggregated version. The basic premise is that a number of activities and related tasks within certain components may be requested more than once and therefore can be combined under appropriate MPC component structures, thus giving a single model for the whole enterprise. For example, the demand management component of phase 1 (Fig. 3.5) may identify a similar forecasting method for two different make-to-order product groups. Interpretation may play a significant role here and therefore it is suggested that a manufacturing knowledge base may be used (Kochhar, 1992) or created to assist the designer in this work. The customization process is summarized in Fig. 3.6. The designer may have a number of options that can be adopted to execute the global tasks of the three MPC system phases; this could be a pragmatic process assisted by a knowledge base of local constraints.

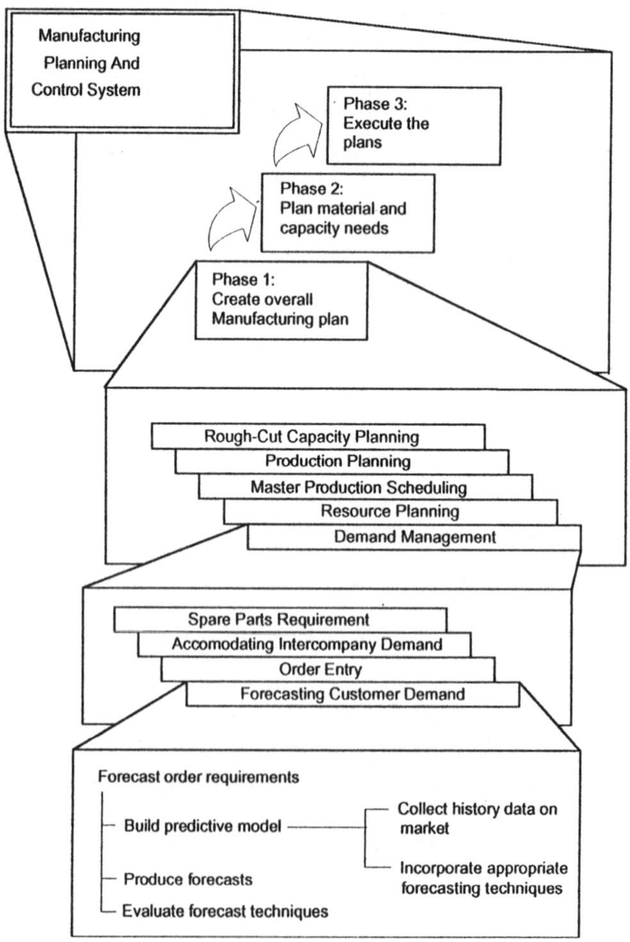

Figure 3.5 Example MPC system configuration.

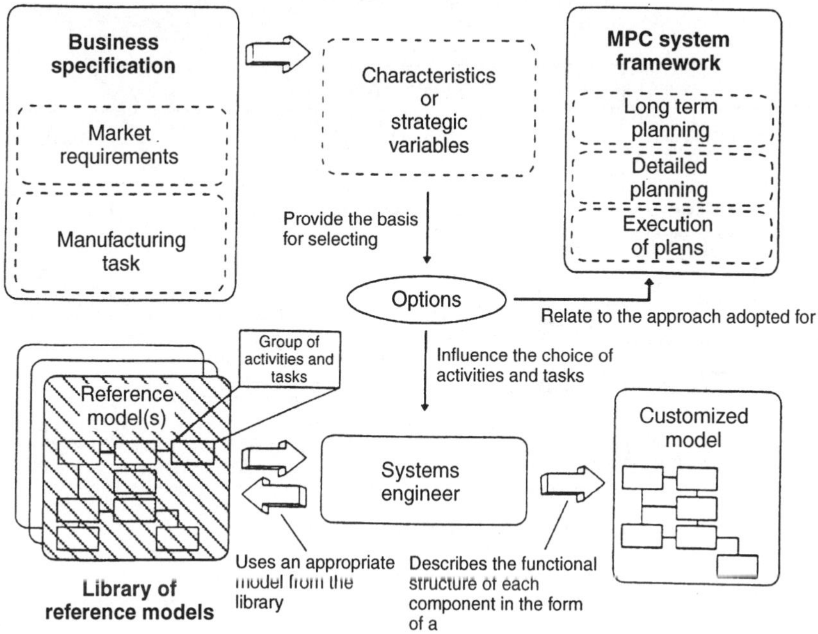

Figure 3.6 Matching requirements to design options.

3.5.5 Concept design

Having established the optimized version of the required MPC systems configuration for all products, a number of tools can be employed to provide representations and descriptions of different domains as shown in Fig. 3.7; the employment of representational tools will facilitate implementation. The overall process is described in the following section and is based on the results of an industrial application of the methodology. The following tools and methods have been used to develop both MPC decision making and information support domains.

1. The IDEF0[1] structured method has been used to develop functional models; these models are developed from the descriptions of the tasks and activities of the optimized conceptual model of the MPC system. IDEF0 is a structured design method and experience has shown (Banerjee, 1987; Banerjee and Al-Maliki, 1988) that it provides a sound basis for developing a functional architecture of systems and

[1] ICAM (Integrated Computer-Aided Manufacture) Definition method Zero. There are several ICAM Definition methods, developed under a US AirForce sponsored programme, which can be used to produce manufacturing, information and other architecture of systems.

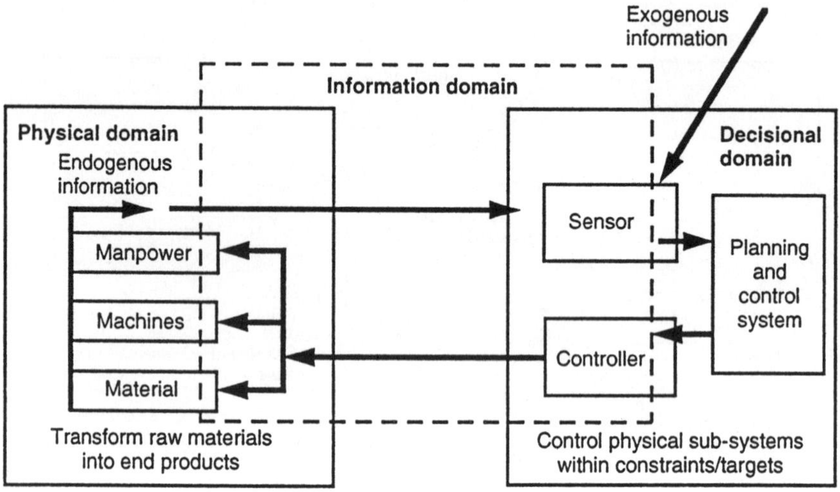

Figure 3.7 Representations of different domains.

domains which can be used as a basis for developing other domain models.

2. Information models and databases can be developed by initially converting IDEF0 models into DFD (data flow diagram) models and then using an appropriate CASE (computer-aided systems engineering) tool (Zhou, 1993). The DFD models are required for information systems design.
3. An MPC decision model, including its hierarchy and time horizon relationship, can be established by using the GRAI method (Roboam *et al.*, 1989). GRAI (Groupe de Recherche en Automatisation Integree) is a methodology developed at the University of Bordeaux, France. IDEF0 functional models provide a logical platform to develop a time horizon decision grid, the GRAI grid, that shows clearly the relationships between various decisions of an MPC system.

Figure 3.8 shows the possible connections between the representational tools. IDEF0 conversion to DFD can be done with ease by adding external entities, namely sources and destinations, and data stores to each model by extracting information from glossary and functional notes of the model (ICAM, 1981). In this development work the outlined IDEF0–DFD conversion process has been used very successfully and an example of this is shown in Figs. 3.9 and 3.10. GRAI grid development from IDEF0 models requires care as some understanding and interpretation of hierarchical functions, their relationships and respective time horizons, are necessary. It is suggested that the top-row decision functions are extracted from the top-level IDEF0 models and other rows and column details are determined from the other functional models. There is not an automatic process to link these tools at present that can facilitate this work.

3.6 MPC SYSTEMS DEVELOPMENT

The outlined methodology has been applied in several industrial organizations for the development of integrated MPC systems. It has been observed that many manufacturing systems are concerned with the production of both components and assemblies for more than one product group. The physical facilities of these systems are designed and organized according to individual business needs to provide the most efficient and effective manufacture of end products. As discussed earlier, MPC systems are 'dependent' on the organization of production facilities and therefore, before a MPC system can be designed, one has to study the particular physical organization of the production facilities of the system. For instance, a cellular system of component manufacture has long been established as an efficient method of component production, but the effectiveness of the system as a whole depends on how well the assembly of end products is carried out to satisfy the environmental needs. This can often lead to conflicts of objectives between the component and assembly production units and the designer must resolve these issues satisfactorily at the customization

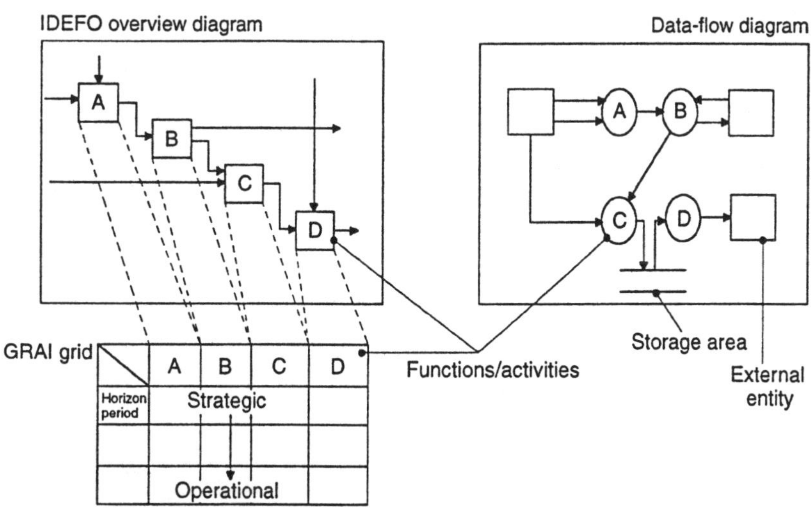

Figure 3.8 Integrating different representational tools.

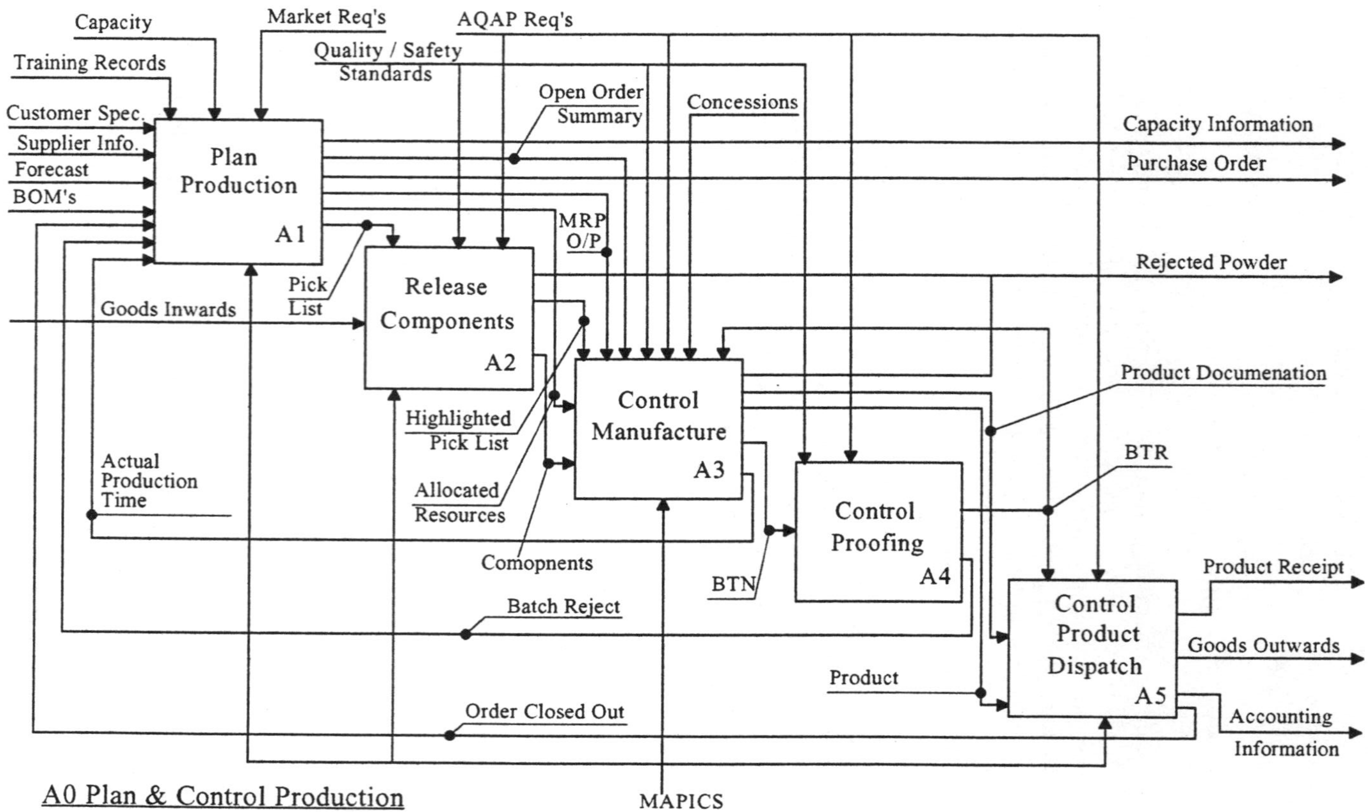

Figure 3.9 A0 plan and control production.

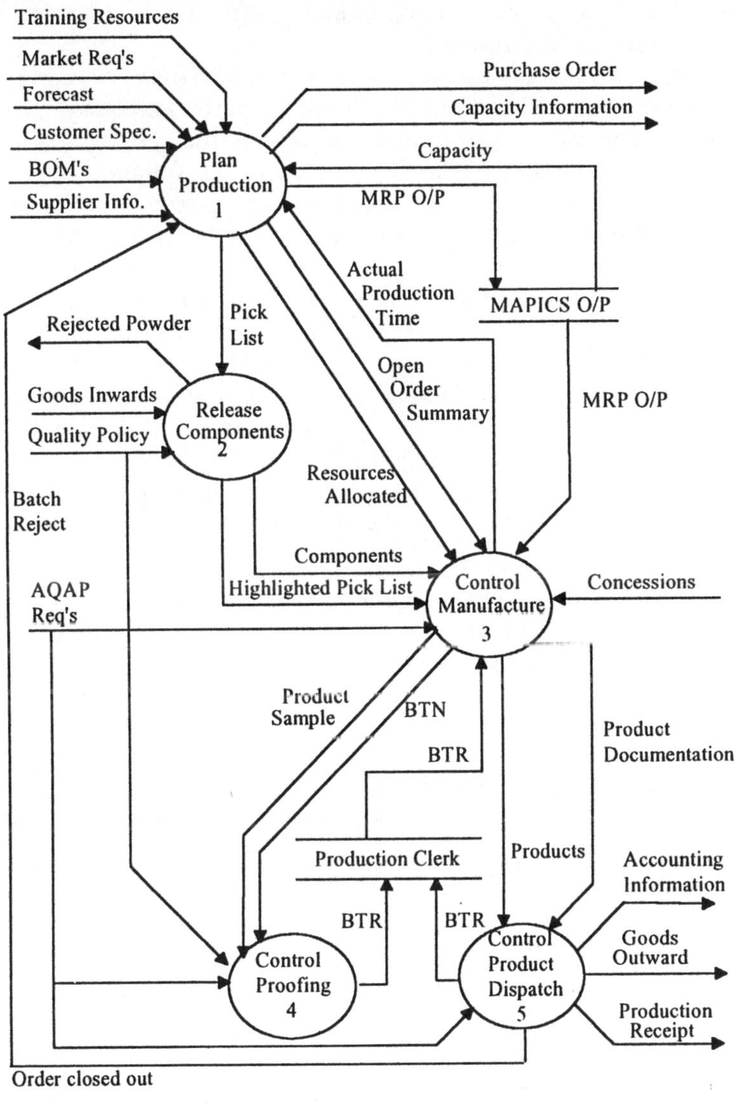

Figure 3.10 IDEF0–DFD plan and control production.

stage. Therefore several conditions may need to be considered when an optimized MPC system's configuration is developed and these include:

- A manufacturing system may consist of one or more component production cells and all operations on a particular component may be performed in one cell.

- It is possible to produce a component in more than one cell with varying degrees of efficiency.
- Each cell should be relatively independent when planning and controlling the production of allocated parts and may provide minimum communication with other parts of the system.
- Production of parts or assemblies in a particular cell are not product group dependent; in other words, parts and assemblies are produced in the most efficient cell. Cells are not identified with any one product group.
- All activity planning tasks and resources are within the control of each area of operation, i.e. within the manufacturing unit or cell.
- Supply of parts for assembly of end products may come from more than one component production cell and from different vendors and therefore there is a need for material flow control and coordination.
- All component production requirements will be driven by assembly schedules and therefore an assembly cell can make demands on the component production cells in the short term for maximum effectiveness of the system.

These conditions, as found from the analyses, must be considered carefully in order that the conflicts between the component and assembly production areas may be resolved. The main aim of the MPC system is to plan and control manufacturing activities, within the constraints of the system, in such a way that products can be delivered on time in the most cost-effective manner. The customization process is concerned with this aspect of the design process. This approach is free of any influence from methods such as MRP and JIT and is based on the actual needs of the system as determined from its business analysis. However the above, in all the cases examined, points to a hierarchical and decentralized control structure. It is not a new concept but the determination of the hierarchy and the relationship among various production units are based on a very different set of parameters and analysis process, as discussed above, and are not influenced by any particular method or theory.

3.6.1 Control system

The next stage is to resolve two other issues, namely the control structure and the decision-making mechanism (procedure). The control structure of the system is concerned with decision-making tasks and functions for various time horizons, their interrelationships and position in the hierarchy. The GRAI (Roboam *et al.*, 1989) method is used here to develop a decision grid and it is based on a functional analysis (IDEF0) of the system (Banerjee and Kourouklis, 1993) as shown at the concept design stage of the methodology (Fig. 3.2). The GRAI grid is a very convenient method for resolving conflicts and at the same time it is a valuable aid in establishing a common understanding of the required control structure. In one case example it was found that a particular assembly unit was responsible for meeting two different types of market demand for the same product group, namely make-to-order (MTO) and assemble-to-order. Almost all orders for the assemble-to-order market are made from standardized parts held in the system as intermediate stocks and wherever possible standard parts are also used for the MTO market. Figure 3.11 shows a conceptual model of this situation. The decision-making centre Control Operations in the Control Manufacture column is responsible for these activities on a weekly basis. A single arrow represents a major information flow between decision centres and a double arrow indicates a decision frame between two or more decision centres. A decision frame represents a group of decision centres that must all participate in completing a decision. Within a frame the partly processed results from one decision centre must be sent to another decision centre for completion.

The mechanism, on the other hand, concerns the specification of the internal procedures, methods and rules to be used to carry out the tasks as defined by the control structure. Figure 3.12 shows a simplified model of an MPC system, derived from functional analysis of a system and its GRAI grid. As the figure shows, the physical facilities for the

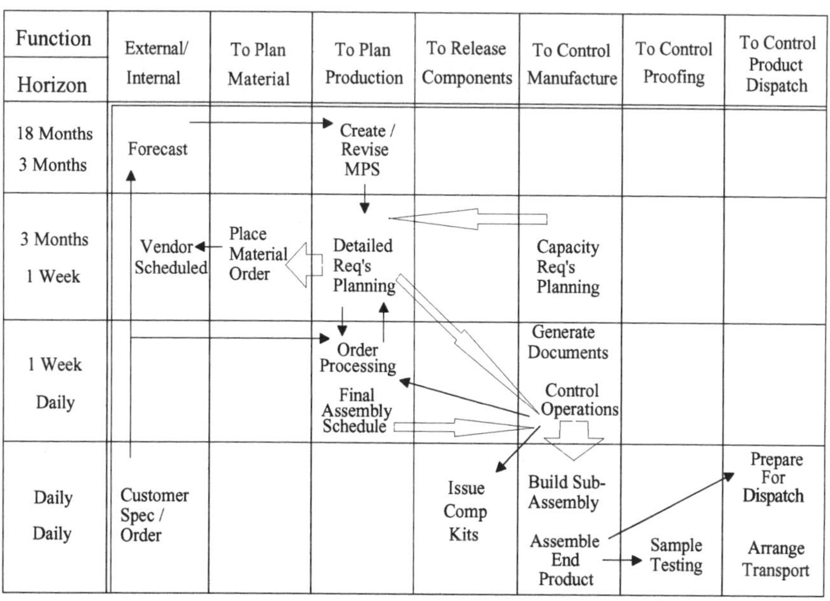

Figure 3.11 Conceptual system – GRAI grid.

manufacture of products in this case consist of several component and assembly production cells. The decision-making mechanisms used are also indicated within brackets in each block. These mechanisms are either standard software packages or specially developed programs that suit the decision-making needs of the system. The ability of the system to carry out effective decision-making often depends on how well these decision-making responsibilities are allocated and coordinated. The control structure should not only identify all control tasks or functions, each with a limited set of planning and control responsibilities, but also must clearly specify their interrelationships. The mechanism required to coordinate the execution of various control functions within the structure can then be specified with relative ease. However, the mechanisms needed to perform the tasks of each control function satisfactorily can be delegated to the owners of that function: how to do something can be best determined locally rather than be recommended from elsewhere. For example, one of the responsibilities of the Cell

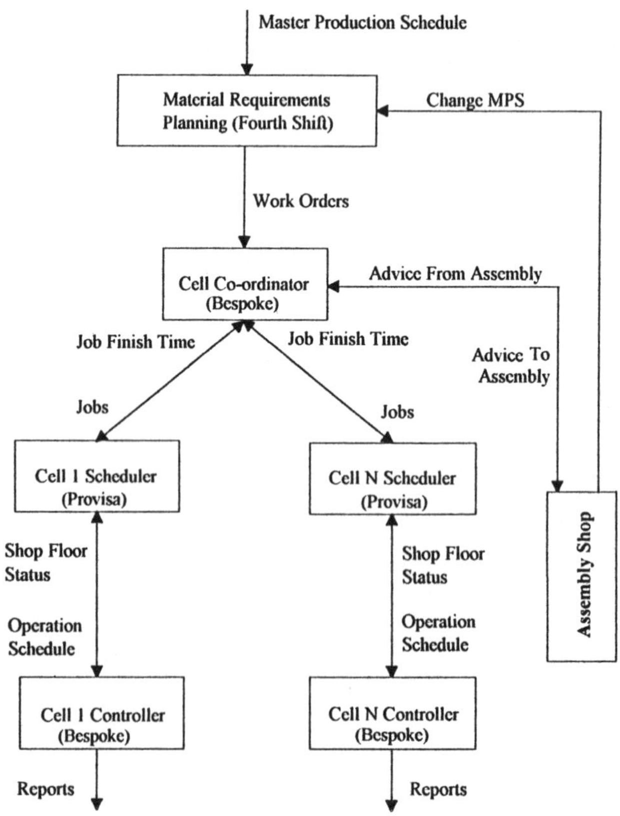

Figure 3.12 Simplified model of an MPC system.

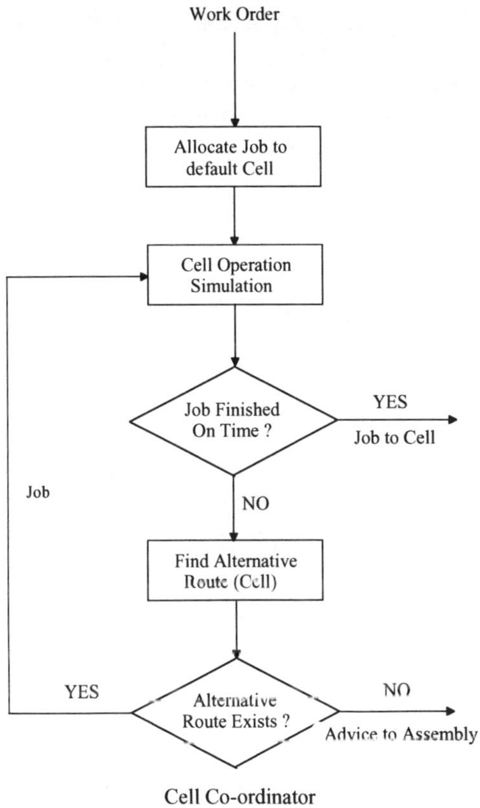

Figure 3.13 Interaction between units.

Coordinator in Fig. 3.12 is to select an appropriate cell to manufacture an order for a part. To do this, the Cell Coordinator may have an internal mechanism that assists the function of identifying the appropriate cell; this internal mechanism may be a decision rule. On the other hand, the selected cell for the part order may decide, using its own mechanism, how and when it is able to satisfy the demand and feed this information back to the Coordinator; here, the Cell Coordinator is concerned with the material flow control between the units. The Cell Coordinator evaluates the finish time and decides whether to consider an alternative cell for manufacture of the order to meet the end-product assembly requirements. If none of the component production cells is able to deliver the order on time for the assembly of the end product, the Cell Coordinator will inform the Assembly shop about the delay. The interactions between these units are shown in Fig. 3.13. The Assembly shop or the Coordinator for the area may reschedule its activities accordingly within the planning period. Since all these decision-making tasks are carried out within the specified time horizon, local adjustments are possible in order to achieve best schedule adherence. This structure has evolved as a result of the customization process and in this type of situation ensures the most effective material flow and manufacturing unit control. In an extreme case of delay the overall plan for the assembly of a product may have to be changed, as indicated in Fig. 3.12, by the feedback loop 'change MPS' from the Assembly shop.

Once a 'job' is inside a cell, it is the responsibility of the cell to plan and control the movement of the job. Every time a cell receives a part order, it reschedules all its operations by considering all factors including the prevailing constraints of the cell. The local scheduler is ideally placed to consider all options and optimize its operations, rather than being given a start time by some other part of the control system. The rescheduling frequency could be an hour, a shift or a day depending on the actual situation. The purpose of the Shop Floor Control (SFC) system is to execute the operation schedule of the Cell Scheduler and monitor the unit status including its work-in-progress to ensure adherence to schedules. It collects and feeds back status information to the Cell Scheduler. It also accumulates relevant statistical data for machines and overall cell performances which may be required for other management decisions such as replacement of a machine and acquisition of additional capacity. It is the responsibility of the SFC system to react to any unexpected disturbances to production in real time. This may be accomplished by adjusting the operations schedule using dispatching strategies or decision rules. The SFC and the cell scheduler together form a scheduling strategy: the Cell Scheduler schedules jobs to individual machines; the local dispatching methods, which are embedded in the SFC system, react to a real situation by adjusting the job queues of a machine.

The above process is free from any influences imposed by many proprietary methods such as MRP systems and it rejects the artificial lead time of manufacture – a management parameter that does not represent reality, used for calculating order release time for jobs. In this process component production cells are given the finish time of the job which is required for end-product assembly.

The coordination mechanisms of various functions are determined by the control structure which is entirely based on the physical organization of manufacturing facilities and the known business objectives of the company. A hierarchical control structure, which decomposes the whole complex plan and control tasks into a number of interconnected smaller ones, appears to satisfy the needs of this type of manufacturing organization. The relationship between two levels in this control structure is a cooperative one and not a master and slave relationship. In a master–slave relationship all the slave's activities are dictated by the master level and the slave has no other option but to comply with the order release data rather than the due date. In this case the cell coordinator is concerned with whether an order could be finished on time by a cell or not, rather than with the release of an order to the cell with a fixed start date. The lead time for the manufacture of an order in the cell in question will depend on its workload and resource availability during a planning period. It is therefore up to the cell to decide, based on a true assessment of its status, when an order could be completed. The coordinator can then decide whether to assign the order to the cell or look for an alternative option such as another cell, by comparing the cell order delivery time and the due date for the assembly shop. On the other side, the assembly activities are planned according to the end-product requirements, but constrained by the availability of components and purchased parts. Effective coordination between the assembly shop and the component production areas through one or more coordinators at this level is therefore very important in order to achieve an acceptable schedule adherence. Through the coordinators it is possible for the assembly shop(s) to establish the availability of parts for production. This helps the local decision makers to adjust their relevant schedules or manufacturing sequences within the appropriate time frame. Effective communication at this level is important to avoid schedule disturbances in order to meet the end-product delivery dates. Any delays in receipt or production of a part can be communicated and the relevant decision made locally without having to await a decision from a higher-level control function such as a centralized MRP system.

The main function of the coordinators is to monitor each other's requirements continuously in terms of end-product demands and to take appropriate measures accordingly within a defined time period or horizon. The objectives are to avoid shortfalls and control the work-in-progress inventory by being more responsive to the actual needs at this level of operation; these tasks are to be performed within the guidelines or schedules as determined at a higher-level control function. It is recommended that tasks such as local schedules, process planning, quality and shop floor control etc. are undertaken locally for maximum efficiency and effectiveness of the system.

3.6.2 Decision-making mechanism

The method of design as outlined above has provided functional models and a decision-making structure appropriate to the needs of the company. Having defined the control tasks and structure, mechanisms are required to support each decision-making task. An information support system is also required to provide relevant data to be used in the mechanisms. A number of proprietary software tools were investigated as mechanisms to support these decisions. This is a difficult task, as the support needs of each decision have to be matched with the available services provided by the commercial packages; the cost of software is also a consideration. In this example, Fourth Shift MRP and PROVISA scheduling packages, marketed by Fourth Shift Corporation and AT&T Istel respectively, were chosen for certain tasks as shown in Fig. 3.12. However, functions such as coordination and shop floor control did not seem to exist in most of the packages examined and therefore needed to be developed to support all elements of the control structure; bespoke programs were therefore written and interfaced with the selected commercial packages. An example of the structure of the decentralized control system for component production is shown in Fig. 3.14. The hierarchical control structure, its functions at different levels and the type of data required for each of these functions are shown. Depending on the complexity of the control structure it is possible to consider manual methods or procedures as mechanisms for tasks that cannot be supported by commercial packages.

The next stage is to develop an integrated database for the mechanisms of the control system, as indicated in the methodology. This part of the design is in a relatively well-developed state and many methods and systems can be used. In this case the design was carried out using a proprietary CASE tool and a relational database, namely TEAMWORK and ORACLE, marketed by HP and Oracle Corporation respectively. The following steps summarize the processes used at this stage:

- Translate functional models (IDEF0) into DFDs. This could be done simply by adding elements such as sources, destinations and storages to the IDEF0 diagram as outlined in section 3.4.5. Fig. 3.15 shows an example of this for the Cell Coordinator model, in which inputs (*I* and C) and outputs (*O*) of the model represents the sources and destinations. *M* represents mechanisms required to perform a function, an input from within the system.
- By using a CASE tool, an entity–relationship diagram (ERD) can be developed. The ERD is a structured conceptual schema tool (Martin, 1988) and is the heart of the database. An ERD model consists of entities and relationships between entities. It also shows attributes of each entity and keys; the values of attributes or a combination of attributes and their relevant keys can be used to uniquely identify every entity instance of an entity type. Figure 3.16 shows an ERD of the Cell Coordinator with a list of its entities and their attributes. Once developed, an ERD can be mapped into any database management system (Howe, 1983). The ORACLE database management system was used in this case.
- Structure charts can be developed for program control mechanisms of logic. A structure chart (SC) is a tree or hierarchical diagram that defines the overall architecture of a program by showing the program

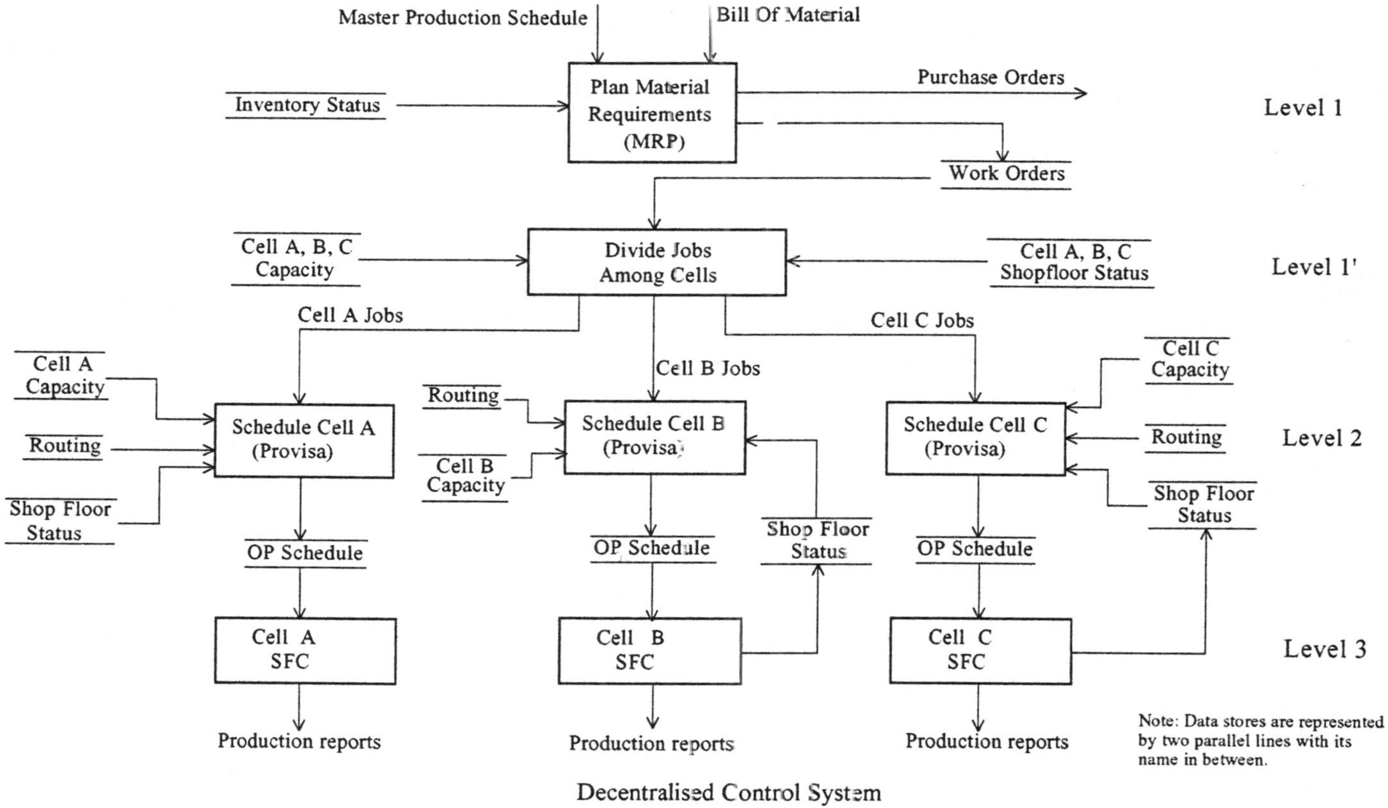

Figure 3.14 Decentralized control system.

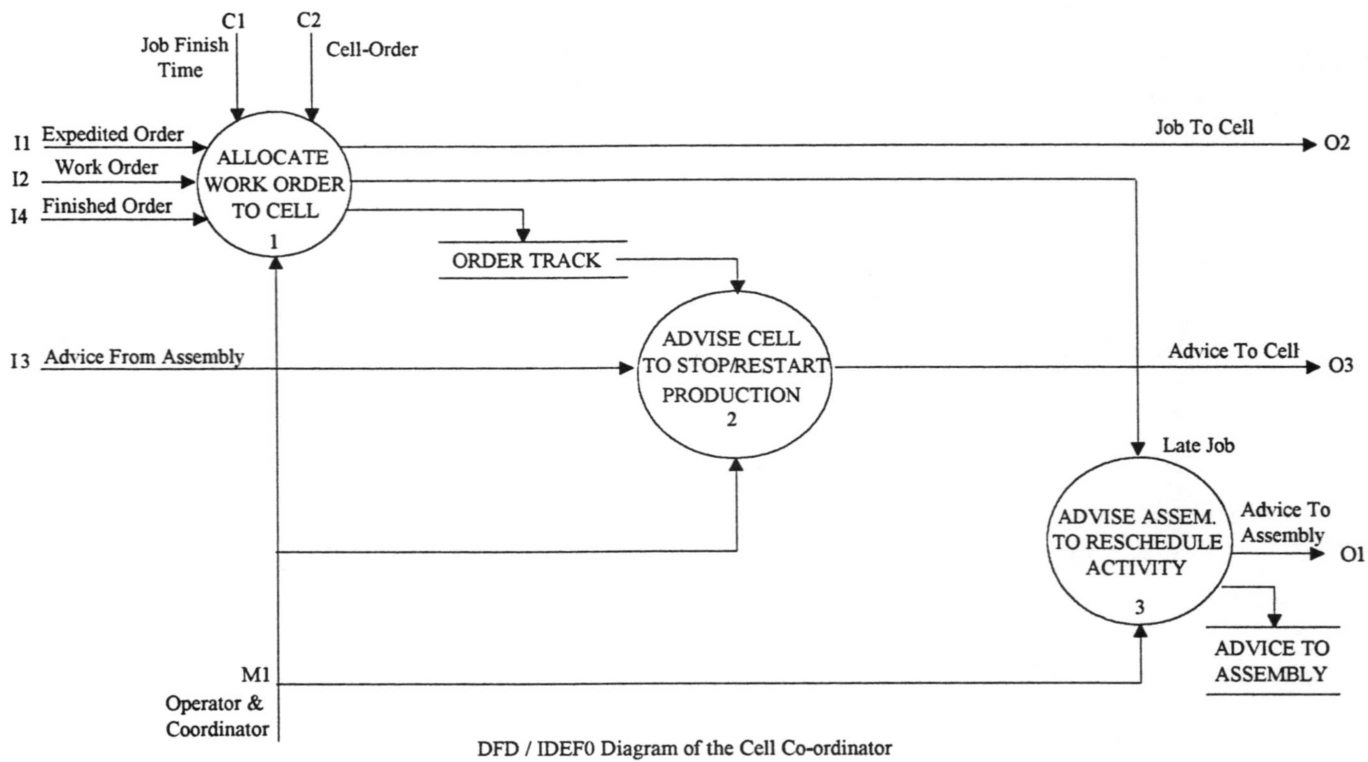

Figure 3.15 DFD/IDEF0 diagram of the Cell Coordinator.

Entity Relationship Diagram and entity attributes of Cell Coordinator Data Model

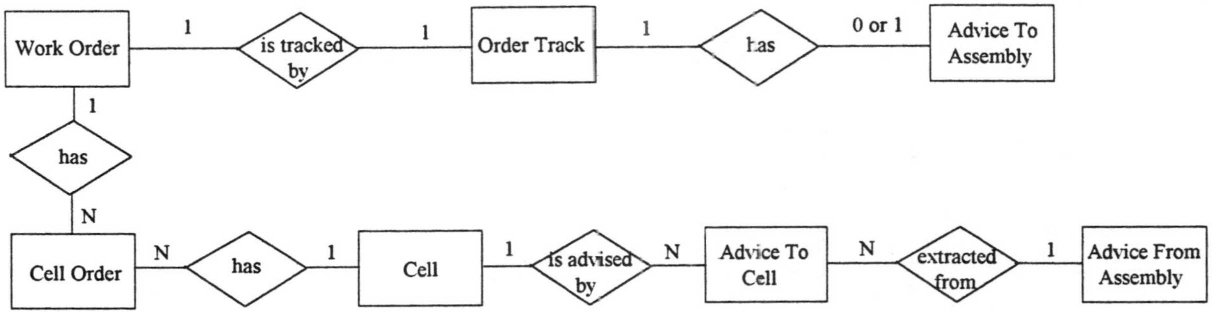

WORK ORDER
- Work Order Number
- Part Number
- Order Quantity
- Start Date
- Due Date
- Order Status

ORDER TRACK
- Work Order Number
- Part Number
- Cell Name
- Finish Date/Time
- Allocated Flag
- Start Date
- Due Date
- Order Status
- Late Flag

CELL ORDER
- Work Order Number
- Part Number
- Cell Number

CELL
- Cell Number
- Cell Name
- Cell Description
- Cell Manager

ADVICE TO ASSEMBLY
- Work Order Number
- Date
- Advice

ADVICE TO CELL
- Cell Number
- Date
- Advice

ADVICE FROM ASSEMBLY
- Sequence Number Of Advice
- Date
- Advice

Figure 3.16 Entity-relationship diagram and entity attributes of the Cell Coordinator data model.

modules and their interrelationships (Martin and McClure, 1988). This program structure must be known or developed before the system can be coded by the programmer. It describes the logical control structure of the system and therefore provides the systems programmer with a view of how the system works; this is needed for the implementation of the mechanisms of the system. An SC consists of modules, arrows connecting the modules and data/control couples transferred between the modules. Figure 3.17 shows an example of an SC for the Cell Coordinator which depicts the relationships between the data internal processes of the function. The details of the internal controls such as allocate cell orders and advice assembly shown in Fig. 3.17 have been described earlier in section 3.5.1. The arrows here show the type of data or information used by the Cell Coordinator to fulfil its functions such as allocation of jobs and advising assembly about availability of parts.

The higher-level modules in the SC diagram call lower modules to perform a task by providing the necessary data; the procedural details can be coded by the programmer from the description of these tasks. In an IDEF0–DFD the database of the system is represented in the form of data store (Fig. 3.15), but by the time the SC is developed the database should be in the form of ERD or a set of normalized relations. This makes it possible to describe the relationships between the data and the processes in the SC as clearly and in as much detail as necessary.

The process of developing this part of the system requires specialist knowledge and resources, even though many automated tools in the form of CASE tools exist in the market place. For a complex system it could be time consuming; however, once implemented, future updates to the system necessitated by any changing circumstances can be incorporated with relative ease. A detailed description of this part of the development, including file transfer translators for different packages used as mechanisms can be found in Zhou (1993).

3.7 CONCLUSIONS

The methodology proposed here caters for an integrated system whose main purpose is to provide effective and efficient coordination and communication between various decision-making functions within a control structure, so that the stated goals of a company can be achieved. It is not influenced by any one theory or method but based on a true understanding of a system's needs and its physical structure. Such a design process is free from such bias, as is often found in most available methods or tools. The approach considers tools and methods that match

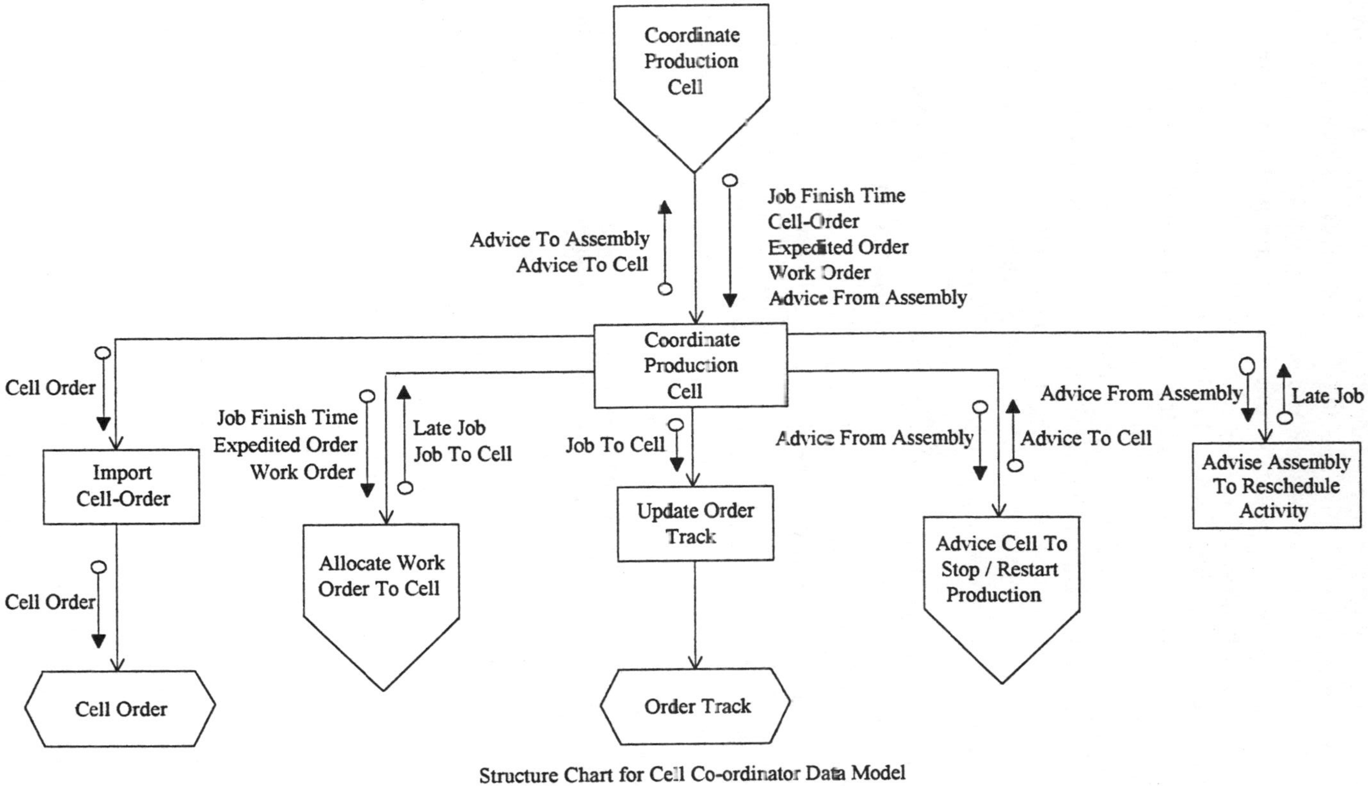

Figure 3.17 Structure chart for Cell Coordinator data model.

closely to the needs of the functions in the control structure. Extra effort is therefore required to interface the selected tools for effective coordination of material flow and manufacturing unit activities. Once implemented it is relatively easy to update the system and/or enhance its effectiveness by way of a periodic auditing and monitoring process. Industrial dynamism and its effect on the systems control structure can be evaluated with relative ease. This is because the systems designers have a very precise knowledge of the systems elements and their task relationships. Use of methods such as MPR and JIT are based on control needs of products to satisfy the environmental requirements. The case example illustrated in this chapter uses MRP for advising what may be required in the upcoming periods but the actual production in the various production units are scheduled on the basis of short-term product assembly needs: a pull type of system coordinated by the Cell Coordinator.

However, this work also indicates several other needs that could greatly assist the design process and reduce the development time to a more reasonable scale. These include:

- availability of a library of reference models for different types of manufacturing systems;
- a better or an automated means for selecting tools for decision-making mechanisms rather than the pragmatic processes used in this exercise. Further work is therefore needed which will enable the designers to match the service provisions of the commercial tools with the service requirements of the controls;
- other methods of information systems design such as object-oriented methods or neutral database systems using languages such as EXPRESS (ISO, 1991; Reid and Banerjee, 1996) for better systems modularity and dynamic reconfigurability;
- extending the methodology to include the design of organizational structure, once other structures have been established.

The methodology described here may require commitment from senior management as the development time and resource requirements can be substantial. Without such provisions the design may at best provide some short-term benefits only.

REFERENCES

Ansoff, I. and McDonnell, E. (1990) *Implementing Strategic Management*, Prentice-Hall International.

Banerjee, S.K. (1987) *Application of Structured Design Techniques in Advanced Manufacturing Technology*. Proceedings of the 2nd Int. Conf. on Computer-Aided Production Engineering, Edinburgh. Mechanical Engineering Publications Ltd., pp. 245–54.

Banerjee, S.K. and Al-Maliki, I. (1988) A structured approach to FMS modelling. *Int. J. of Computer-Integrated Manufacture*, **1**(2), 77–88.

Banerjee, S.K. and Kourouklis, A.P. (1993) *Integrated MPC Systems: Design and Implementation Issues*. Proceedings of Int. Conf. on Engineering and Production Management, Mons, Belgium.

Carrie, A.S. and McIntosh, R. (1992) UK research in manufacturing systems integration, in Pels and Wortman (eds) *Integration in Production Management Systems*, Elsevier, Amsterdam, pp. 323–36.

CIM-OSA (1991) *AD 1.0 – Architecture Description*. ESPRIT Consortium AMICE, 489 Avenue Louise, B14, B-1050 Brussels, Belgium.

Dar-El, E.M. (1993) *Achieving World Class Manufacture – WCM, TQM or TPQM*. Proceedings of 12th Int. Conf. on Production Research, Laapeenranta, Finland, pp. 111–20.

Duimering, R.P., Safayeni, F. and Purdy, L. (1992) Future Factories and Today's Organisation. Proceedings of 8th Int. Conf. on CAD/CAM, Robotics and Factories of the Future, Metz, France.

FORCAST (1991) Report on the Eureka collaborative programme. University of Coventry, Dept of Manufacturing Engineering, Coventry, UK.

FORCAST (1992) Feasibility study report. The Cimulation Centre, Chippenham, UK.

Gorranson, H.T. (1992) Dimensions of enterprise integration, in Petie, C.J. (ed.). *Enterprise Integration and Modelling*, Proceedings of the 1st Int. Conf. on Enterprise Integration and Modeling Technology (ICEIMT), The MIT Press, Massachusetts, USA.

Hall, G., Rosenthal, J. and Wade, J. (1993) How to make reengineering really work. *Harvard Business Review*, Nov–Dec, 119–31.

Higgins, P., and Browne, J. (1990) The monitor in production activity control systems. *Production Planning and Control*, **1**(1), 17–26.

Hill, T. (1993) *Manufacturing Strategy: The Strategic Management of Manufacturing Function*, 2nd edn, The Macmillan Press Ltd, London.

Howe, D.R. (1983) *Data Analysis for Data Base Design*, Edward Arnold, London.

ICAM (1981) ICAM Modelling Manual, IDEF0, IDEF1, IDEF2, ICAM Program Library, AFWAR/MLTC, Wright Patterson Air Force Base, Ohio 45433, USA.

Ingham, H. (1971) *Balancing Sales and Production: Models of Typical Business Policies*, Management Publications, British Inst. of Management.

ISO (1991) *ISO 10303–11: Industrial Automation Systems and Integration – Product Data Representation and Exchange – Part ii: Description Methods*, The Express Language Reference Manual.

Jacobs, S. and Kethers, S. (1994) *Improving Communication and Decision Making within QFD*. Informatik V, RWTH Aachen, Ahornstr. 55, 52056 Aachen, Germany.

Kochhar, A.K. (1992) Development of knowledge base to assist the selection of manufacturing control systems. SERC Report, Dept Manufacturing and Machine Tools Division, The University of Manchester Institute of Science and Tech., PO Box 88, Manchester M60 1QD.

Kieffer, J.P. and Thiel, D. (1992) *Industrial Dynamics and Simulation of Production Systems: Towards the Building of Computer-aided Decision Systems in Organisations of Production Investment and Management*. Proceedings of 8th Conf. on CAD/CAM, Robotics and Factories of the Future, Metz, France.

Martin, J. (1988) *Systems Analysis and Design: A Structured Approach*, Prentice-Hall International.

Martin, J. and McClure, C. (1988) *Structured Approach: The Basis for CASE*, Prentice-Hall International.

Miles, R.E. and Snow, C.C. (1978) *Organisation Strategy: Structure and Process*, McGraw-Hill, New York.

Reid, S.P. and Banerjee, S.K. (1996) Creating an enterprise infrastructure using EXPRESS. To be published in *Int. J. of Computer-Integrated Manufacturing*.

Robinson, A. (1990) *Modern Approaches to Manufacturing Improvement*, Productivity Press, Cambridge, Massachusetts, MAA02140, USA.

Roboam, M., Zanettin, M. and Pun, L. (1989) GRAI-IDEF-MERISE (GIM): Integrated methodology to analyse and design manufacturing systems. *Int. J. of Computer-Integrated Manufacturing Systems*, **2**(2), 82–98.

Will, A., Bertrand, M., Wortman, J.C. and Wijngaard, J. (1992) *A Framework for Production Control and Materials Management*. IFIP G.7 Working Party Conference on Integration of Production Management, Eindhoven, Netherlands, p. 10.

Williams, T.J. (1994) *The Generalised Enterprise Reference Architecture and Methodology (GERAM) – A Proposal from the IFAC/IFIP Task Force for a Potential Universal Standard*. Proceedings of the European Workshop on Integrated Manufacturing Systems Engineering (IMSE '94), INRIA, Grenoble, France.

Zhou, M.L. (1993) Integrated production planning and control systems design. Unpublished PhD thesis. Dept of Design Manufacture and Engineering Management, University of Strathclyde, Glasgow, UK.

CHAPTER 4

Analysis and management approaches for complex production systems

P. Massotte

ABSTRACT

This chapter provides a global and modern approach to the complexity existing in many production systems. It describes studies and solutions that have been developed in an industrial context to control and manage this complexity.

Different recent works have shown that well-known models used for the analysis of human behavior, in economics and in production distribution, contain unsuspected regimes of 'deterministic chaos'. The aim of this chapter is to exploit these findings and to study some complex behaviors in manufacturing and assembly shops. It also shows how deterministic chaos could be produced by a human decision-making process: the equations derived from decision rules applied to a complex manufacturing system lead to chaotic behavior in a realistic region of parameter space. Finally, a methodology together with associated tools has been implemented in order to define the nature of a production system, and to highlight and validate characteristics of far-from-equilibrium systems.

Such target systems viewed as deterministic chaos cannot be controlled through conventional management systems. Some experiments were conducted to determine how to manage and control them; the analysis of different classes of problems in such complex production systems enables the right approach to be induced with the relevant management rules to be implemented for improving control. A comparative evaluation of their prerequisites and usefulness is presented and relevant recommendations are also given.

4.1 INTRODUCTION

Until now, only the quantitative aspect has been considered when describing and modeling a system. The Galilean principle, on which this approach is based, consists of determining main parameters and variables of a system, then defining and measuring their values and numbers to study it. These numbers are linked together by mathematical formulae; they are organized into equations. Solutions to these equations make it possible to predict the future. This principle assumes that the system is predictable and that there is absence of ambiguity.

This is in agreement with the deterministic approach, as defined by Laplace (Field and Golubitsky, 1992) which stipulates: 'If we know, at a given time T, the location and the speed of each particle in the sky, then we are able to know all the future of the Universe'. Applying this principle to the planning and scheduling fields, we could say: 'If we know the location and process status of each part in a production system, then we are able to determine the future and evolution of this manufacturing system, based on the bills of materials, the routings and the technical description of the product or process'.

Unfortunately, if we consider a multi-product, multi-process production shop involving thousands of steps and operations, with hundreds of parts, associated with a particular management system and subject to various inputs, we cannot calculate and predict what will happen, in time, what the **Work In Process** (WIP) will be, the location and the status of all the parts, etc. Consequently, we are not able to determine the future and the dynamic evolution of the production system. This problem is due to the **volume** of information and **interactions** we have to handle, the difficulty of finding the right **algorithm**, and also the application of the principle of uncertainty as defined by Heisenberg (Davis, 1992). These concepts outside the traditional realm of production control and management systems are considered as unusual and complex by people working in the industry.

We must therefore first define the meaning of complexity, as encountered in any production system. The different types of complexity are:

1. **Structural** complexity, which is related to the volume and the structure of the data we manipulate.
2. **Ill-defined** complexity, due to the nature of the algorithms involved in the management of the production system.
3. **Behavioral** complexity, which involves new concepts concerning chaos and fractals and very well fitted to the above considerations.

We will also indicate how to check for the presence of a given type of complexity, how to live with this new paradigm and how to deal with it. Finally, we shall study the impact and consequences of each type of complexity on the behavior, management and control of a production system. Since the aim of this chapter is to describe our experience in detail, instead of focusing on the techniques themselves, we will discuss in depth their implementation. We shall also describe results obtained with several optimization techniques based on operations research and artificial intelligence, and we shall see how they fit with different kinds of complexity. We shall then give a few recommendations.

4.2 COMPLEXITY: DEFINITIONS AND GENERALITIES

First, we have to define the meaning of the term 'complexity'. Complexity is a basic characteristic of a system: it expresses either the inability to describe a system, or to calculate a solution, or to deduce its behavior from the knowledge we have about the behavior of each of its parts. Also, we have to point out that a complex system is different to a complicated one. A complicated system is difficult to understand and to execute: it is confusing and the only way to study such a system is to simplify it and to split it up in elementary parts.

These two concepts are not really considered in production systems. People do not generally make a difference between complex and complicated systems. At present, all of them are identified by two main measures: the **volume** and **diversity** of the products. A typology, which is well known to the specialists, has been defined based on these criteria. However, because of new economic constraints, changes in the structure of production systems, introduction of new technologies, the evolution of management systems and variation in demand, etc. the previous typology is no longer applicable. These systems must be capable of evolving. They are governed by some physical laws, and comprise fundamental constraints and interactions. For these reasons, a new approach based on dynamic complexity and **uncertainty** (Kieffer, 1987) is more appropriate and has been successfully proposed to different production management specialists in industry. It is not our subject here to discuss this new typology or the differences between complex and complicated: we will only focus on the concept of complexity.

4.2.1 Preliminary considerations

For many years, industrial systems, and more generally production systems, have been evolving continuously to include more and more complexity. Initially they were related to the assembly or the manufacturing of simple products based on elementary operations. Later, these were organized and combined together to form more complex products. To produce such products, a great number of sophisticated operations are then necessary: their bills of materials (BOM) and associated routings become longer and incorporate more variety.

For instance, in a typical semi-conductor manufacturing line, many local, complex routings are involved in thousands of operations. The production system is multi-product (with several hundred part numbers) and multi-process (each one related to a family of products), and it is difficult to describe or to model such a system with general algorithms and conventional programming. Their complexity is high, even without considering any changes in priorities that might disturb the system and prevent the achievement of good balance in the flow of parts. In these systems, the complexity is a function of the structure of the involved objects and cells, and of their associated procedures. In the case of a study using modeling and simulation, complexity will be expressed by the cost of building up the model: length of coding, number of formulae, execution time, etc. To illustrate what is meant, let us take some examples:

- In an MRP system, when the structure, capacity or sophisticated local rules are not considered, the complexity is low and the scheduling system is governed by simple programs. However, each time an assumption is changed, numerous calculations or simulations must be performed in a process called **try for fit.**
- In a **pull** system, the management of production is basically simple, but the implementation of the strategy involves, at the same time, a flow of information associated with the physical flow of products. Inventories being low, interactions between cells are quite important and this requires specific conditions which apply to stable and steady production systems.
- Nowadays, more and more mixed push-cum-pull strategies are implemented. When they are not well defined and applied (for instance, in the case of inconsistent clustering of the process, or when buffers and 'kanban' are not well determined), unexpected events can happen and the work-in-process (WIP) is subject to many variations.

For all the above reasons, we have to reconsider the concepts of complexity and not limit them to a mathematical or functional point of view.

4.2.2 Different types of complexity

In fact, in our context, any production system may be subject to three kinds of complexity and phenomena:

- structural complexity of the system itself,
- ill-defined complexity due to the algorithms to be used,
- dynamic complexity resulting from the behavior of the system, sometimes leading to deterministic chaos.

The following examples can illustrate these complexities.

Structural complexity

Structural complexity, as related to the structure of the products, comes about for different reasons:

- A single product may contain parts at several levels in the bills of materials, themselves structured as a tree. Problems in such particular workshop are related to the synchronization of the parts flow and to the large volume of various technical data to be managed.
- Parts manufacturing, in flexible production systems, requires the scheduling of several common resources and the organization of the means involved.
- A large computer is an assembly of several thousand components. In such a production system, as was the case in an IBM manufacturing plant, many models could be built from a few replicated basic components, because of the structure of the product, the size of the bills of materials (BOMs), and the addition or removal of specific or similar features; about 30 000 different combinations and models can be planned. This results in a very high degree of personalization.
- This complexity can also be related to the nature and the size of the production system in which different kinds of relationships exist between the cells. In a manufacturing plant, the shop is often organized similarly to a BOM; based on tree structures, the cells will be clustered in consistent sectors and the routing of the products, simple for some, may be very complex for others. Nevertheless, complexity can vary according to the constraints of synchronization and also according to the complexity of the products (quantity of assembly levels, links, etc.).

In each example, we have few steady 'states' for the target system under study, but computations needed for this type of model are prohibitively complex.

Ill-defined complexity

There is not always an accurate and reliable numerical model, or the model used is only partially reliable. In process control, for instance, reasoning may have to be based on partial or incomplete information. This type of problem is also encountered in economic modeling or in diagnosis, where the degree of uncertainty increases. Here, the systems become much more difficult to understand and hence to reason on. Solving these problems on a computer is a problem: algorithms require a different kind of modeling and their resolution involves new ways of programming. This complexity is also encountered in planning and scheduling problems where algorithms are NP-complete. Many calculations are required to find the solution and new approaches based on different algorithms or heuristics have been implemented to reduce the impact of this problem. Another way to make precise and relevant statements is to abstract the support set until consistent associated relations are shown (the principle of incompatibility by Zadeh (Graham and Llewelyn Jones, 1988)).

Dynamic complexity

The complexity, which was first structural and algorithmic, becomes progressively behavioral. This is due to a simple fact: very often, in planning and inventory control for instance, only the static aspect is considered and the focus is on finding better scheduling functions. Introducing dynamics, and bringing attention to the interactions and relations existing between the elements of the system, can highlight different kinds of characteristics. A production system, even with few cells and subject to a simple strategy, may be subject to unexpected events or an unusual behavior: something simple may well lead to something complex and unpredictable. In this case, the complexity involves either instability or a large number of different 'states' that the system may have. Here, complexity is associated with the ability of a system to display a long range of coherence in space and time, and to undergo transitions between different and various states.

4.2.3 Comments

The third type of complexity described above is not sufficiently taken into consideration in most present production systems. Several computer science-based techniques are used to manage the controlling and monitoring of such systems. However, simplification (also called **reductionism**, and generally based on statistical and multivariate data analysis), and the analytical approach (an area of **constructionism**, with optimization methods) may fail when faced with the difficulties of scale and complexity; many production systems are difficult to model. Consequently, the analysis of their behavior and predictability is limited or difficult with current tools, techniques and technologies.

In this sense, we shall apply to the field of scheduling the principles defined by P. Anderson (Varela *et al.*, 1974): 'The behavior of the complex aggregates of elementary particles cannot be understood by extrapolation of the one related to a small number of particles. At each level of complexity, new characteristics appear and they have to be studied and modeled through quite different and specific ways in order to understand the behavior of the whole system'. This is in contradiction with the commonly used approach, which consists of developing submodels to describe and analyze the local behavior of a system, then assembling all the submodels to study the global behavior (as a result of the interactions between the different local behaviors). As we shall see later, focus is needed not only on the functions and tasks required by the production system, but also on the interactions between these functions.

We cannot satisfactorily describe a complex system as an assembly of objects or components associated with local and elementary laws. A different approach, based on behavioral complexity, must therefore be designed, implemented and applied. New ways to model a complex system have been developed; these are based on cognitive and natural sciences and they themselves depend on the new tools used in information systems. Computers are the most significant tools in this area. They are considered as the 'tools of complexity' and they present a new way to solve complex problems, to highlight the relationships between the different kinds of science, to propose new approaches in problem solving, to differentiate between a simple or a complex system and to determine whether a system can be simulated or not.

In the rest of this chapter, different types of complexity are examined to explain how they can be handled in a real industrial world. Moreover, achievements and results obtained in some production systems are detailed.

4.3 MANAGEMENT OF STRUCTURAL COMPLEXITY

One of the main discoveries in natural sciences is that each entity of a system is managed by simple rules. These entities interact together to determine a global behavior. In this case reductionism is a basic principle applied to a production system. The difficulty is how to combine:

- simple and basic physical or mathematical laws;
- large numbers of various, and more or less complex, relationships;
- complicated initial conditions.

This is also typically the case of production systems in industry. Very often their structure is quite simple, but they are subject to lack of control and difficult to optimize. For instance, planning problems can be hard to solve when there is no possible decomposition of tasks or splitting of responsibilities. This is also the case of scheduling problems; the strategy consists in splitting a single and difficult problem into several, hopefully easier, subproblems. In addition, many problems in planning and scheduling, can be viewed as 'nearly decomposable' (Simon, 1981). Under these conditions, the difficulty lies in handling the data (Paul, 1993).

We know that the human brain is able to process simultaneously five to nine chunks of information. This means that each time a large number of variables that are difficult to measure are involved, the level of complexity will be high. Here we are not talking in terms of 'algorithmic' complexity but in terms of 'computational' complexity. This is why computers are at the core of this problem; the computational complexity will depend on the 'time' required to make a calculation and to find the solution to a specific problem. In this area many approaches are in progress both in mathematics and information systems to improve performances for calculations (Massotte *et al.*, 1991).

Most computer architectures are based on von Neumann principles, which are sequential, and all the programs and software packages developed in computer sciences exploit this characteristic. Now, as a result of the downsizing effect, users and scientists have designed new algorithms and computer architectures (clusters of workstations and parallel processors) to improve the performance of complex production systems.

As an example, MRPII systems have been implemented in some assembly lines, for a whole set of products, at the plant of IBM Montpellier (France). A few years ago, running the complete planning programs required several days; thanks to the parallelization of some software packages, the run time was decreased to a few minutes. This was an important step which made possible the use of the 'try for fit' process and a new reasoning based on 'back-orders'. It was then possible to develop new capabilities in production control and management, to back up more conventional approaches.

In addition, next generations of production systems will require high levels of flexibility and adaptability. This is due to major changes in customer demand: customers require more personalized products; they also require them sooner, and at a lower cost. In order to recover from unstable situations and to allow the reassignment of finished products and the reuse of components, **implosion** techniques have been developed; while MRP is a top-down replanning, implosion is based on bottom-up replanning, a complementary approach to MRP (Scardigli, 1994). Experience has shown that this approach is very useful in backing up MRP and recovering from unexpected situations.

Basically, implosion uses simple algorithms. However, because of the presence of multi-level constraints, specific directed graphs like **trees**, are useful for representing this class of problems, which are NP-complete. Different search strategies have been explored. Some of their properties have been highlighted and very efficient search methodologies combining breadth-first and depth-first procedures have been successfully implemented.

In the light of the previous considerations, structural complexity is mainly related to computational difficulties. Concentrating on practical and quite conventional mathematical methodologies enables us to solve most of the performance problems. Also, the implementation of an appropriate information system architecture, combined with parallel engineering, still gives an improvement in performance. In this section, we have discussed neither stochastic nor statistical methods; in this matter again, the reasoning is quite straightforward and simple. Such a view is not surprising; as we have mentioned several times, complexity is included in the data (amount and structure), not in their processing. Techniques such as group technology and search-based approaches in decision trees are often sufficient. In most cases, the situation can be improved algorithmically, bearing in mind that the computer is at the heart of many production control systems, processing and storing information. Again the most important attribute is speed, or performance, not only in processing algorithms. The main item involved is the processor itself (storage, communication and control). Here, parallel or distributed processing is essential. Nevertheless, solving a problem requires formulating the algorithm in formalisms that are compatible with advanced computer applications.

4.4 MANAGEMENT OF ILL-DEFINED COMPLEXITY

As we have seen, ill-defined complexity is caused by the combinatorial nature of the algorithms used for determining the solution to a given problem: solutions require a huge number of calculations. However, advances in computer science have allowed heuristic and even iterative approaches to be implemented. They are often used to reduce the impact of this type of complexity and represent a new paradigm in solving industrial problems: they are based either on the modeling problem itself or on the replacement of the right algorithm by a simpler and equivalent one or by heuristics. The following are some alternatives which have proved effective in real situations:

- simulation, which is used instead of numerical optimization;
- knowledge-based systems (KBS), which are in fact a new way of programming based on logical and declarative approaches;
- constraint satisfaction, again instead of optimization and mathematical programming;
- neural networks, which, as with simulation and KBS, give real-time adaptive control through decision support systems (DSS) (instead of open loop decision making);
- genetic algorithms and simulated annealing, which are generic computer-search approaches to avoid formulating the problem with conventional algorithms.

As we shall see, each alternative has both advantages and limitations.

4.4.1 Emergence of simulation

Reasoning in uncertainty with partial or incomplete data is a very complex task; mathematical models are appropriate for capturing data or performing data analysis and finding a solution. Statistical and Bayesian approaches have been used, and some tools have been developed as a result with encouraging results. Unfortunately they remain difficult to implement because of end-user skill problems, level of abstraction, etc. The most pragmatic, user-friendly and efficient solution consists of building and running a **simulation model** in order to obtain all the necessary data and evaluate the performance of the real system. Analysis of the results is performed manually by the decision maker. Simulation techniques are much better than numerical optimization for 'exploration' of the parameter space (Davis, 1991) rather than for the 'exploitation' of a continuous parameter space where variables are neither complete nor independent. They make better use of knowledge of a domain, they do not require so much awareness of problem formulation and algorithm selection and require less computational effort.

When the expectations are not realized (which means the objectives are not reached) the decision maker uses his/her own expertise and know-how to change and adjust the values of the model parameters. Then, through an iterative process, the model can determine the best configuration or scheduling of the real system.

However, the tuning of real systems has received little attention in the past despite its importance to the success of the scheduling system. For many reasons, such as time constraints and organizational problems, there is no room for a time-consuming evaluation of a new shop floor configuration or new policy. This demonstrates that there is a need for the manufacturer to have a versatile tool in order to simulate, possibly on a real-time basis, every new configuration of a flexible production system. Every time a significant change and a new situation occurs, simulation provides a complete set of data to the user or to the decision maker. This enables him/her to know and to understand the behavior of the system, then to analyze the solution. For instance:

- the addition, deletion or modification of the existing material or human resources used in a production system;
- production commitments to the customer, in terms of quantity and cost;
- the ability to deliver a new or specific product on time.

Simulation is well accepted by the operators in the production area. In fact, with currently available tools, modeling is clearly understood and easily readable; operators and technicians have a real impression of controlling and managing the model, which is a simple but true representation of their production system.

Considering the differences that may exist between the 'ideal' model and the real world, simulation must be used to understand a production system, to make a diagnosis and highlight some specific points, to bring trends to light and to validate solutions.

Simulation is a typical example of reductionism: to be efficient, it must be a simplification, not just a slight abstract of the real world. A 100% representation of a production system would be useless because:

- it would generate more noise than interesting results;
- since initial conditions and assumptions would be applied to dynamic behavior, it would be unable to give valid predictions or help with planning.

A process is improved by iterations: the aim of simulation is to evaluate the process, and successive analysis–action–validation steps are necessary if improvements are to be made. This can take a long time. Automation of the decision and action determination process is then required, using a DSS; this has been tested in a real production system and we now describe the techniques which can be used for such an implementation.

4.4.2 Decision support systems and knowledge-based systems

An easy way to develop our approach, and to describe the concepts behind such an approach, is to take an example; we will consider an integrated decision support system called LMA (Line Manager Advisor). It has been designed for improving the process control and scheduling of a complex workshop (jobshop approach) (Beaudoin and Massotte, 1990).

LMA performs the simulation and analysis of a production system, then 'common sense' reasoning on the results, as carried out by any specialist in scheduling. This is a whole array of important tasks which requires a great deal of specialized knowledge and know-how that most people do not possess and can only be performed by experts in scheduling and planning. Programs to perform these tasks are based on 'expert systems' or KBS: they make up one of the main techniques implemented in LMA.

The architecture of LMA is briefly described in Fig. 4.1. Three main functions can be identified, as follows.

Model generation

An automatic model building system was designed to permit dynamic modeling of the production system. It is based on information coming from technical files such as:

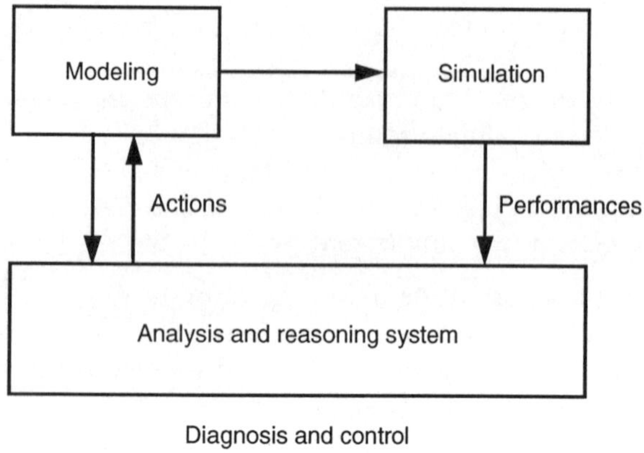

Figure 4.1 Main architecture of a Line Manager Advisor.

- product specifications;
- routings;
- bills of materials;
- process description;
- equipments availability.

This part is highly knowledge-intensive and uses predefined knowledge related to:

- the combination of tools necessary to build up a product;
- a description of the elements relating to each product/process and how they have to be used;
- how to simulate the behavior of specific cells.

As these tasks require different forms of knowledge, there is no unique representation such as production rules that can be used so an object-oriented approach is applied. This gives real advantages: quick modeling and easy alterations during line monitoring. The basic object of the model is a node (workcell): it is characterized by state variables and behaviors (procedural attachments). The model is built automatically from a library of nodes which are connected according to the characteristics of the flow of parts.

Simulation module

The associated simulation system is based on the same techniques. First, the model is initialized using a history file, then it is exercised using a discrete event-driven, object-oriented, simulation system; the parts or products are moved from cell to cell according to priorities and queueing disciplines.

The simulation results are displayed in graphs and tables to give the performances of the production system (work-in-process (WIP), turn-around time (TAT), throughputs, waiting files, task-times, etc.). The data may be static or dynamic and they are recorded in a standard history file (with the numerical results of the previous modeling sessions) which can be processed by the diagnosis module.

Diagnosis and control module

The aims of this diagnosis module are to interpret the results of the simulation and to determine the best corrective actions in terms of balancing parts flow and shop floor performances: definition of scheduling policies, adjustment of production programs, reassignment of priorities, analysis of bottlenecks and reassessment of resources. Generally, considerable time is spent by the decision maker in investigating the system, collecting data, designing the model, analysis and programming decision rules. The contribution of artificial intelligence (AI) techniques and the use of an integrable KBS shell is therefore crucial. All these techniques are used in an integrated environment to:

- provide facts about entities and results,
- assist in model buildup, algorithm selection or statistical analysis,
- assist the user in decision making and explaining their rationale, etc.

To enable flexible, fast and dynamic model generation, update and simulation, a consistent environment based on object-oriented programming techniques was preferred. The knowledge-based part of LMA is encoded in an object-oriented shell environment. The complete application is available on a workstation, in a generic form; it can be dedicated to any assembly shop.

Results and consequences

As illustrated above, KBS is often the basis of any DSS. Most of the work consists of enumerating the possible scenarios used by the decision maker. This requires the definition and the formalization of the operating rules when faced with various possible situations, performance results and bottlenecks.

In our example, which is an assembly shop of electronic parts, about 60 different production rules were implemented in the knowledge base related to the analysis module. All the knowledge and know-how devoted to the scheduling was heuristically organized; plausibility factors were also assigned to some items relating to some specific priorities and sequences.

It was demonstrated that the KBS approach is a very effective support for a decision maker in the area of controlling and optimizing the TAT and the WIP, and managing the scheduling of the production system. Also, it was very easy to upgrade the knowledge base to accommodate new experiences obtained in the manufacturing line; they were directly collected, modeled and introduced into the KBS by the end-users themselves. However because the KBS approach is based on sequential search strategies, it is limited to well-bounded problems.

Building a complete and integrated decision system from a model, as it is done with LMA, is a new challenge in the field of monitoring and control. In addition, such an application is viewed by the decision maker as becoming the reality itself, instead of just being a tool to study, analyze and improve reality. Consequently, the whole decision process is changed and new expectations can be considered.

As we have said before, the optimization process using this simulation and KBS approach is iterative. Examining how the different parameters are tuned, several observations can be made:

1. Because of the wide range of conditions, signal types, symptoms and scenarios, any system tuned to a finite amount of data will eventually encounter a new situation for which its performance is substandard. The reasoning system, based on a KBS, has to be regularly updated. This requires some skill. If the system cannot be improved based on this experience, then it will repeat the same mistakes in the future. As an example, if we consider a scheduling system tuned to low throughputs, as soon as it encounters a high throughput it will fail.
2. To be considered as adequate, the system must subsequently have learning capabilities to handle new conditions. Two types of learning are generally involved. First, experience shows that real situations highlight conceptual bugs in the underlying management rules and algorithms; corrective actions are taken by the developer who changes the production rules in the KBS or modifies some algorithms in the DSS. The second type of learning concerns the 'parameter tuning', whose settings greatly influence the performance of the real system. However, it is difficult to evaluate the interactions between parameters and to determine the volume of data necessary for a set of parameters as well as the sensitivity of a change.
3. Adaptation: Optimization greatly depends on the organization of the real system. Some approaches and algorithms are valid in specific situations, under some conditions, but do not perform well with another type of shop. This adaptation problem requires the use of 'general domain' production rules and algorithms in the DSS. It has to be complemented with 'strategies' specific to the context and to the type of real system under study.

With regard to the problems and weaknesses encountered, several other approaches have been investigated in order to improve the situation: constraint logic programming (CLP), neural networks, simulated annealing and genetic algorithms. These approaches, also called **computer-search** approaches, are detailed in other chapters of this book.

We shall therefore just briefly review some of them to explain how they have been implemented in real production systems. We will also discuss the typology we established on the basis of the results obtained.

4.4.3 Constraints satisfaction programming

Many problems in planning and scheduling can be viewed as problems of 'constraint satisfaction' in which the goal is to find the proper production program within limits on time, costs, parts and materials. Constraint-satisfaction problems do not require new tree-search methods of their own; they can be solved using search strategies we have already discussed, and which lead to numerical optimization or KBS. They can be viewed (Rich, 1983) as a form of unimodal hill climbing (i.e. there are no local maxima other than a global maximum). Each state in the problem in which hill climbing is performed represents the set of possible solutions that have not yet been ruled out. Progress is defined as moving toward a state in which the set of possible solutions is smaller.

Constraint solving can be widely used in production engineering and knowledge representation, because it allows great expressive power. More specifically, production control systems can be easily designed and developed over well-understood domains such as sets, Boolean expressions, integers, rationals, real numbers and so forth. These domains have natural algebraic operations associated with them. These operations may include set or graph intersection, disjunction or multiplication. Moreover, these domains are associated with predicates to express the 'constraints'; they are typical representations of how a specialist in scheduling acts, or reacts, when faced with problems he/she has to solve on a daily, or real time, basis. They can include various forms of equality, inequality, graph isomorphism, etc. In this framework, we developed and implemented some applications to determine this sequencing of multi-products for a given set of equipments (each product having its own operations instructions and priorities, according to various demands).

We first analyzed and selected different domains for which constraint-satisfaction methodologies were most appropriate. Moreover there is a proliferation of extensions to KBS languages, to model related problems; among them we can mention logic programming languages such as PROLOG. All these languages are very useful, but they have their own limitations.

The popularity of logic programming stems from its unique nature, which combines expressive and computational power with simple and clear semantics. The semantics of PROLOG, which are operational and declarative, are both simple and elegant and coincide in a natural way. However the semantics of the resulting logic programs are defined within the context of the Herbrand Universe (the set of all possible terms that can be formed from the constants and functions in a given program). In this universe, only those terms which are syntactically equivalent can be combined and unified together. A problem arises when two terms are semantically identical but syntactically distinct. For this reason, a more general framework, called the Constraint Logic Programming (CLP) scheme (Lassez, 1987; Fron, 1994) was used for developing languages more powerful than traditional logic programming languages.

We started implementing several applications using the CLP® language, which is an extension of PROLOG. This experimental implementation of the CLP paradigm was carried out on a UNIX workstation. Other applications have now been developed using the final version of an IBM PROLOG extension, called CLAIRE. It is clear that constraint solving permits a more natural representation of complex problems and gives solutions with good performances. Even if CLP can be considered as part of numerical optimization (NO), it is a powerful paradigm, easy to implement, and very useful for process optimization.

In conclusion CLP can be seen as a new way of programming, able to improve drastically the productivity of the scheduling process in a production system. CLP provides a general framework in which powerful languages can be created. A complex problem can easily be formalized, avoiding the restrictive and limited formalism required by conventional mathematical programming methods. In addition, such an approach is highly parallelizable and is thus very attractive.

Some studies have been carried out in this framework, in the Advanced Technologies Group of IBM, which gave very promising results (Paul, 1993). In the field of scheduling in flexible shops, problems were described more easily and could be solved on a cluster of PS-processors to shake off combinatorial constraints. To quote M. Lassez (1987): 'KBS and logic programming opened a new era in computing. Now, constraint logic programming should let us develop an exciting wave of applications'.

4.4.4 Neural networks

Neural networks are directed graphs consisting of 'nodes' (which are processes, sub-processes or tasks), and 'arcs' called links. A state variable is associated with each node. A real-valued weight is associated with each link, and a real-valued bias is associated with each node.

Transfer functions determine the state of each node j. They take a given form as described in Miller *et al.* (1990): $F(S(W_{ij}.X_i) - PS_j)$, where PS_j is the bias of the node j, W_{ij} the weights of its incoming links and X_i the states of the nodes connected to it by these links. F is either a sigmoid or a step function.

Only feedforward neural networks, whose topology has no closed loop path, have been used to monitor a production system for scheduling purposes. Input nodes are submitted to a pattern (a vector of real values representing the status of the real system: WIP, TAT, utilization rates, throughputs required, buffer sizes, strategies, etc.). Output nodes with no arcs away from them are assigned to the determination of the class of the real system. All other nodes are hidden nodes organized in several hidden layers. Such a network is fully connected.

The network is used as a classification tool. When the input nodes are set, all other nodes in the network can set their states as values propagate through the network. The output nodes indicate which typical class of situation the system is in and what kind of corrective action or tuning must be applied to the real system to improve its performances. Results are obtained instantaneously and provide the decision maker with useful information about the situation and the actions to be applied to the production system.

However, the adequacy of the solution depends on the values of the weights as well as the structure of the neural network. Therefore, each time a new situation is encountered, the network needs restructuring or, more simply, parameters have to be retuned through learning.

Neural networks are an efficient technique when an extensive history is available about the different situations encountered in the production system and the relevant configurations and corrective actions applied to it. Training and learning is then possible. Generally, we used back-propagation techniques in our solutions. However, several problems were encountered during the implementation stage of this approach: it is difficult to identify and classify the different types of situations. Too many 'new' situations were regularly encountered and the technique was not flexible enough to learn, adapt and train the tool to unplanned situations. More sophisticated neural networks and training features have been studied (Muller, 1993) to improve the situation, but no significant results are yet available.

In addition, expertise in planning and scheduling is often difficult to elicit, to model and to distribute to each class of situation. People in manufacturing are not yet ready for such new techniques as they require high technical skills. Consequently, fully automated adaptative control techniques are not yet ready and mature enough for a broad usage in industrial production systems.

4.4.5 Genetic algorithms and simulated annealing

Genetic algorithms have been successfully used to optimize the scheduling and assignment of parts to different machines or equipments in a workshop. The main advantages are as follows:

- This approach is a general one and the formalization of the scheduling problem does not require complex abstraction. Modeling is relatively easy and consists of a general format with some variable components chosen from a limited domain of solutions.
- They are able to find nearly global optima in complex spaces. As stated before, the search space is multimodal and this situation generally leads hill-climbing algorithms to get stuck in local optima. In some specific cases, simulated annealing (Beziat and Massotte, 1989) was also used: very good results were obtained, specifically for parts assignment and plant layout. This technique led to the development of a tool called 'PLOOT' (Plant LayOut Optimization Tool). Nevertheless the applicability of genetic algorithms seems at present to be more suited to scheduling and assignment problems.
- Discontinuities (this point will be discussed in detail later) are a major problem for mathematical optimization based on well-known techniques such as mathematical or linear programming. This is due to the use of discrete variables (parts, equipment, etc.) and to the implementation of just-in-time (JIT) scheduling techniques which require batch orders (according to the level of acceptable inventories, called **Maximum Allowable Quantities** (MAQs), of parts and final products). Genetic algorithms do not require any form of smoothness of 'linearization' for rule threshold or parameter optimization.
- As for simulated annealing, genetic algorithms are very fast at finding global optima. This is important in jobshop scheduling; the decision maker always requires a 'good' answer (not the 'best') to his/her problem, as soon as possible.

Simulated annealing is derived from statistical physics and has been successfully applied to determining and optimizing the scheduling of multi-product and multi-process production systems (Michalewicz, 1992). However, it can be considered as a specific application of genetic algorithms, so it will not be discussed any further in this chapter.

More generally, simulated annealing and genetic algorithms are derived from natural sciences and statistical physics. Even though they are based on a random approach, it has been demonstrated that they converge to a global optimum. However, an acceptable approximation, near the optimum, can be effectively calculated in a short time; decision makers in industry are generally satisfied, as they require a fast and acceptable solution. This rationale is very important and these methods are able to give an answer to the problems involved, which are difficult to model through conventional exact methods. For this reason, these approaches can be considered as being better than analytical or iterative methods.

4.4.6 Summary and recommendations

As seen above, a decision maker can be faced with different situations according to the nature of the process and the knowledge he/she has about it and about the parameter space. Experiments showed that hybrid techniques are necessary to achieve good results in process optimization.

Most of the applications we have developed use simulation of the target system, to understand and analyze its performance. Optimization by iteration is easy and more user-friendly. However, to **tune** a target system automatically, a more sophisticated approach combining several techniques such as simulation, expert systems and numerical optimization must be designed. For instance, KBSs are used when the search for the optimal solution is domain dependent. As soon as the knowledge of the domain becomes incomplete, hybrid applications based on numerical optimization and simulated annealing, or even genetic algorithms, seem to be the best solution for a good decision-making process. In fact, the combination of different optimization techniques depends on the level of knowledge we have in the field of 'causal relations' and 'process knowledge':

- Causal relations specify parameter space conditions under which changing specific design variables causes an output parameter change.
- Process knowledge specifies the rank, direction and magnitude of changes impacting the design variables.

Usage of the techniques is summarized in Fig. 4.2. The horizontal axis indicates the level of knowledge (low, medium, high) about the causal relations. The vertical axis, in the same way, indicates the level of knowledge and information about the process.

For instance, when the process effects are well known and no causal relationships are established between the variables, multivariate data analysis is first necessary before tuning the real system with a KBS.

Process Knowledge	(low)	Causal relationships	(high)
(high)	Data Analysis		KBS
		KBS + G.A + N.O	KBS + N.O
(low)	G.A + N.O		Simulation

Figure 4.2 Usage of different optimization techniques.

In this way, when the structure and the role of variables is known, though data are incomplete and the process uncertain, simulation can be used in conjunction with KBS and numerical optimization.

4.5 DYNAMIC BEHAVIOR OF A PRODUCTION SYSTEM

This section deals with **chaotic dynamics**. It is a relatively new era of mathematics applicable to complex dynamics fields which has not very often been studied in production systems. Here, we focus on theoretical approaches to determine the possible situations we may encounter.

Chaos does not result from large amounts of data or from the large size of interconnected production systems. Simple systems may result in chaos; a small change in a control variable can turn an orderly flow of parts from a relatively simple system into a highly complex chaos of vortices. This is why, very often, behavioral complexity results from simple mathematics.

Chaos is a seemingly random process generated in some nonlinear dynamical systems that are extremely sensitive to changes in initial conditions; in this sense chaos means 'without order'. However, in chaotic dynamics, chaos is fully deterministic and based on fixed rules that involve no elements of chance; the future is determined by the past. Thus, 'deterministic chaos' means 'predictable unpredictability' or 'orderly disorder'. This situation is common in industry and the behavior of many phenomena which was considered as random can be explained in terms of simple deterministic laws (Gleick, 1987). In fact, the term 'deterministic chaos' refers to the irregular output of a deterministic system. Deterministic chaos is characterized by three properties:

1. The system is **sensitive dependent** on initial conditions.
2. In very stable conditions, the value of the output is fixed or hops back and forth between two limits (periodic) or more than two limits (quasi-periodic). When the system seems to be out of control, it is made up of a set of **unstable orders**; in fact the value of the output is a seemingly non-repeating erratic. Its behavior cannot be predicted.
3. **Mixing** which means that a chaotic time series will eventually (even after a long waiting time) come as close as is required to any value within an allowed range of the output.

To apply these concepts to a production system, we will consider it as a dynamic system. This dynamic system is represented by a set of equations, states (the state describes the system and its conditions) and dynamics (which specify the rules of how the state evolves with time, along an 'orbit'). The states converge towards 'basins of attraction', which are regions of small dimension in the state-space that attract nearby orbits, called 'strange attractors'.

The two main causes of chaos in a production system are described below; some examples will be discussed to illustrate how the chaos may appear. We have to point out that this study is based on the analysis of mathematical models. The most difficult task will be applying these concepts in the real world, to highlight the chaos in a production system and to exploit its properties.

4.5.1 Initial conditions necessary for chaos

The existence of deterministic chaos is not necessarily caused by a large amount of data. Simple systems, governed by a few non-linear rules and very sensitive to initial conditions will be unpredictable: they are generally governed by Lorenz attractors.

There follows a description of two very common characteristics of complex systems, which are prerequisites for any chaotic behavior. They are called sensitivity and non-linearity.

Sensitivity

Sensitivity is a characteristic present in many closed loop systems involving either feedback loops or amplifiers. It translates the reactivity of a production system into disturbances.

For instance, if we consider a production system in which work-in-process (WIP) is denoted by X and the turn-around time (TAT) by Y, X and Y will be very sensitive to repair and to any feedback. To illustrate, we first consider a simple operation (or a set of simple operations) as described in Fig. 4.3.

In this example X_0 is the initial demand, and D depends on the yield ($D \in [0,1]$.

The WIP is expressed by: $X = X_0/(1 - D)$.

The TAT, called Y, compared with the raw process time, Y_0, follows the same rule and becomes: $Y = Y_0/(1 - D)$, when it has stabilized after a given transient period of time.

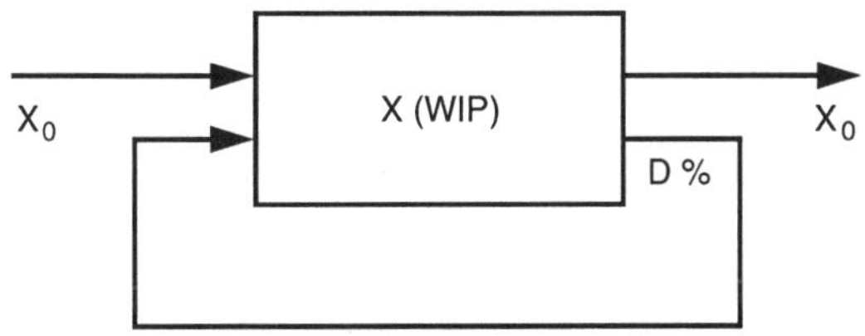

Figure 4.3 Production system with a feedback loop.

As we have already seen, when D varies and is near to '1', the WIP, X, becomes very 'high', and the feedback is equivalent to an amplifier. The system is overloaded. Nevertheless, the throughput remains constant.

Nonlinearity and discontinuities

A system cannot easily be studied through modeling. Here we use analytical modeling to calculate the evolution, or 'curve', of the system, and to predict its behavior. When possible, the corresponding parameters and variables will be expressed through linear equations or through linear relationships. In this case, for instance, the variables X and Y at a given time T, are used in proportional relations to the power of '1', and they have to be combined linearly in a formula. In reality, the variables are sometimes multiplied together or divided.

A set of such equations may not have a possible 'algebraic' solution. Because of this, we are not able to predict the curve of the system, nor therefore its sensitivity or reaction to some initial conditions.

Here are some examples:

1. In a manufacturing line, the throughput can be expressed by the ratio: $Z = F(X/Y)$, where X is a quantity of product and Y is a time duration.
2. To implement the continuous flow manufacturing (CFM) concepts, the demand, called E, is initially equal to X_0. The quantity of parts/products sent from the previous operation is a combination or a selection of the two values (see Fig. 4.4): $X_{01} = \text{Min}(X_1, X_2) = Q_{min}$ or $X_{02} = \text{MAQ} = Q_{max}$.

   The information related to X_0 is determined when the WIP reaches a given quantity called the maximum allowable quantity (MAQ).
3. In a very flexible manufacturing system, some nodes are fairly autonomous and they are controlled or managed through behavioral rules such as: $X = F(X, X^2, \ldots)$.

   Such conditions and management rules are **nonlinear** and they make the management system more complex.

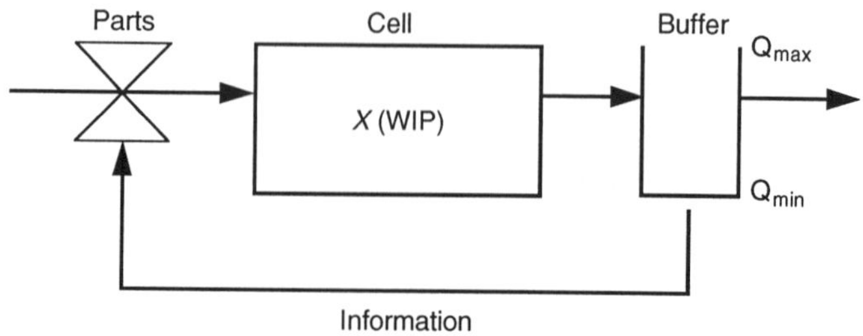

Figure 4.4 Discontinuities in a 'pull' production system.

4.5.2 Qualitative analysis of a chaotic system

Let us start with an example. In a flexible shop, or manufacturing line, some cells can perform several different operations, and the parts are sent back to an identical or different cell in the manufacturing line. Similarly, in many complex assembly lines involving several hundreds of operations, some of them are replicated several times along the process and could be performed tens of times on the same product. Basically, we are faced with a process subject to **nonlinearities** and **feedbacks**. The state variable, or the set of state variables, is called X.

The aim of this section is to analyze the evolution of the state variables of such production systems. We will adopt a dynamic point of view, in order to establish a relation, or 'rule', to generate the new value of X_{n+1} (for instance the WIP and/or TAT) from the previous value X_n.

Let us consider the simple closed loop system depicted in Fig. 4.5, which describes a nonlinear behavior linking the steady state value X to the control parameter C, at stages (n) and (n + 1). Here, (n) is the index representing the number of times parts flow through a given cell (or a set of cells). Sometimes, this index is called **number of passes**.

The evolution of X is influenced by the variation of some parameters present in the problem which can be modified by the external world. We call these entities, or problem parameters, control parameters. They are denoted C.

The mathematical expressions describing how a set of state variables X evolves with time are given by the equation:

$$dX/d(t) = F_n(X, C)$$

When this nonlinear function $F(X_n, C)$ acts as an amplifying factor and, under some conditions, it may lead to **chaos**.

Whatever the form of F_n, when a steady state is reached, this corresponds to the solution of $F_n(X, C) = 0$, since the rate of change of the state X is zero.

These relations impose certain restrictions, and many physical systems cannot be reduced to mathematical formulae. However, we will apply them to part of a process regarded as more typical or generic from the mathematical point of view.

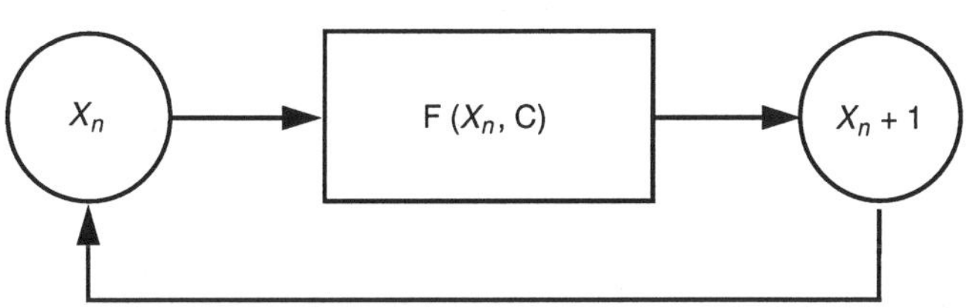

Figure 4.5 Evolution of a dynamic system.

$F(X_n, C)$ corresponds to a dynamic system which successively propagates an initial X_0 value at the points $X_1, X_2, \ldots X_i \ldots$ This sequence of points is called the path, or the 'orbit', of X_0. If the path is ordered, it is a case of **ordered dynamics**; if not, it is described as **chaotic dynamics**.

After several successive backward loops or recyclings, in the production process, the quantity of parts present in a cell is defined by: $X_n = (1 + R)^n.X_0$, where X_0 is the size of the initial batch in the cell, and $R = D/(1 - D)$.

Growth of X is exponential. However, in reality, this growth may not occur for several reasons:

- The processing capacity of a cell is limited to a maximum value X_m.
- The yield, leading to R, is controlled, and may change over time or through circumstances so that it is generally improved: the rate always drops from R to a lower value, when the population approaches X_m.
- The resources, and consequently the effort, assigned to a manufacturing system are sized according to the 'criticality' of the situation. There is 'autoregulation' of the system. This introduces the concept of **self-control** which we will not develop further here.

In practice, whenever the growth of the population becomes too high, some control systems are activated, through the MAQ, through **dispatching rules**, or through **bottleneck management**: the solutions offered may vary according to how these control systems cope with chaos. The three examples are described in more detail below.

Limitation by the maximum allowable quantity (MAQ)

The MAQ is defined at cell level. It enables regulation or limitation of the flow of parts in a shop. The input of parts in the cell is limited by a buffer control approach. When a cell has its maximum number of parts, the upper limit of WIP is reached, and parts from the preceding cells are not accepted. This WIP is denoted by X. Its evolution is mathematically represented by:

$$X_{n+1} = ((1 - A) + R).X_n$$

where: $A = C_1.X_n$.

Here, C_1 depends on $1/X$, and it is defined so that input is '0' when X is equal to X_n. When the limit is reached, the WIP becomes:

$$X_{n+1} = X_n + X_n.(R - X_n/X)$$

Then, for the steady state, by appropriate alteration of the units of measurement, the model is simplified, to give the following quadratic equation:

$X = (1 + R).X - X^2$

The graph of the trajectories (phase plane) is shown in Fig. 4.6. As we can see, the behavior of the cell is a deterministic chaos. This situation appears as soon as $R \geqslant 2$, that is to say for a yield lower than 50% ($D \leqslant 0.5$).

Symmetry-breaking **bifurcations** related to Fig. 4.6 can be calculated. We will limit the calculations to the first two:

- The first bifurcation occurs when $F(X) = X$.
  In this case, two solutions are possible: $X = 0$ and $X = R$.
  Then $F'(0) = 1 + R$ and $F'(R) = 1 - R$.
  With $R > 0$, only $X = R$ is valid.
  Bifurcation is reached when $abs[F'(X)] = 1$ that is to say when $R = 2$ and $X = 2$.
- The second bifurcation is given by
  $F^2(X) = X \rightarrow -X^2 + (R + 2)X - (R + 2) = 0$
  then $X_1 = ((R + 2) + ((R^2 - 4)^*.5)/2$
  and $X_2 = ((R + 2) - ((R^2 - 4)^*.5)/2$

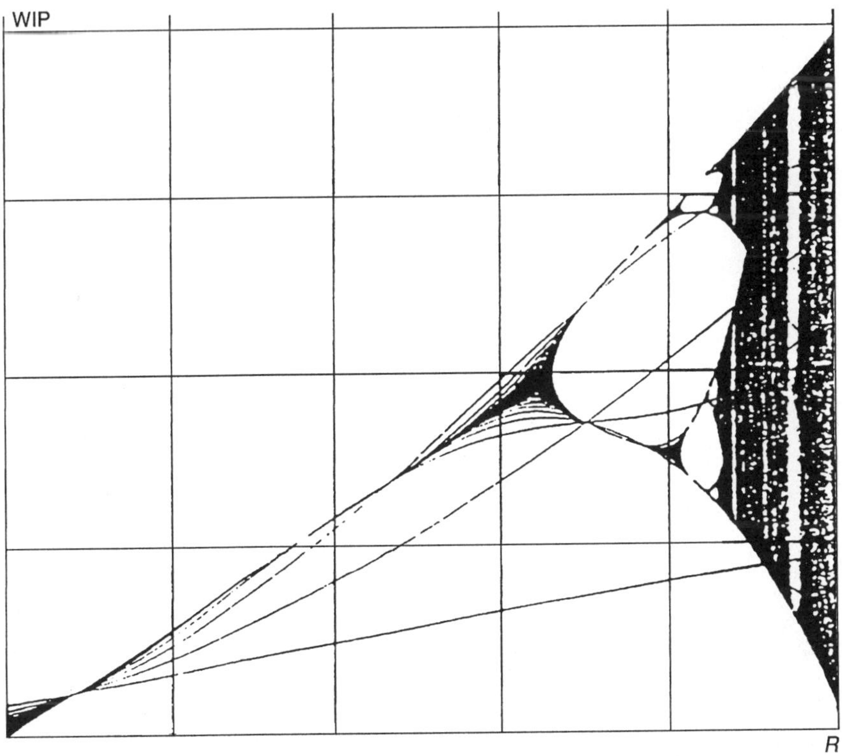

Figure 4.6 Phase plane trajectories (with MAQ).

We can easily deduce that: $R = 2.45$, $X_1 = 0.517$ and $X_2 = 2.931$. This therefore applies to manufacturing lines where some sets of operations are duplicated several times in a process.

Limitation by 'dispatching rules'

In order not to affect the main flow of parts, the dispatching rules may be changed, giving lower priority to recycled and repaired parts in the cell. This approach is important when the TAT of the critical products is to be reduced. Unfortunately the distribution of TAT could be statistically flat.

Here, the variable growth rate R, used in the previous mathematical expression, will be limited by a threshold and is now replaced by: $R - C_2.X_n$, where $C_2 = R/X$. This indicates that the growth of the population becomes zero when:

$X_n = X_m$ (X_m being the upper possible limit of X).
Thus: $X_{n+1} = (1 + R - C_2.X_n).X_n$
When C_2 is replaced by its value, we obtain: $X_{n+1} = X_n + R.X_n(1 - X_n/X)$

As stated above, the formula yielding a stationary state value becomes:

$$X = X + R.X(1 - X)$$

The representative phase plane graph is shown in Fig. 4.7.

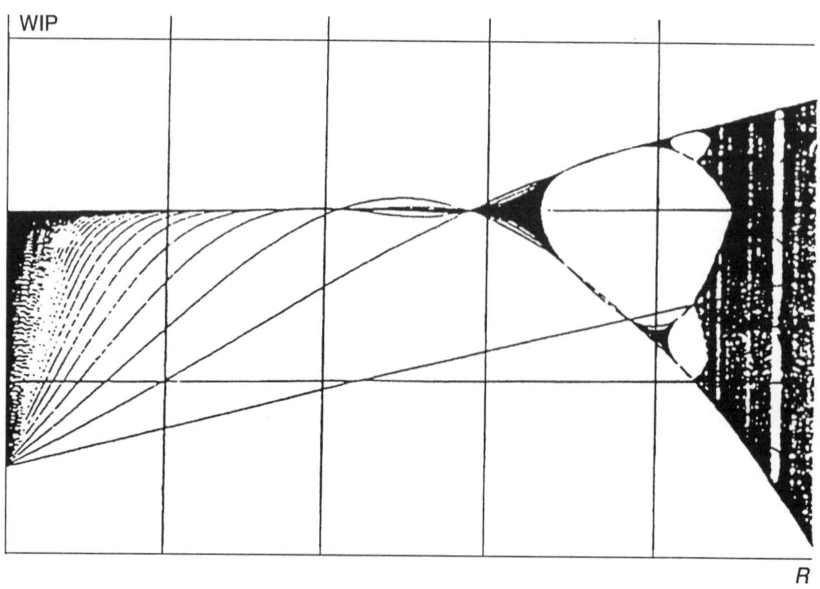

Figure 4.7 Phase plane trajectories (with dispatching rules).

Again, the behavior of the cell shows deterministic chaos:

- The first bifurcation: when $F(X) = X$, then $X = 0$ or $X = R$.
  We can define: $F'(0) = 1 + R$ and $F'(R) = 1 - R$.
  Consequently: $R = 2$ and $X = 2$.
- The second bifurcation is obtained with $F^2(X) = X$.
  Then: $-X^2 + (R + 2)X - (R + 2) = 0$ gives X_1 and X_2 as before.
  With $R = 2.45$, this leads to $X_1 = 0.619$ and $X_2 = 1.197$.

These formulae are similar to the very well-known equations defined in 1845 by P.F. Verhulst (Berge *et al.*, 1988).

Bottleneck management

In some recent manufacturing and assembly shops, such as in semiconductor plants, the process comprises hundreds of products, described by different routings, each one involving hundreds of operations. The process is often repetitive because of the similarities of some operations: masking, exposure, cleaning, burn-in, etc. The equipment at these stages is often sophisticated and expensive. Consequently, they are limited in quantity. Those hardware limitations explain why bottlenecks occur. Generally they are solved through the management of the product flow, in the shop itself.

The common way to handle such a bottleneck is to apply the following strategy. 'If a given operation is a congestive one, never leave it unloaded. As soon as the queue length is below a threshold, load the shop with the products having the highest processing time at the bottleneck, even without a demand!' As a consequence, the performance of the production system will be quasi-optimal.

If X_b and T_b are the values of the WIP and of the threshold at this bottleneck, respectively, the value of the total WIP, over time is defined by

$$X_{n+1} = X_n + K(T_b - X_b) - X_b, \text{ where } K \text{ depends on } (1/X_b).$$

This formula can also be expressed by:

$$X_{n+1} = X_n + C(1 + X - T/X)$$

where C is a constant, acting as a control parameter.

In a stationary state, this leads to the more general formula

$$X = X + C/X(X^2 + X - T)$$

Here again, because of the form of the formula, the resulting behavior of the WIP will be chaotic.

Comment

The studies described above consist of modeling the process and simulating it on a computer. They are theoretical: it must be noted that time and space are discrete and that the chaos revealed by the simulations is a result of a virtual world built in the computer. In this sense, this approach is highly **constructionist** and the conclusions we draw have to be interpreted carefully. This fact is reinforced by the fact that most of the studies presently available in industry are based on mathematical models. For this reason it is always necessary to check the presence of chaos in a real production system before trying to exploit its properties. That is the aim of the next section.

4.6 ANALYSIS OF THE BEHAVIORAL COMPLEXITY IN A PRODUCTION SYSTEM

The purpose of this section is to validate the concepts and experiments described above, where chaotic behavior was theoretically demonstrated in a production system by modeling simple cells. In fact, real production systems are an assembly of many simple cells. Modeling a complete one is complicated: the cells to be considered are interdependent and the mathematics are not sufficiently precise to model such systems and describe their behavior. Moreover, we intend to verify and to extend the results previously obtained, in industry.

To validate our assumptions and to prove the presence of such chaotic behavior in a real production system, experiments were conducted and several tests based on observations were performed. For this purpose, a set of simulation tools, image analysis techniques and tests of hypothesis were set up in order to:

- explore the production system and check for the presence of any chaotic phenomenon;
- measure and visualize this phenomenon in the system under study.

Such a validation is very important: if a production system is chaotic, it is subject to behavioral complexity. Then, many interesting properties can be applied and exploited in order to improve its management system.

4.6.1 Detection of the chaos in a production system

First, several mathematical methods can be used to detect the chaotic behavior of a dynamic system. Most of them are qualitative and are able to determine, with a high level of confidence, whether the system under study is chaotic. The methods we implemented and experimented on real systems included:

- spectrum analysis – we developed a Fast Fourier Transform (FFT) program to analyze a large set of observed data;

- phase graph analysis (with a 3-D visualization tool called GALAXY, implemented on a RS/6000);
- Poincare map analysis;
- Sugihara May test.

All these approaches can describe the characteristics of a production system which may be subject to chaos. However, our experiments, did not confirm clearly or prove the chaotic type of the dynamic system. Best results were obtained with FFT; time series representing daily deliveries of TCM products were analyzed and nearly continuous spectra were observed.

Second, to improve the mathematical behavioral analysis of the dynamic production system, a different and quantitative approach was implemented. It is based on **Lyapunov** exponents. This approach, thanks to specific and precise calculations, makes it possible to predict the status of a dynamic system by measuring the deviation of the trajectory near the attractor.

The formulae and calculation principles will not be given in detail as they are quite complex. The presence of chaos is detected by the analysis of some parameter values called Lambda and Sigma (Berge *et al.*, 1988). Suffice to say that the condition of a system can be specified by the value of the two exponents:

- **If** Lambda and Sigma are positive, **then** the system is chaotic.
- **If** Lambda and Sigma are negative, **then** the system is steady (non-chaotic).

Considering the complexity of calculations to determine the values of the Lyapunov coefficients, two methods were implemented:

- use of the set of differential equations describing the system, when no real data are available;
- use of a significant set of data coming from observations related to the system.

Reliable analysis of a production system requires a large set of data (about 10^D to 30^D, D being the dimension of the attractor). For our experiments we were not able to collect a reliable set of data large enough to validate our assumptions. The IBM plant in Montpellier, France, is involved in the assembly and testing of TCM modules. In this case, the information system collects and stores the quantity of parts produced every day in the production system. As the life time of a product is around three years, the time series consisted of around 1000 values. Under these conditions, it was very difficult to confirm reliably the evidence of chaos in the production system.

Third, in order to obtain information and results from the production system concerned, one solution consists of modeling then simulating the system. As pointed out before, simulation is a good and realistic way to represent complex systems with many more details and behavioral rules than would be possible with mathematical methods. Again, two approaches have been defined:

- The Dynamic Simulation Language (DSL) is an IBM product. It is a high-level language enabling continuous simulation of a production system. DSL is based on the resolution of differential equations, and consists of many functions representing different behavioral methods in a dynamic system. DSL is fully adapted to transient analysis and it was possible to simulate all the production systems we intended to study.
- NETSIM stands for NETwork SIMulation; it is a tool developed at this laboratory to model and simulate a network. Each node of the network is a piece of equipment or a cell and the nodes are linked together by a procedure which defines the management methods between two cells, the values of the parameters etc. Netsim is an object-oriented, user-friendly tool. It is written in Smalltalk.

DSL and NETSIM are considered as very useful for generating large sets of data that are well fitted to different types of production systems. These data are the inputs for the Lyapunov program, to calculate the exponents associated with the dynamic system under study.

Figure 4.8 shows some outputs obtained with NETSIM: it was verified, as in the real system and according to the values of the control parameters, that there were cycles with 16, 24 and 32 periods of time. The impact of such behavior is already significant; it was possible to visualize simple periodic systems. When many oscillations are superimposed, it is difficult to say whether the system is periodic, quasi-periodic or chaotic. Very often people considered it as a random phenomenon and which leads to a demand for a more constrained management system (which is a mistake). This is why it is important to determine exactly the kind of behavior taking place.

4.6.2 Validation of different approaches and methodology

All the experiments we conducted showed that determining the nature of a dynamic system may require several studies with the various tools we have developed. We found it possible to define relationships between the tools and associated methods, according to the characteristics of the observed data and also depending on the structure of the production system.

This can be summarized in the diagram depicted in Fig. 4.9. This methodology has been successfully used in several studies we have recently conducted.

Each time the quantity of observed data was not large enough,

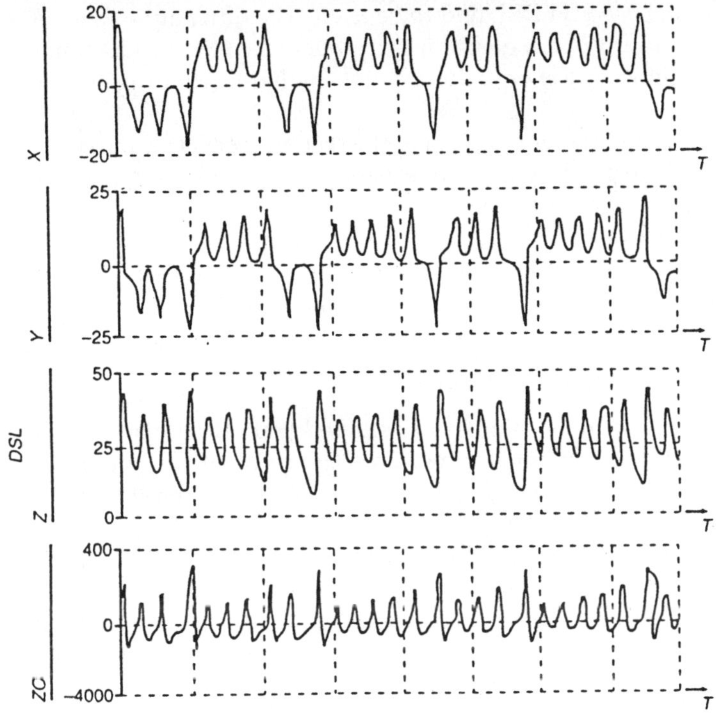

Figure 4.8 Simulation of a flexible production system, with NETSIM.

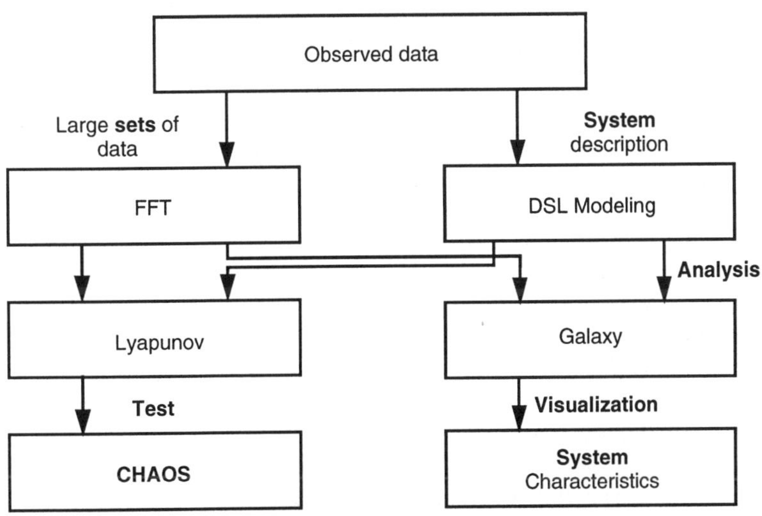

Figure 4.9 Methodology for the analysis of a non-linear dynamic system.

simulation based on DSL had to be used. The difficulty was to select the most significant parameters for the modeling of the production system. The definition of the equations is not the hardest part of the modeling process.

FFT is a good tool for analyzing the response spectrum. However, this is only a help. Proving the existence of chaos must be based on Lyapunov exponent testing.

Finally, visualization is interesting as a qualitative tool, to demonstrate how the variables of a dynamic system are arranged and to represent a periodic trajectory in pseudo-phase-space coordinates.

4.6.3 Results and recommendations

Several production systems were studied theoretically and experiments were conducted over a long period of time. We were able to check the existence of chaos in some complex assembly lines, based on Lyapunov exponents.

However, in several production systems, analysis of WIP showed that it was largely impossible to predict future events or determine any cycle in WIP variations. Mathematical modeling of such systems demonstrated chaotic behavior. In that case, the demand remained constant and the variables considered were the WIP and the throughput. The control parameter was the size of the **kanbans**. These characteristics were selected as they are easily visible. Moreover, a partial display of the attractor is sufficient as a first approach to validating assumptions.

Experimention and observation never demonstrated unambiguously that chaotic behavior existed in the production systems, for the following reasons:

- The yield which involves feedback varies with time.
- The number of state variables depends upon the complexity of the dynamic system, which changes constantly because of modifications in the process and routings.
- As explained above it was difficult to collect large sets of data. In many cases we had a 'strong belief' in the existence of chaos, but we could not state that assertion as a fact with a high level of confidence.
- A simple approach consisted of highlighting on bifurcation: we observed a doubling of the number of states each time the control parameters were changed (first, one stable steady state, then 2, 4, 8 . . .) till non-predictable variations took hold. As seen above, it was possible to test for the evolution of the system towards chaotic behavior. However reliable and regular measurements are difficult to make and some caution.
- When checking the real sets of data, some anomalies were detected due to, for instance, the introduction of random events, like failures, which disturb the set of observed data.

To conclude, the mathematical approach indicated that chaos occurs in the production systems under study. The validation of such assumptions was difficult, but we have a strong belief that such behavior exists quite often and that assumption will be taken into account in the design and implementation of future management and control systems. Also, complementary studies have still to be conducted in real systems in order to verify if the aggregation of different chaotic cells forms a chaotic system: we do not know yet if the behavior of macro-systems can be similar to the behavior of micro-systems. Currently it seems that macro-systems can be subject to compensation effects, as in quantum theory, leading to global behaviors different to those we have in mind.

4.7 IMPACT OF THE BEHAVIORAL COMPLEXITY ON MANUFACTURING CONTROL SYSTEMS

4.7.1 Introduction

The main characteristic of a chaotic system is related to its unpredictability: the orbit of a strange attractor, and therefore the state of the system, is multidimensional and it is difficult to calculate and to determine the value of the variables and predict how they will evolve. Consequently the production system is non-predictable and we cannot anticipate how it will react. As a result, the approaches generally used for managing and controlling a production system do not apply and must be reconsidered.

The aim of this section is to study briefly two families of control systems widely used in industry and often described in the CIM (computer-integrated manufacturing) field. Their characteristics will be highlighted and some management rules will be deduced from these characteristics.

This classification will help the decision maker in choosing the type of strategy that has to be applied in any given situation. However, as the objective of this chapter is not to develop sophisticated control systems, we shall just describe main principles.

4.7.2 Case study 1: A 'planifying' control system

First, let us consider workshops conventionally controlled using an MRP control system. We consider that such scheduling systems are 'planifying' because a predefined quantity of parts/components is 'pushed' at the input of the workshop according to a given production program (the meaning of 'planifying' is stronger than 'planified'). Also, let us suppose they contain feedback loops of information. Different shops are connected together to make up a complete facility. The production system under study is submitted to a variable input, expressed by the demand (DM), the nature of which is either stochastic or deterministic chaos.

Such a production system thus resembles the one we studied in the Computer Manufacturing Division in IBM Montpellier–France. The problem consists of analyzing the behavior of this complete production system. For this purpose we used **industrial dynamics** techniques to study the interactions and the behavior of the different workshops:

CHIP ⟷ CARD ⟷ BOX.

Analysis based on dynamic modeling

The model was first developed using DYNAMO (System Dynamics Modeling language by Pughs–Roberts Associates), then submitted to a step function, in terms of demand (DM), at **box** level. The purpose was to analyze the consequences of the variations on DM at chip and card level and the evolution of the WIPs.

The equations used to process the flow of information were simple. The parameters which could be varied were: the size of the buffers, the 'rates' or throughputs, and the delays (response times) introduced in the formulae.

Figure 4.10 illustrates all the results obtained: when the demand 'DM' changed, the WIP was subject to decreasing variations. Nevertheless, for some values of the control parameters, these variations persisted in time, and in some cases, an oscillation occurred or an increasingly oscillating curve. In these conditions, the system is said to be '**pumping**'. In addition, these oscillations, due to feedback loops, are either amplified or absorbed according to the values of the different parameters. Although the system was modeled with simplified equations, it was difficult to control the system in a stable way and we could not firmly demonstrate that it was chaotic. We think this is due to the so-called decoherency and it has still to be checked.

System submitted to a chaotic demand

Now, let us consider, as the input stimuli, a real set of values for DM (see Fig. 4.10). Contrary to appearances, this initial demand is not 'random': for several reasons that we will not explain in detail here, the vector DM issued from a history in the BAT shop has a '**memory effect**' of previous events and situations. In fact, DM is the result of a deterministic demand modified by several and successive simple laws, fits and starts, generated by different decision makers. In this study, we did not have sufficient available data, whatever the product, to demonstrate firmly that the DM vector is chaos; nevertheless there is a 'strong belief' to support such an assumption.

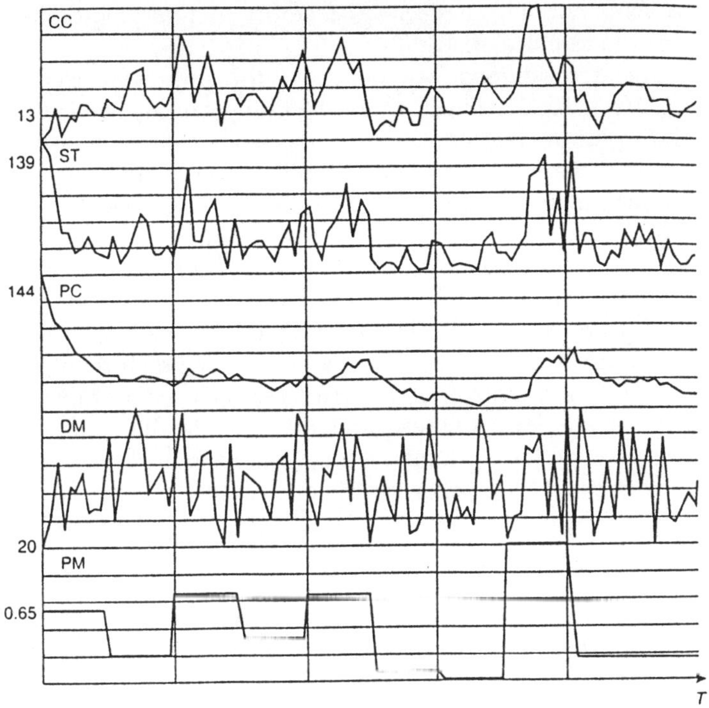

Figure 4.10 Behavioral analysis of a production system.

The curves showing the evolution of the different parameters are shown in Fig. 4.10. The production is represented by the curve (PC). We may observe that the evolution of the order file (CC) and of the WIP (ST) follows the input (DM). However, according to the quantity of buffers, their size, the delays, etc., the results are either smoothed, attenuated or amplified (pumping effect). When the buffers are too big, and corrective factors in the feedback loops too high, the system may diverge: it creates and amplifies the noise. Then the resulting inventory is submitted to abnormal and unfortunate variations.

With a more planifying system, which means that we require the system to follow more constraining and stronger rules, expected results may not be satisfactory because the production system is a 'dangerous system' with unexpected evolutions. In some cases, the resulting performance parameters may evolve by 'fits and starts'.

Again, a control system is not generally designed to process chaotic demands: such a condition degrades the overall performance of the production system. Thus, a planified system (designed to be submitted to very stable inputs and a predefined production program) with tight control rules is not able to react satisfactorily to a varying demand like this one or to structural disturbances.

Under these conditions, we observed the only way to limit the pumping effect, or the effect of the resulting chaos, is to react with **noise**: the solution is to introduce unexpected disturbances in the inputs. In this case, we associate a 'white noise' to the regular linear demand. Introducing 'noise' and uncertainties into the control parameters and values of the calculated inputs compensates for or even eliminates the pumping effect as well as the **coughs** and their unexpected or unwanted effects. Obviously, such a strategy seems surprising: production control managers usually try to smooth the inputs of the production system to eliminate large disturbances and to take into consideration its limited capacities. What we recommended is then to add 'white noise' to these smoothed data to eliminate the effects of behavioral complexity. Experiments have also to be conducted to evaluate what would happen with 'chaotic noise'.

Consequences

Most of the time, highly controlled systems (**planifying systems**) are of little use: they have too many constraints, which are difficult to manage, and are subject to uncertainties. They do not regulate the production system: variations and variances in performance can go out of control and it is difficult to monitor the complete system. There are two obvious questions:

- What organization can we propose?
- What type of modeling needs to be designed to analyze a seasonal or varying demand?

Whatever the level of complexity embedded in the model of such a system, we cannot represent everything in a model without generating noise. As written before, a simulation model just enables us to determine trends and macroscopic effects. Also, the manager of a shop is always introducing '**bad noise**': he/she modifies, permanently, in an unplanned way, the priorities of the products to be manufactured, according to situation changes or contradictory requests; he/she creates further disturbances and 'coughs'. As a result, the capacity of a manufacturing shop being limited, some parts are penalized more than others, which are prioritized; this cannot be taken into account in the model. Consequently, we have to keep in mind that the 'buffer' effect is always increased.

The only way to monitor such a system is to build a simple model, with small buffers, with an 'output follower' control system to correct the inputs and define overall strategies and guidelines. Finally, in order to integrate the 'history' (memory effect) of the inputs and results, the model must have smoothing and moving average capabilities, completed with a good sequencing system, at cell level only.

4.7.3 Case study 2: A flexible manufacturing system

Let us consider a flexible manufacturing system: multi-product, multi-process, with duplicated equipment, feedback loops, etc., with several and complex routings (Fig. 4.11).

Such production systems, related to electronic components, were modeled and simulated using RESQ (RESQ means RESearch Queueing Package: a discrete event simulator language developed by IBM). The resulting data were analyzed, as described above, and we concluded that they may have a chaotic behavior.

In this system, each node, or cell, in the network (production system) has its own control system with dispatching and behavioral rules. Chaos is due to the interactions between the cells. This may happen when there are only a few cells: oscillations created by the 'calls' of the buffers are propagated from one cell to another and vice versa. In practice, this results in the presence of many possible states for each cell. When this phenomenon is amplified and subject to nonlinear functions, it becomes chaotic. This is due to the 'chaining effects' of the physical and logical structure of the production system. These variations in the WIP, among the different cells of the shop, with time, are called a **caterpillar effect**.

Under these conditions, we cannot predict the behavior of the system, where the parts will be, and when. Some specialists recommend the use of discrete event simulation, but, again, such a resulting model, which could be not very complex, cannot integrate all the parameters and assumptions: it is not suitable for management purposes. Simulation can only give trends and approximate strategies.

In terms of effects, the cells have their own 'elasticity', and the disturbances are reduced because they are smoothed by themselves. With small buffers, the adaptation of the outputs to the inputs will be

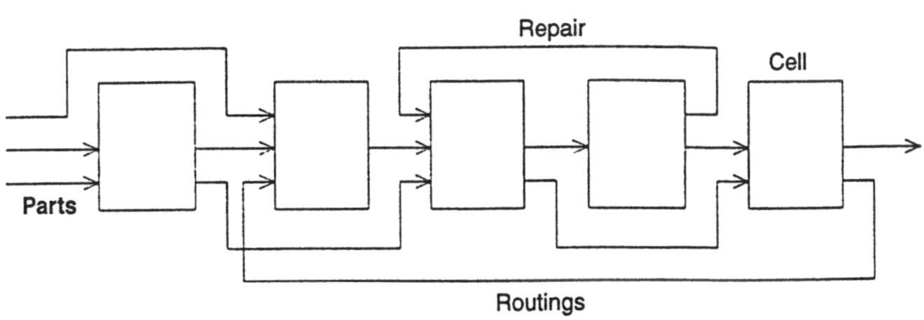

Figure 4.11 Example of a flexible manufacturing system.

quite fast. Using NETSIM, which is another kind of simulation tool (see above), we were able to define the best size of the buffers. Then, one effective method was to leave the system 'free': as a production system is such as a programmable network, it can regulate itself thanks to self-organization effects.

Here again, a model must be built up which is conditioned by the inputs or, according to inputs, is able to determine which global strategy should be applied in the production system. Experience shows that simulation is more appropriate and easier to use than an analytic approach.

4.7.4 General comments applicable to assembly shops

In any case, the reliability of the results depends not only on the quality of the inputs (data and stimuli), but also on the validity of the model built for the study (level of abstraction and structure). All the approaches we have studied are highly reductionist and modeling the problem to get the right solution is the critical point.

Behavioral complexity is mainly due to the importance of interactions between the agents present in the production system. This is why simple deterministic functions may generate **chaos**. For this reason, we recommend concentrating more on the interactions taking place in the production system, rather than trying to improve the functions themselves, through scheduling.

In the same way, order can emerge from disorder. Moreover, chaos and fractals lead to the same mathematics, which affect self-organization and autonomy. These concepts have a direct effect on the control (i.e. the management) of the production system, and these properties have to be further analyzed for a better exploitation. This has already been done in cellular automata: solutions can be generated from simple and communicating entities which can evolve quickly with time, towards single, periodic, quasi-periodic and strange attractors. Some properties may emerge from such networks and can be applied to scheduling with regard to specific situations. It is a very promising approach.

As for management and control, a completely different approach is needed. When faced with a complex system, we can conclude that:

- Reducing the values of some control parameters leads to a more stable system. It is under control but becomes inflexible, and changing its state requires a lot of effort and resources.
- Choosing the appropriate values for the control parameters will position the system in the area of a 'low' chaos. It will be easy to move the system into the right part of Figs. 4.6 and 4.7. This means that the system is very flexible and multiple states can be reached quickly, according to demand. This principle is to be used in self-organization:

it is a way of getting self-maintaining and self-reconfigurable production systems.

- Finally, as such production systems are very sensitive to initial conditions and to the value of the control parameters, new concepts have been defined to develop inventory and control systems managed through their performances and their outputs rather than their control parameters. This approach is a very promising one.

Without detailing here the activities in progress, we can stipulate that:

- Further work is needed to apply autonomy to scheduling. This concept is directly derived from the studies undertaken in the field of multi-agents systems.
- Self-organization characteristics of complex production systems have to be more precisely defined before being applied in industry.
- We have still to understand how self-organization controls interacts with evolution and selection principles.

In addition, as these concepts have an impact on skills, education and social organization (Warnecke, 1993), the development of multi-skill approaches is required in this field.

4.8 CONCLUSIONS

Algorithms, behavioral analysis and, to some extent, the management of complex target systems have been investigated in this chapter. More precisely, an in-depth study was carried out in the field of ill-defined and dynamic complexity.

To summarize, most production systems can be subject to deterministic chaos, called the **caterpillar** effect. Under such conditions, conventional management and control systems do not apply. This is why we have developed a methodology for identifying such behaviors.

Since an environment is difficult to control and the target system is unpredictable, we cannot force it to react as we wish and a new paradigm, together with an appropriate new control and monitoring approach, must be used.

We cannot cope with complexity by more complexity, that is to say, with a management system too detailed and complex: the more **complex** a target system is, the more **simple** its associated control must be.

Conventional approaches are generally based on the **products** (flow of parts, scheduling with a given set of capacity limitations, resource constraints) although the emphasis must be put on the **process** (tuning, balancing): a given release of the parts cannot be managed without taking into consideration the process itself, or the coordination of actions within this process.

Instead of being monolithic with a complete set of embedded functions, the management system should be layered. In this chapter we have not developed these concepts but we found that two application levels are sufficient: the first higher one gives the global guidelines to be applied to the target system, the second enables local control to be performed at microscopic level. It is characterized by a certain rigidity, leaving some whirlpools at microscopic level. The resulting models have to be modular and, more specifically, the lower control levels must be able to adjust to evolving situations: local rules will take into account the tasks and characteristics of the related cell or workcenter.

The microscopic level must be analyzed in depth and a precise and local scheduling should be developed accordingly. Scheduling, sequencing and/or dispatching should take into account real situations subject to uncertainties and non-linearities (ill-defined complexity). As conventional analytical techniques are not sufficient, approaches should be used which are based on simulation in which AI and genetic algorithm-based methods are combined.

Flexible and efficient control of a target system can be achieved through such concepts. They will be the basis of future integrated manufacturing systems.

Some of the above concepts have been successfully implemented in an application called Line Management Advisor (LMA). This application is an example of the tools developed at IBM Montpellier–France by the Advanced Technologies Group, to perform dynamic scheduling of a manufacturing shop.

Our experience shows that solving complex problems requires new and specific approaches; as a production system is subject to several variables and stimuli, several types of complexity are present at the same time in the production system. This requires the combination of different techniques in a single solution. Such solutions are called 'hybrid applications'; they involve different skills or problem-solving techniques which are not always available at the same time and in the same area. For this reason, production managers must review their organization, skills and training programs.

Until now, only static aspects of production systems have been generally considered. Behavioral complexity introduces a new paradigm and a more dynamic view of production systems is needed. It is necessary to emphasize **interactions** and coordination of functions within rather than the **functions** themselves of a production system. Also, inventory and control systems based on the analysis of the 'results' rather than the definition of process control parameters can be satisfactorily implemented. However, a general approach associated with an appropriate methodology has to be refined.

This is why studies should be undertaken into self-organization principles in production systems. In the future, flexibility and adaptability of production systems will be achieved through autonomous and independent agents interacting together. Also, properties of chaos will be used to implement more adaptive production systems. In this framework the behavioral complexity which is described above is of the utmost importance.

GLOSSARY OF TERMS

AGV	Automatic guided vehicle
AI	Artificial intelligence
BAT	Bond assembly and test
BM	Bill of material
CFM	Continuous flow manufacturing
CIM	Computer-integrated manufacturing
CRP	Capacity requirement planning
DM	DeMand
DSL	Dynamic Simulation Language
DSS	Decision support system
FCS	Floor control system
FFT	Fast Fourier Transform
FMS	Flexible manufacturing system
GA	Genetic algorithm
KBS	Knowledge-based system
KEE	Knowledge engineering environment
LMA	Line manager advisor
MAQ	Maximum allowable quantity
MCS	Manufacturing control system
MRP	Material requirements planning
NC	Numerical control
NO	Numerical optimization
PLOOT	Plant layout optimization
RESQ	Research queuing package
SMS	Shop management system
TAT	Turn around time
TCM	Thermal control module
WIP	Work in process
WMS	Workstation management system

REFERENCES

Beaudoin, J. and Massotte, P. (1990) *Line Manager Advisor and Object-oriented Approaches*. European Conference on Productivity, Boehlingen, Germany.

Berge, P., Pomeau, Y. and Vidal, C. (1988) *L'Ordre dans le Chaos*, Hermann, Paris.
Beziat, P. and Massotte, P. (1989) *Plant Layout Optimization (PLOOT): A tool using Group Technology and simulated annealing*. MICAD, Proc. 8th Int. Conf., pp. 87–106.
Davis, L. (1991) *Handbook of Genetic Algorithms*, Van Nostrand Reinhold, New York.
Fron, A. (1994) *Programmation Par Contraintes*, Addison-Wesley.
Gleick, J. (1987) *Chaos, Making a New Science*, Viking, New York.
Intelligent Manufacturing Systems I (1994) Inter-regional Meeting on Information Technologies and Integration in Manuf., Dubrovnik/Helsinki.
Kieffer, J.P. (1989) 'Typologies de Systemes de production'. *Revue Française de Gestion*, November.
Lassez, C. (1987) 'Constraint logic programming. *BYTE*, August.
Massotte, P., Paul, C. and Robert, D. (1991) *Parallelisation of Applications in CIM: Results and Criteria for Further Applications*. High Performance Computing, Montpellier, October.
Michalewicz, Z. (1992) *Genetic Algorithms + Data Structure = Evolution Programs*, Springer-Verlag.
Miller, T., Sutton, R.S. and Werbos, P.J. (1990) *Neural Networks for Control*, The MIT Press, Cambridge, MA.
Muller, J.D. (1993) La Perception Structurante. PhD thesis, ENSAE, Toulouse.
Paul, C. (1993) Parallelization of applications in manufacturing. PhD thesis, Laboratoire d'Informatique, Nantes, France.
Rich, G. (1983) *Artificial Intelligence*, McGraw-Hill.
Scardigli, C. (1994) Implosion: Gestion des stocks par la replanification amont. PhD thesis, Laboratoire de Genie Industriel, Grenoble.
Simon, H.A. (1981) *The Sciences of the Artificial*, MIT Press.
Varela, F.G., Maturana, H.R. and Uribe, R. (1974) 'Autopoiesis: The organization of living systems, its characterization and a model'. *Biosystems*, **5**, North-Holland, Amsterdam.
Warnecke, H.J. (1993) *The Fractal Company*, Elsevier.

FURTHER READING

Arecchi, F.T. (1985) 'Caos e ordine nella fisica'. *Il nuevo saggitatore*, Soc It. Di Fisica, **1**(3), 35.
Banker, R.D., Datar, S.M., Kekre, S. and Mukhopadhyay, T. (1985) Costs of products and process complexity. Carnegie Mellon University, EDRC report, grant 8522616.
Balmes, R. (1989) Flexible Line Simulator (FLI) – functional specifications. Technical report 79070, IBM Montpellier–France.
Beaudoin, J. and Massotte, P. (1990) Line Manager Advisor (LMA). Technical report 79079. IBM Montpellier–France.
Chevalier Milhe, N. (1994) Ordonnancement d'un systeme de production. PhD thesis, LIRMM, Montpellier, France.
Devlin, K. (1988) *Mathematics: The New Golden Age*, Penguin.
Giard, V. (1987) *Gestion de la production. Calcul economique*, Economica, Paris.
Greene, J.H. (1987) *Production and Inventory Control Handbook*, McGraw-Hill.
Haken, H. (1978) *Synergetics*, Springer-Verlag.
Haynes, B.R. and Billings, S.A. (1992) 'Qualitative analysis of chaos in nonlinear data systems and system identification'. *Journal of Systems Engineering*, **2**(2).

Lesourne, J. (1987) 'Un systeme autopoietique: le marche'. *AFCET – Revue de Systemique*, **1**(2).

Lesourne, J. (1989) 'The OR contribution to strategy formulation in a turbulent environment'. *European Journal of Operations Research*, **38**, 286–9.

Massotte, P. (1993) *Behavioural Analysis and Management of Complex Production Systems*. FUCAM 1993, Int. Conf. on Ind. Eng. and Prod. Manag., Mons, Belgium.

Midoux, N. (1985) *Mecanique et Rheologie des Fluides en Genie Chimique*, Lavoisier, France.

Mosekilde, E., Larsen, E.R. and Sterman, J.D. (1989) *Coping with Complexity: Deterministic Chaos in Human Decision Making*. TIMS 29, Conf. on Intell. Manuf. Sys., Osaka, Japan.

Pao, Y.H. (1989) *Adaptive Pattern Recognition and Neural Networks*, Addison-Wesley.

Perez, J.C. (1988) *De Nouvelles Voies vers l'Intelligence Artificielle*, Masson, Paris.

Proth, J.M. (1992) *Conception et Gestion des Systemes de Production*, PUF (Presses Universitaires de France).

Sterman, J.D. (1988) 'Deterministic chaos of human behavior: Methodological issues and experimental results'. *System Dynamics Review*, **4**.

Temmyo, T., Hasegawa, M. and Matsuka, H. (1988) *Distribution Control in Manufacturing Cells*. IBM Tokyo Research Laboratory, Internal Report (IROS 1988).

Titli, A. (1979) *Analyse et Commande des Systemes Complexes*, AFCET, Cepaudues Editions, Paris.

Tse, E., Lee, H. and Bevan, W.U. (1987) *Computer Integrated Manufacturing Enterprise Management Systems*, Springer-Verlag.

Zeigler, B.P. (1987) 'Hierarchical modular discrete-event modeling in an object oriented environment', *Simulation*, **49**(5), 219–30.

CHAPTER 5

Loading and scheduling a hybrid multiproduct plant via a rule-based planning system

A. Artiba

ABSTRACT

In this chapter, a production system organized in parallel multiproduct manufacturing lines is presented and the objectives of its management system are stated. Then, a classification of the essential constraints that characterize this type of system is given. Also, a classification of the different pieces of knowledge and data necessary for the loading and scheduling development is presented.

A prototype developed for the Baxter Company is described and presented. The results according to the given evaluation criteria are discussed.

In the last part, an architecture of a multi-model system is presented (incorporating the simultaneous use of several models: optimization methods, heuristics, expert systems and simulation tools) for loading and scheduling. It illustrates the perspectives and the aim of our research work. This chapter is a report of the author's actual experience in the design and implementation of a production control system.

5.1 INTRODUCTION

Currently manufacturing companies are faced with markets which demand a great variety of products of high quality. These products have a shorter and shorter life cycle and increasingly smaller delivery times. These companies must therefore make their production system more and more flexible, reduce the costs related to production and storage and react rapidly to fluctuations in demand. For these reasons, companies follow two technological axes:

- an increasingly higher degree of automation;
- more and more sophisticated computerized management techniques.

The development of production schedules is an extremely important task in industry, although this is an area where a theoretical approach has not yet provided a general solution because the specific features of practical examples are both numerous and varied. The research work presented in this chapter concerns loading and scheduling of production systems organized as multiproduct parallel production lines. This type of production system is common in industry, notably in the pharmaceutical, chemical, cosmetic and brewing industries, and could also be extended to other types of manufacturing systems.

The multiline/multiproduct problem is a composite case, i.e. it has aspects of both parallel machine and flowshop, and this type of problem has received little attention in the literature so far. In addition Gupta (1988) has shown that this so-called hybrid flowshop problem is NP-complete. For this reason, human expertise is the key to the solution and is based on:

- experience;
- decision strategies and global company objectives;
- heuristics.

The knowledge-based approach is well suited to integrating these three points; it can simulate human expertise and reason in each specific situation in order to take the appropriate decision.

The loading and scheduling of parallel multiproduct manufacturing lines are, at this moment, open problems (Artiba, 1994). In addition, there is no commercial software package able to tackle efficiently this type of problem (Artiba, 1990; Musier and Evans 1990, Van Dam *et al.*, 1992).

Real mutual benefits can be achieved between expert systems and operational research (O'Keefe, 1985): 'The knowledge engineer can benefit from the experience of OR in model building; expert systems will use OR techniques. For the OR scientist, the expert system is another tool in the tool-kit. Integration with other computing systems will lead to DSS (Decision Support Systems) able to draw on a knowledge base and to reason with the user.' Muller *et al.* (1987) illustrate some similarities between the OR and expert system approach and point out that OR should take advantage of the new techniques provided by expert system technology to tackle complex production scheduling problems.

In this chapter, a production system organized in parallel multiproduct manufacturing lines is presented and the objectives of its management system are stated. Then, a classification of the essential constraints which characterize this type of system is given. Also, a classification of the different pieces of knowledge and data necessary for the loading and scheduling development is presented.

The knowledge acquisition phase is discussed and a classification of rules is given. Each kind of rule is illustrated by an example. Then, the scheduling algorithm is presented and illustrated by a demonstrative example in order to highlight the methodology used. A prototype developed for the Baxter Company is described and presented. The results according to the given evaluation criteria are discussed.

In the last part, an architecture of a multi-model system is presented (simultaneous use of several models: optimization methods, heuristics, expert systems and simulation tools) for loading and scheduling. It illustrates the perspectives and the aim of our research work. This chapter is a report of the author's actual experience in the design and implementation of a production control system.

5.2 DESCRIPTION OF HYBRID PARALLEL MULTIPRODUCT MANUFACTURING LINES

The parallel multiproduct system consists of parallel lines using shared and peculiar resources. Each line produces a certain group of products. The lines in a parallel multiproduct system are interdependent; this interdependence can arise from:

- product characteristics;
- shared resources;
- production line balancing.

In the present case, the discussion is limited to systems which are:

- semi-continuous;
- where products may be transferred during manufacture from one line to another;
- without intermediate storage areas (buffer zones).

At this stage in the description of the procedure, it is useful to illustrate this kind of production system with an industrial example, which will be referred to throughout the remainder of the text. The shopfloor presented is concerned with the manufacture of pharmaceutical products used for perfusion at the Baxter company in Belgium.

The manufacturing procedure for these particular products can be conveniently divided into four distinct steps (Artiba, 1990): the preparation of the solutions; their analysis; the filling of the bottles or sachets with the solution and their sterilization (Fig. 5.1).

The analysis of starting materials and mixed solutions requires the use of instruments which form part of the common resource pool. Thus, the mixing tanks (considered as transforming machines) are shared between

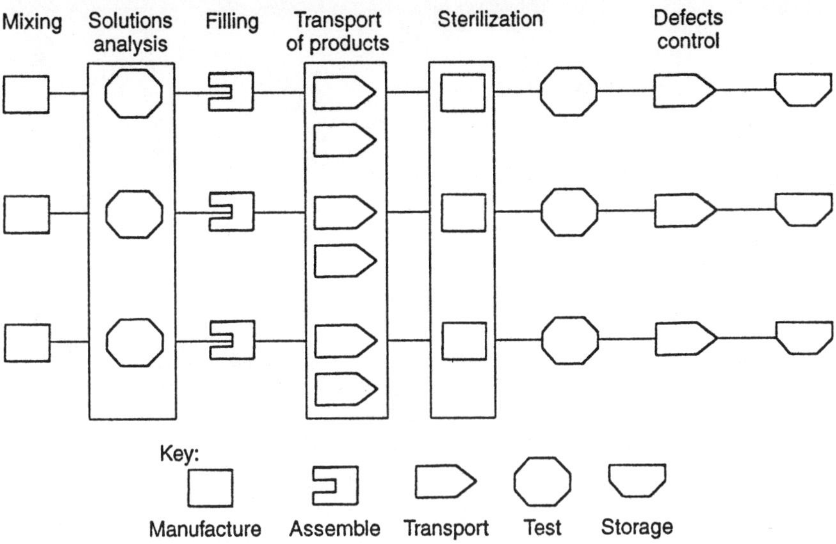

Figure 5.1 Pharmaceutical manufacturing process.

different solutions to be handled on different lines. A similar situation exists for other resources: printing machines, sterilizing trays, autoclaves, sealing machinery, prepackaging machinery, operators, etc.

Figure 5.2 presents the SADT (Structured Analysis and Design Technique) model of the main operational activities of the Baxter manufacturing process (pharmaceutical department – perfusion materials).

The first activity (solution mixing) consists of filling the (appropriate) mixing tank(s) with distilled water (or de-ionized water), adding the components (following the bill of material: BOM), mixing the solution and analyzing it by the laboratory. The analysis time may vary widely from one solution to another.

The second activity (line set up) concerns the preparation of the production line: cleaning of the line machines (with distilled water at a defined temperature and for a certain time, following the BOM), assembling the filters, controlling (automata) the connection between the line and the (right) mixing tank and flushing the line with the solution.

The filling activity includes filling of plastic bags (for three production lines: V1, V2 and V3) or glass bottle (for one dedicated line), over pouching (put the bag with the solution into a protection sachet), printing and arranging the filled bags into the specific type of sterilizing tray.

The sterilization activity deals with the choice of the qualified

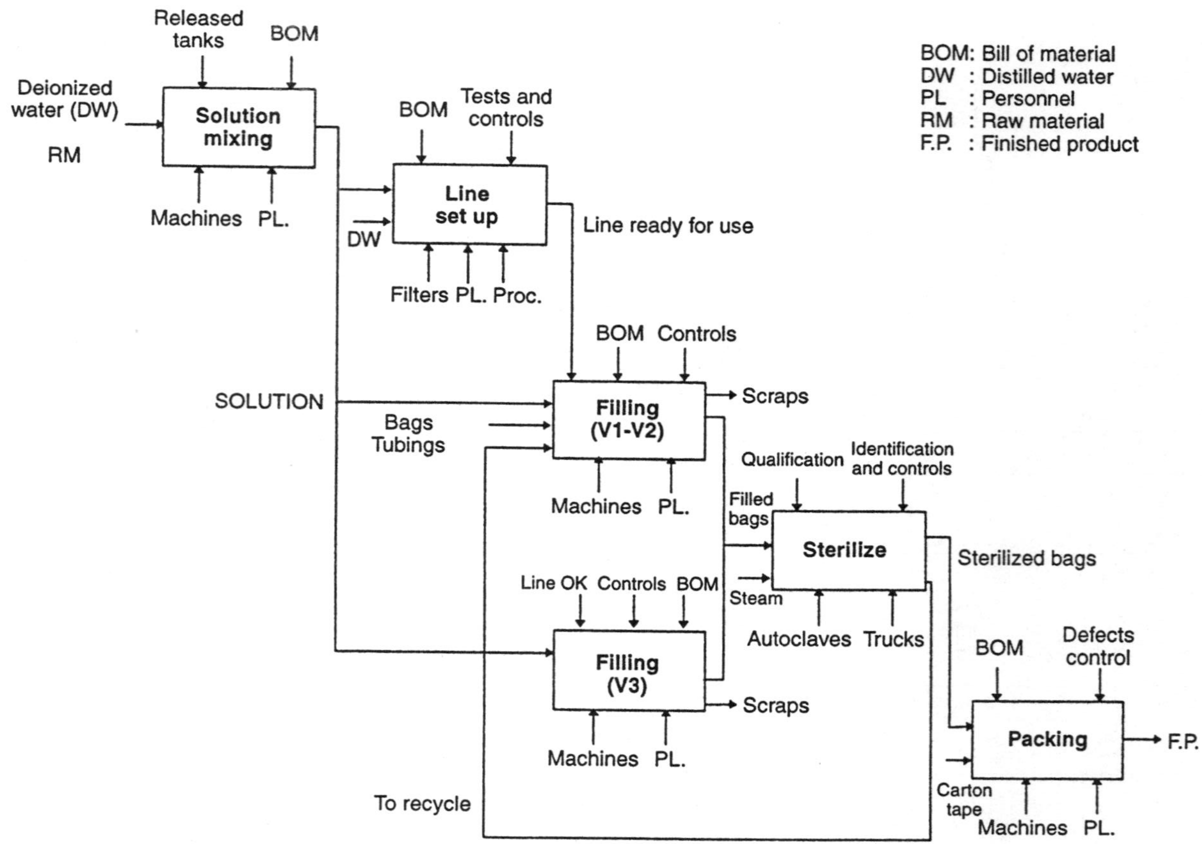

Figure 5.2 SADT model of the manufacturing process.

autoclaves, the arrangement of the trays into the (right) autoclaves, the sterilization process itself and visual controls (to recycle if any defect).

The packing activity consists of defects control, arrangement of perfusion bags (or bottles) into the appropriate carton (size, language, etc.), inclusion of prospectus and palletization.

5.3 OBJECTIVES OF THE PRODUCTION CONTROL SYSTEM

The planning/scheduling function consists of programming in time the production sequences on different lines, in the presence of various constraints linked to the product, the process, and the order situation, bearing in mind the overall objectives of the company. The problem is rendered very complex by the large number of constraints, the product diversity and interdependence, the dynamic and shifting nature of the production environment, as well as by conflicts between some of the stated objectives.

The principal objectives are:

- satisfaction of all the strong constraints in order to develop a workable schedule;
- satisfaction of the maximum number of weak constraints. The acceptability of a schedule is assessed in terms of the weak constraints which it can take into account. The success of the schedule is measured in terms of the degree to which these constraints can be respected;
- minimization of production costs;
- minimization of stock costs;
- quality control (as a function of the choice of machines);
- optimization of resource use;
- adherence to delivery schedules.

The user (planner) can suggest to the system some hypothesis to answer what-if types of questions (using the system as a simulator). The planner can use the system to generate several schedules (according to his/her modifications) and then chooses the preferred one.

5.4 CLASSIFICATION OF THE CONSTRAINTS

This section is devoted to the listing and classification of a set of constraints on the production system described previously.

We make the distinction between **strong** (or imperative) constraints, i.e. those which must be observed in all cases, and **weak** (or conditional) constraints, which may be relaxed, though this could result in a less efficient scheduling.

Depending on the particular production unit, some constraints which are classified as weak, herein could be designated as strong (e.g. delivery deadlines).

5.4.1 Constraints linked to the products

The type and nature of a product limit the use of some resources and the choice of production sequences. We give some examples below:

- production periods: the manufacture of some products can be restricted to some periods (strong);
- availability of raw materials (strong);
- products sequence: some products cannot be manufactured in the same sequence on the same line (strong);
- simultaneous production on parallel lines: some products cannot be manufactured at the same time on parallel lines (strong);
- resources capacity: the capacity of a resource may depend on the type of products using it. For instance, for the product P_i, the resource R_j cannot exceed 80% of its capacity, whereas for the product P_k the use of the resource capacity can reach 100% (strong);
- minimum quantity (strong);
- maximum quantity (strong);
- production quantity has to be expressed in a multiple of batch size (strong).

5.4.2 Constraints linked to the process

These constraints concern the use and the availability of the resources such as:

- the efficiency coefficient of a resource (weak);
- the minimal rate of use of a resource: a resource cannot be used under a certain rate of use (for example 20% of its capacity) (strong);
- the maximum rate of use of a resource (strong);
- the duration of resource assignment (strong);
- the type and the number of available resources (strong);
- the set-up times (strong);
- the calendar of production for each line (strong).

5.4.3 Constraints linked to the product orders

These constraints concern the quantities and the delivery dates of each product:

- yield quantities (weak);
- due dates (weak);
- the state of the inventories (weak).

5.5 KNOWLEDGE ACQUISITION AND CLASSIFICATION

In this section, we present and discuss how the acquisition phase has been carried out in collaboration with different experts and managers of the Baxter Company. We highlight the difficulties encountered during this period (which took about six months) and describe the different steps. We then present the knowledge representation formalism used and give a classification of rules, each type of rule being illustrated by an example.

5.5.1 Knowledge acquisition

Knowledge acquisition is a non-trivial task in the industrial world, especially if this world is completely unknown to the knowledge engineer. This was the case for the current study. The knowledge engineer can acquire the knowledge by means of surveys, by recordings, by studying books or by observing people at work, but no analysis is complete without face-to-face discussions with the domain experts (Hart, 1989). In this case, the knowledge was acquired essentially by means of interviews. This widely used technique allows rapid acquisition of general concepts for the problem at hand (Nassiet, 1987).

The different steps and the essential difficulties encountered during the knowledge acquisition phase are:

- Expert's availability and set up of a work group: Because of their responsibilities in the company, it is difficult to bring together domain experts. In order to solve this problem, meetings were organized by the head of the planning department. The different department heads (production, laboratory, purchasing, sales, quality control and planning) systematically attended these meetings during which the objective of the system, the different phases of development and the programme of participation of each of the experts were defined. The formation of a work group and the policy of monitoring a project is essential for the success of the project. The efficiency of the development team is chiefly dependent on the presence of a driving force in both the industrial section and the knowledge engineers.
- Motivation of persons involved in the project: All the sections which are potential sources of information and knowledge (i.e. different departments) for the production control system must therefore be motivated. This motivation first of all begins with an explanation of the role that each section concerned can play in the development of the system. Then, it is necessary to show each person's gain to this system (e.g. definition of needs and constraints linked to the different departments involved in the development of the system). In our case, the managers of the Baxter Company were motivated and they were

able to transmit this motivation to the different people concerned in order to commit them fully to the success of the project.

- Collection (or acquisition) of knowledge (multi-expert case): The knowledge and information are shared among several experts of the same field who have with different origins and interests. We consulted them in turn in order to gather a synthesized knowledge of the field. Each expert provided information relative to a specific topic. To communicate with experts from different fields, one must become familiar with their technical jargon. We constructed a dictionary of the vocabulary used in the company as we went along the knowledge acquisition phase.
- Knowledge acquisition procedure: The knowledge extraction phase itself concerned the analysis of all the planning strategies which are used by the experts in each situation (known), and the different rules associated with them. Strategies and rules were analyzed in detail to check their accuracy and their completeness. These constitute the kernel of the prototype. The acquisition of knowledge was performed by the development of a complete planning (product sequencing and resource allocation). With this approach, we could identify and characterize the reactions, choices, rules and information used by the experts.

All the sessions were recorded. Between two sessions of knowledge acquisition, the work of knowledge encoding consisted of:

- carrying out a synthesis of information kept,
- structuring the rules, and
- establishing the list of questions corresponding to a lack of knowledge, to ambiguities or to imprecisions encountered in the previous phase.

At the beginning of a session, this list of questions was submitted to the relevant expert. The formalized rules were then either validated or modified by the expert. After these two stages, we proceeded with the acquisition phase. This phase of information gathering and knowledge acquisition demands a considerable and continuous effort from the part of knowledge engineers, during all the phases of development (identification, etc.). The approach chosen ensured the consistency of information and knowledge extracted from the different experts. The next section describes the knowledge representation formalism and the classification of rules.

5.5.2 Knowledge representation formalism and classification

Dynamic-type knowledge is described below: it consists of tactical and strategic rules and expert directives at the different stages of the planning establishment.

Conditions–actions type of production rules are very simple and practical means to represent dynamic knowledge. They allow a very accurate description of the decisions that an expert has to cope with in order, for instance, to carry out a diagnosis or to plan a task. The formalism must represent as accurately as possible the expert's knowledge. In our case, representation using production rules seemed to be the most appropriate one.

To solve a problem, experts explain their strategies and directives in an 'if conditions then actions' form.

As stated in the operational architecture of the system (section 5.9) these rules call external procedures and algorithms for specific aims: priority rules (EDD, etc.), search heuristics, selection algorithms, calculation procedures, resource allocation (bin-packing like algorithms), data validation, presentation of results in desired format, information updating, etc.

We note that the experts express their preferences on potential alternatives (if any) and strategies. They can justify their choice for an adopted solution but never refuse the use of well-adapted algorithms (e.g. resource allocation).

We can summarize the expertise as the expression of objectives and constraints embedded in solution strategies and heuristics. The important task to be achieved by a knowledge engineer is to structure these pieces of expertise knowledge, verify their accuracy, search for (existing) heuristics and algorithms and eventually adapt them for some specific problems to complete (or improve) this know-how. We insist on the fact that when we can apply an algorithm we have to do so and never replace it by a set of rules.

A classification and examples of rules used in our prototype follow.

Analysis rules

These are the balancing rules for the production lines according to different aspects including:

- global load;
- line by line load;
- line by line capacity for the target horizon for each type product.

Example:

if the sum of the load of line 1 and line 2 is less than the corresponding capacities mentioned in the operating plan,
and if the load of line 3 is more than its capacity C mentioned in the operating plan **and** if the load surplus amounts to at least one day of production of one of the two lines,
then produce on line 1 or line 2 with respect to the capacity available.

Management rules

The rules have been classified into several groups.

General rules
Concerning all the lines and all the products.

Example:

if a product is incompatible,
and only a small quantity is to be produced,
and the production period is not mentioned,
and the sum of the analysis times does not exceed the laboratory capacity,
and it becomes badly needed,
then produce it at the end of the day with a filter replacement.

Rules specific to types of products
This knowledge is related to a set of products clustered according to specific criteria.

Example:

if there is at least a five-day-of-production demand for 'small formats',
and no other deadlines concerning other products manufactured on the same line occurs at the same period (five days),
and it happens neither on a weekend nor on a day following a weekend,
then launch the 'small formats' producing campaign during this period.

Rules specific to a family of lines

Example:

if products of 'X format' are produced on line 1,
and chemical solutions produced on line 1 are different from the ones produced on line 2,
then the bag format on line 2 must be different from 'X'.

Allocation rules of critical resources
The expert knowledge about allocation of critical resources are of two types:

- satisfaction
- anticipation

Example 1 – satisfaction:

if there are not enough tanks for a production day,
and the permutation of incompatible products is not satisfying,

then plan the products with the same solutions and the same quantities on line 1 and line 2 (parallel use of tanks).

Example 2 – anticipation:

if a tank is free for a day,
then plan the production of a product requiring a high analysis time (more or less six hours of analysis) at the beginning of that day.

Disturbances resolving rules

These are the rules concerning the planning adjustment after disturbances. The system tries to restrict in a local way the consequences of disturbances, by a permutation of the products to be manufactured on the same week, for instance.

Example:

if there is a lack of a raw material for a day X,
then prompt for the day Y of the availability of this raw material and permute these two days (X and Y).

5.6 DATA MODELING

The structure of the data is organized according to its nature and the context of its use. We break up production data and management data, related to the products and linked to the resources, as illustrated in Fig. 5.3. We note that data information linked to resources includes information about processes.

5.6.1 Production data

Related to products

- identification (name, type, etc.),
- resources used (production line, specific resources, etc.),

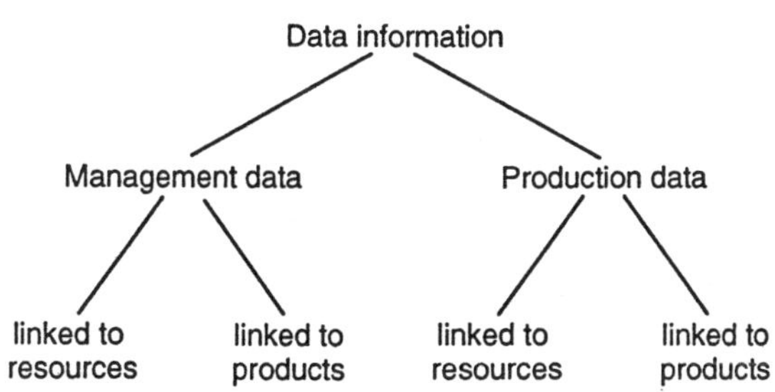

Figure 5.3 Organization of data structure.

- operational data (batch size, production period, etc.),
- data linked to the nature of a product (successor and antecedent products, freeing lead time, etc.).

Related to resources

- identification (resource name, position on the shop floor, etc.),
- data characterizing the use of a resource (number, set-up times, etc.),
- operational data (state, product in process, etc.),
- analysis data (breakdown frequency, average duration of a resource breakdown, etc.).

5.6.2 Management data

Related to products

- commercial data (sales forecasts, minimum coverage of each product, etc.),
- scheduling data: we distinguish between two types of data: the **forecast data** (stock levels of products and raw materials, due dates, etc.) and the **monitoring data** (state of operations available (finished, etc.), state of the progress of work (normal, stopped, etc.), etc.).

Related to resources

- forecast data (production calendar, scrap rate, etc.),
- monitoring data (number of available resources, end of operation, etc.).

The model selected to represent the data is an object-oriented model. The capability of the structured objects to represent static interrelated data leads us to choose this model for the following reasons (Artiba *et al.*, 1989):

- Modularity: the objects are stand-alone independent entities with both their descriptive data and the corresponding procedures.
- Close to the reality: natural representation of the knowledge resulting in increased ergonomy, representation facility, extendibility, performance.
- Possibility to give values by default: reasoning with incomplete data is allowed. This means that for a given attribute (i.e. number of operators dispatched for a production line), if there is no explicit data, then the system looks for an implicit data (if any) which represents the value by default.

In production management, the data can be organized as a hierarchy of objects. Such a hierarchical structure is a tree in which object properties can be transmitted from the root object to the low-level objects (inheritance principle).

Alasuvanto *et al.* (1988) present two production management systems developed with an object-oriented language and give the advantages of using such programming approach. Further details on object-oriented programming can be found in Cox (1986).

Figure 5.4 shows an example of an object representing a specific product with its properties and their values. This object has a name (AAA64) and belongs to the 'products' class. This class inherits attributes (properties) from the 'Production' class (line, specific resources, lot-size, etc.) and from the 'Management' class (sales forecasts, minimum coverage in months, due dates, etc.).

The values of attributes are imported from a database. To each product corresponds an object belonging to the 'products' class. Figure 5.3 shows a Nexpert/Object (expert system shell) screen editor representation. We can associate a procedure (called Meta-Slot in Nexpert) to a specific attribute (or property). To the property 'quantity-to-produce' is associated a demon which is activated when the value of this property changes (in Nexpert it is called if-change Meta-Slot). If the value of this property changes then the level-of-stock and latest-due-date are automatically recalculated for concerned products (which are planned).

5.7 FUNCTIONAL ARCHITECTURE

In this section, the different phases for the development of a schedule are indicated. The functional modules corresponding to each phase are shown in Fig. 5.5.

5.7.1 Net requirements analysis

This module performs three functions:

- determination of yield quantities;
- selection of products to manufacture: this first selection of products is based on minimum and maximum quantities per product and on their latest completion dates;
- production line load calculation: the load of a specific line is determined by the sum of the quantities of the selected products assigned to this line.

5.7.2 Production line balancing

The first phase is the verification of the loads of each production line. This phase determines the overloaded, underloaded and balanced lines.

The second phase is the deviation of some products from the

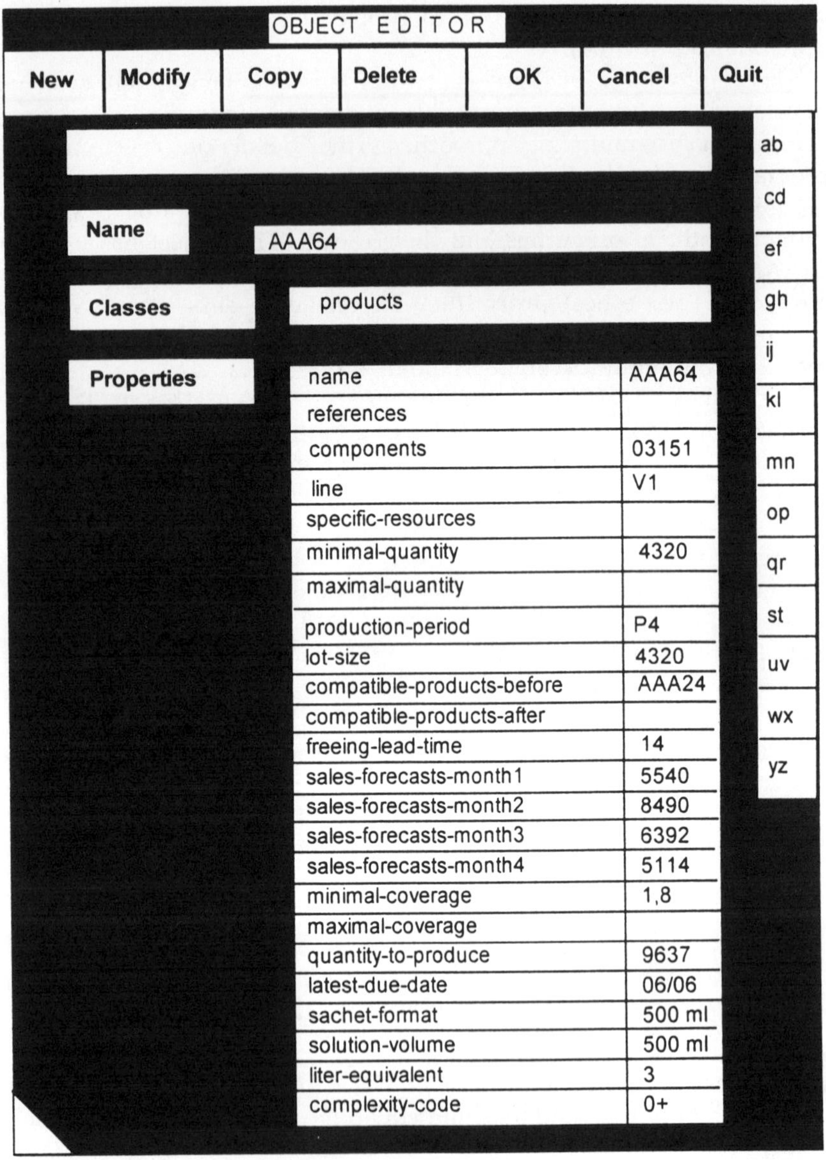

Figure 5.4 Example of a product object.

overloaded line to the underloaded one. The selection and the deviation of products depend on some criteria such as: the priority of the customers for whom the products are destined; the due dates of the products; the characteristics of both lines; the state of the shared resources.

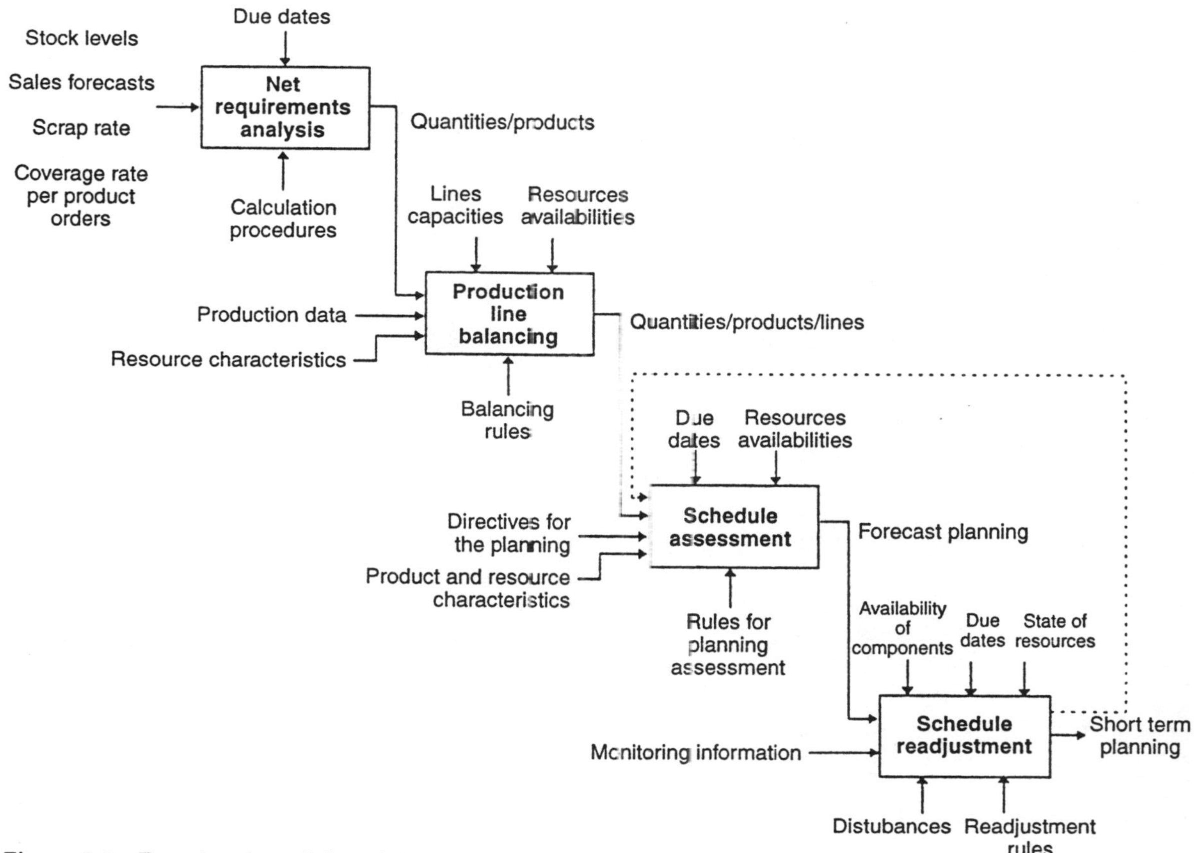

Figure 5.5 Functional modules of the production control system.

5.7.3 Schedule assessment

In a parallel multiproduct production system, the first issue to be considered is with which line to begin the determination of the production sequencing.

The first phase is thus the determination of the priorities of the different production lines. This is based on some dynamic criteria as follows. The line which has absolute or the highest priority is: the line with the highest number of urgent products (products for immediate delivery); the line with the most restrictive production calendar; the line using the highest number of shared resources; the line reserved to a specific type of product.

The production sequencing is largely based on the human planner's expert knowledge. This knowledge represents the tactical and strategic rules used by the human planner to resolve or anticipate the resolution of current and foreseeable problems.

The general strategy of the expert is first to work out a planning which respects the strong constraints (those which cannot be relaxed) and then to try to work out a planning which respects the maximum number of the weak constraints (those which can be relaxed but only by reducing the quality of the planning).

5.7.4 Schedule readjustment

The deviation between the forecasted and the actual situation is determined by comparing them. The production disturbances are detected by the system if there is a deviation between the forecasted and the actual situation; or detected and introduced into the system by a human operator. The different phases of the schedule readjustment module are: disturbance detection; analysis of disturbances; planning correction.

5.8 SCHEDULING ALGORITHM

1. Determination of the average number of products per day.
2. Determination of priorities of lines.
3. Applying the earliest-due-date (EDD) rule to the line with the highest priority (provision of the first product to be manufactured).
4. Construction of a list of products that can be manufactured with the previous one in the same sequence (same line, same sachet format, compatible products, analysis complexity, etc.).
5. Choice of a sequence (constraints: succession, resources, complexity, etc.).
6. Allocation of resources (a bin-packing like procedure).
7. If the sequence is feasible go back to 1, else go to step 5.

Further details on these different steps as well as other aspects of our system can be found in Artiba (1990).

5.9 IMPLEMENTATION AND VALIDATION OF THE PROTOTYPE

5.9.1 Industrial application

The industrial support is the pharmaceutical manufacturing department of the Baxter company in Belgium. This department produces a set of differentiated perfusion materials such as solutions in glass bottles or plastic bags.

5.9.2 Operational architecture of the prototype

Figure 5.6 describes the different components of the prototype.

- Planning data (1): characteristics of products and resourses (database system used: DBASE).

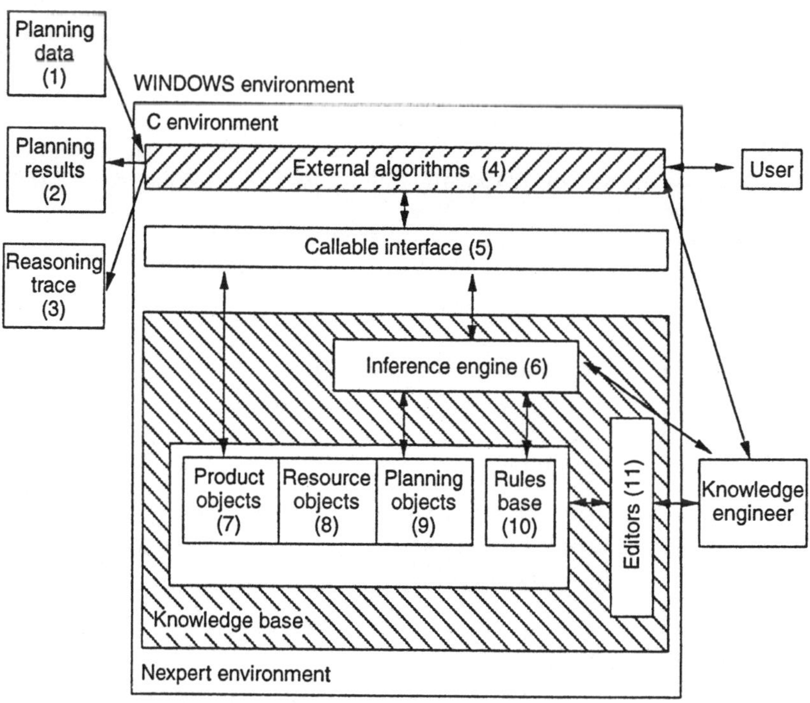

Figure 5.6 Operational architecture of the prototype.

- Planning results (2): stored in text files and present needed information (for each period of a production line: products, quantities, used resources, etc.).
- Reasoning trace (3): to explain the reasoning of the system.
- External algorithms (4): procedures and algorithms written in C language (calculation procedures, search heuristics, data validation, etc.).
- Callable interface (5): library of procedures to integrate Nexpert (expert systems shell) into existing computing environments.
- Inference engine (6): Nexpert inference engine.
- Product objects (7), resource objects (8), planning objects (9): an object describes a product, a resource or one-day work plan.
- Rule base (10): production rules described in section 5.
- Editors (11): Nexpert editors: rule editor, object editor, class editor, etc.

5.9.3 Prototype evaluation criteria

There are some methodologies developed in the literature (Adrion *et al.*, 1982; Ganascia, 1983; Michel *et al.*, 1986; Nassiet, 1987) which submit the case studies simultaneously to the human experts and to the computer system in order to compare both results.

The evaluation criteria were specified in the first phase of the development of the system. We distinguish two types of criteria (Artiba, 1990):

- **Technical criteria**:
  - the number of manufacturing lines and the number of products taken into account,
  - the production quantities per product,
  - the production sequences,
  - the degree of respect of constraints.
- **Operating criteria**:
  - the facility of use of the system,
  - the response time,
  - the facility of access to results,
  - the system integration.

The following example illustrates the scheduling strategies and how some of the constraints have been taken into account. It shows the different steps given in the algorithm.

5.9.4 Comprehensive example

Table 5.1 gives the initial situation and the solution procedure is explained step-by-step.

Table 5.1 Initial situation

Product name	*Batch size*	*Min. quantity*	*Max. quantity*	*Production period*	*Complexity code*	*Back order date*	*Ordered quantity*	*Sachet format*	*Monthly sales forecasts*
P1	1 000	2 000	6 000	1 or 2	4	02/06	3 800	0.5	2 800
P2	2 000	4 000	8 000	4 or 5	1	05/06	4 000	0.5	3 800
P3	1 500	3 000	12 000	–	2	05/06	8 000	1	4 000
P4	1 000	8 000	–	–	1	01/06	55 000	0.5	60 000
P5	2 000	2 000	6 000	4 or 5	4	04/06	4 000	1	3 500
P6	1 000	4 000	14 000	1, 2 or 3	2	07/06	9 000	1	5 000
P7	2 000	8 000	20 000	–	3	03/06	13 500	1	15 000
P8	1 000	8 000	20 000	–	3	07/06	20 000	1	20 000
P9	1 000	4 000	8 000	4 or 5	5	05/06	5 000	0.5	5 000
P10	2 000	6 000	16 000	–	3	06/06	15 000	1	8 000

The matrix given in Table 5.2 shows the compatibility between the different products. The letter B means that the product P1 can be produced **B**efore product P2 without any cost (we say that P1 is compatible before P2). Similarly, the letter A means compatible **A**fter; and blanks means non-compatibility with the two products. We can see that some products are compatible with several other ones and some are not compatible with any.

Table 5.3 shows the capacity of each production period in 'litre equivalent' (the measure unit used).

Table 5.4 shows a developed schedule for five days concerning one manufacturing line. The schedule contains the periods of production of each product and indicates whether there is a filter change or not (due to the succession of two incompatible products). The large dot between P8 and P5 symbolizes a filter change after the production of P8 at the end of period 4 of the third day.

Methodology

The first criterion applied to choose the first product to be placed in the sequence is EDD (earliest due date). In our example, the product P4 is selected. Then, we select the products compatible with P4. These are P1, P2, P3 and P9. As P3 has a different sachet format, it is discarded. The fact that P7 has only P3 as a compatible product is a further reason not to

Table 5.2 Products compatibility

Product	*P1*	*P2*	*P3*	*P4*	*P5*	*P6*	*P7*	*P8*	*P9*	*P10*
P1	–	B		B						
P2	A	–		A					A	
P3			–	B			B			
P4	A	B	A	–					B	
P5					–					
P6						–			A	
P7			A				–			
P8								–		
P9		B		A		B			–	
P10										–

Table 5.3 Output capacities of different periods

Production period	*1*	*2*	*3*	*4*	*5*
Capacity	4000	5000	5000	5000	4000

Table 5.4 A week production scheduling

Period / *day*	1	2	3	4	5
1	P1	P4	P4	P4	P4
2	P3	P3	P7	P7	P7
3	P8	P8	P8	P8 ●	P5
4	P4	P4	P4	P9	P2
5	P6	P6 ●	P10	P10	P10

choose P3 and to try to manufacture P3 and P4 together in the same sequence. Given a choice between P1, P2 and P9, P1 is selected because it is the most urgent (with the earliest deadline). The production period for P1 is the first or the second one. For P4, there is no specific production period. The quantity ordered for P1 is 3800 but the output capacity for the first period is 4000; so 4000 units of P1 are produced in the first period. The ordered quantity (for a month) of P4 is more than one production day. The adopted strategy then is to produce each week a small quantity of P4.

To manufacture two or more products in the same sequence, we verify their solution analysis complexity. This constraint represents the time of the solution analysis and the lab resources used. This analysis time varied between zero (distilled water) and seven hours. A complexity code is given to each product according to its analysis time and the kind of resources used. In this example, P1 and P9 cannot be manufactured in the same sequence, because they both have a high level of complexity (in this case, the sum of the different code values in the same sequence cannot exceed 8).

An important constraint is the average number of products to manufacture per day and per line. In the example, we have ten products for five production days. But there is also a maximum number of products per day to take into account. This number is five for the compatible products (equal to the number of production periods per day), and two for the incompatible ones.

After a product has been placed in the production sequence, the data are updated (compute the latest due date, the quantity to produce, the stock level, resources availability, etc.).

For the second day, the EDD heuristic rule is applied and the product P7 is chosen. P3 is the only product compatible with P7. P7 and P3 are manufactured in the same sequence if there is no product that must be manufactured on this day (i.e. no back order). As the product P7 is compatible after P3, then P3 is produced first. There is no problem with the analysis complexity. The quantities are 9000 litres for P3 and 14 000 litres for P7. These quantities respect the batch size, the minimum and maximum quantity per product.

For the third day, P5 can be chosen. This product is incompatible. Then, we look for the other incompatible products with the same sachet format which must be produced on the same line. The other incompatible products are P8 and P10. For these two products, their quantity and due date are analyzed. As P5 has to be produced in period 4 or 5, it is placed at the end of the day. The products P8 and P10 have the same due dates, but not the same ordered quantities. The most appropriate quantity for the rest of the day is the ordered quantity of P8.

On the fourth day, applying the EDD heuristic rule, P2 is chosen. The list of compatible products contains P1, P4 and P9. Because P1 has already been produced on the first day and its ordered quantity is fully satisfied, it is eliminated from the list of compatible products to P2. The ordered quantities of P9 and P4 represent the output of two production periods. To complete the day, P4 is produced in the first three periods (the sum of analysis complexity, $1 + 1 + 5 = 7$, does not exceed 8).

For the fifth day, P10 is chosen. It is an incompatible product. Its ordered quantity is less than the output of a full production day. Then we look for other products to complete the production sequence. Priority is given to the other incompatible products (to reduce the costs). As in this case there are no more incompatible products to satisfy, we look for the most appropriate compatible product (ordered quantity, list of compatible products, production period, analysis complexity, etc.). In the example, P6 is selected.

The sequencing of the other parallel lines is similar to this one, but the constraints must be taken into account due to the lines with the highest priority (which are sequenced before). These constraints represent products exclusion and the restriction on shared resources.

5.9.5 Analysis of results

The results obtained according to the different criteria given above can now be analyzed.

Evaluation in relation to technical criteria

We have studied a production plan for 107 products split up as follows:

- 61 products allocated to line V1;
- 32 products allocated to line V2;
- 14 products allocated to line V3.

The prototype takes into account all the products handled on the three parallel lines V1, V2 and V3 for line balancing. The development of the plan for V1 and V2 allows us to explore the fundamental problems which the scheduler has to face.

For products produced in large quantities (several days of production per month), a fluctuation of ±10% can be tolerated in the quantities produced (tolerance level). However, for products produced only in small quantities (approx. ½ day per quarter), production must be complete. The results produced by the prototype respect these constraints.

The production sequences themselves are assessed according to three criteria:

1. **The number of changes in the sachet format per month**: A change in the sachet format from one day to another requires a retooling operation for the pre-packing and sealing machines. The products sharing the same sachet format on the same line should be grouped together on successive days. Table 5.4 shows that the production of day 2 and day 3 corresponds to the same sachet format (P3, P8, P5, P7). We can observe that this criterion is less important than that of the due date. This is the reason why we change the sachet format from 0.5 for the first day to 1 for the second day.
2. **The number of changes of filter each day and each month for each production line**: This feature is a most important test of the quality of the production plan. Filter changes, necessitated by a sequence of incompatible products, lead to additional set-up costs. This interruption in production can be avoided by planning the filter change during mandatory pauses. The number of filter changes proposed by the prototype for the trial months is less than that resulting from a manual plan. In our example, two changes of filters are unavoidable. This is explained by the systematic use of the product compatibility table.
3. **Sharing and evening out of production difficulties over the planning horizon**: It is necessary to spread out the laboratory workload (solution analysis), and that of the retooling teams, over the month. The aim is to plan days of uniform difficulty rather than 'very difficult' or 'very easy' days. Measures of this parameter are the number of complex products handled per day on the different parallel lines, the number and times of filter changes, and the overtight scheduling of resources (e.g. plan to use a resource at 10.00 h which should become available at 9.55 h). This third point is difficult to assess, because it takes into account three criteria which may be contradictory in some cases (e.g. whether to plan for several complex but compatible products or invoke filter changes). In our case, the objective of cost minimization is the guiding one.

The analysis complexity code used in our prototype allows a partial solution of this problem. The different sums of complexity codes for the five days are: (Table 5.4 for products sequences and Table 5.1 for complexity code) 5 (4 + 1), 5 (2 + 3), 7 (3 + 4), 7 (1 + 5 + 1), 5 (2 + 3).

The constraints mentioned have all been integrated into the prototype and we have checked the validity of their inclusion during the prototype development. Delivery times have been respected for all the products considered by the prototype. The system calculates, in a dynamic fashion, the latest possible end dates after each insertion of a product in the sequence. The planner does not have to recalculate these dates, which are thus generated by the system.

Examining these different criteria, which we have labeled technical criteria, demonstrates the feasibility and reliability of the system and the precision of the reasoning, but does not prove that the model is complete or optimized.

Evaluation relative to operating criteria

We have questioned the experts and the system users about the flexibility of the prototype, the access time and the ease of use of the results.

Ease of use

The planner is an engineer familiar with computer-aided tools and with the use of spreadsheets for calculations, database techniques for access to data such as stock inventory, source material stocks, etc. held on a central system. He finds the system, with a single executable program, easy to use, but indicates that for simulation:

- the data access should be improved by using a DBASE compiler with simple and ergonomic displays;
- the data access and system start should be more user-friendly (use of a root menu at power-on, use of F-keys);
- we should develop a module to explain the reasoning method for users. At this stage of the prototype development there is a complete record of the reasoning process, but not in a form easily accessible to non-initiates of Nexpert.

Response time

The plan for two lines for a fortnight is determined in about five minutes. The prototype runs on a PC-AT COMPAQ-386/20 with four megabytes of core memory. The system is already at the limit of its capacity, and an upgrade in term of performance is desirable (e.g. PC with UNIX or OS/2, use of a 486/100 processor, etc.). These upgrades would compensate for the inevitable increase in the response time for use with more lines and products.

At present, one person is fully occupied with the planning and scheduling of four lines. It can be estimated that the time saved in the development of the production plan would liberate the operator from ±80% of the present workload.

Access to results
Access to the results is made easy by the use of any text editor. For the sake of better ergonomy, an improved interface is desirable.

System integration
The system as it stands has been or can be easily integrated with the other aids available in the company. In addition the knowledge base can be imported on any system (Macintosh, workstation, VAX, mainframe, etc.). The different modules are written in C.

5.10 CONCLUSION

In this chapter, we have presented a production system organized in parallel multiproduct lines and have provided a classification of the essential data and constraints characteristic of this type of system.

This type of production system is common in industry, notably in the pharmaceutical, chemical, cosmetic and brewing industries, and can perhaps also be extended to other types of manufacturing.

The results obtained show that an expert system approach is well adapted for this kind of problem. The solutions obtained are feasible but we cannot guarantee that they are optimal. The system prototype presented is a first step towards a more intelligent system, integrating simulation of product and resource selection criteria as a function of dynamic objectives. Thus the system must determine which criteria to select as a function of each situation. This is a step-by-step authentication. The integration of a simulation module therefore improves the quality of the scheduling generated by the system.

The production plans established from real data have proved the adequacy of the management system for the problem of the loading and scheduling of multi-product parallel lines.

At the level of the prototype integration, it can be observed that the Nexpert/Object development tool is very powerful, but would require a large investment if one is to develop beyond the stage of the writing of a few simple production rules.

As far as the hardware is concerned, the system would require more than the capacity of the PC-AT currently used. To some extent this limit is also due to the operating system used and Nexpert/Object which occupies a large part of the memory.

For further improvement of the system the man–machine communication aspects must be enhanced, and a more powerful system with more memory, or a workstation, used.

The results obtained show:

- the interest and the feasibility of the approach used as an aid to the loading and scheduling of multiproduct parallel lines;
- the appropriateness of the proposed architecture for the problem studied, integrating a line-balancing phase before the plan calculation;
- the semantic richness of the approach, and its conviviality for the modeling and representation of different system entities.

The hierarchy of selection criteria for products and resources and the way in which the various constraints are taken into consideration are established in a static manner in the management system prototype.

To create a dynamic hierarchy, simulation tools must be included in order to present several solutions from which the best can be selected. The progressive evaluation of the proposed solutions responds to the needs of an interactive plan construction. This allows the operator to query the solutions proposed by the system, and to suggest modifications. The evaluation and testing of these modifications calls for simulation techniques. The same is true for the readjustment of a plan following perturbations. For known and solved perturbations the corrections can be efficient and already introduced into the system in the form of rules for treatment of perturbations. However, the system can always be confronted with an unknown perturbation, or a known one for which there is no easy solution. For this type of perturbation the system must look for the best possible solution by comparing the solutions generated by a dynamic hierarchy of management criteria.

5.11 CURRENT DEVELOPMENTS AND RESEARCH AXIS

This final section of the chapter highlights the extension of our development of production control systems. It describes the reasoning behind a multi-model system and presents its architecture. This system is under development but the multi-model approach has been validated (Artiba *et al.*, 1994; Aghezzaf *et al.*, 1994).

The OR heuristics offer the advantage of providing optimal (or near optimal) solutions for well-structured scheduling problems. The AI formalism allows us to capture all the details of real world constraints; it also allows us to represent and manipulate the human scheduler's knowledge.

Kusiak (1987) proposed a tandem architecture: knowledge-based system and optimization algorithms. This approach is powerful when the scheduling problem to be solved fits the stored optimization algorithms. A framework combining simulation and expert systems for production planning and control was proposed by Falster (1987). He highlighted the advantages of combining these two approaches.

> The heuristic problem solving frameworks that have emerged from the field of AI can be seen as complementary to the analytic technique produced by OR. These frameworks can provide a basis for exploiting knowledge of model assumptions, parameters, set-up, and applicability to (1) make existing OR techniques more accessible and usable to an end user, or (2) opportunistically exploit a collection of analytic/heuristic procedures as appropriate during the planning/scheduling process. (Smith, 1992)

In Fig. 5.7, we propose an architecture of a multi-model based system for loading and scheduling problems. This system combines an expert systems technique, object-oriented approach, simulation, algorithms and heuristics.

Simulation is used to validate the generated schedule in order to evaluate the impact of the decisions taken on the global system. The simulator can also be used by the planner in an interactive way to validate his or her changes to the work plan. It can be used to answer what-if types of questions. For example, the sales service may ask the planning service if a certain quantity of an urgent product can be produced by a given date.

The numerical results of simulation describe the new state of the system under consideration. These results serve to calculate the performance measures of the schedule (status of jobs, resources, etc.).

After each decision is made, the new status of the system is analyzed. According to this analysis phase, some directives are generated to guide the building of a new local system on which a new decision will be made. Depending on a built local system, an algorithm or heuristic will be applied to solve the given local problem and simulate its impact.

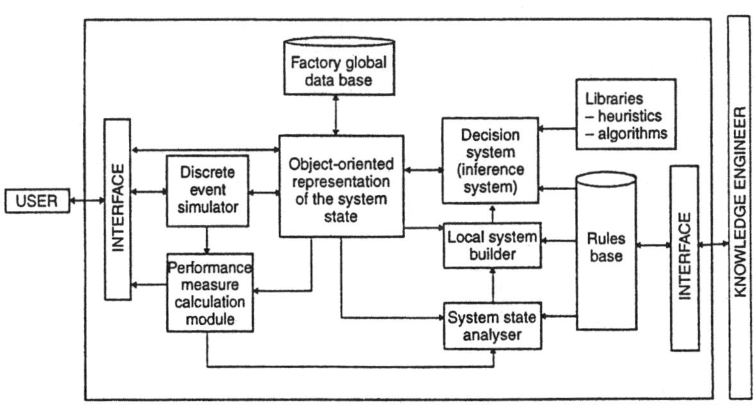

Figure 5.7 Open system architecture of a multi-model loading and scheduling system.

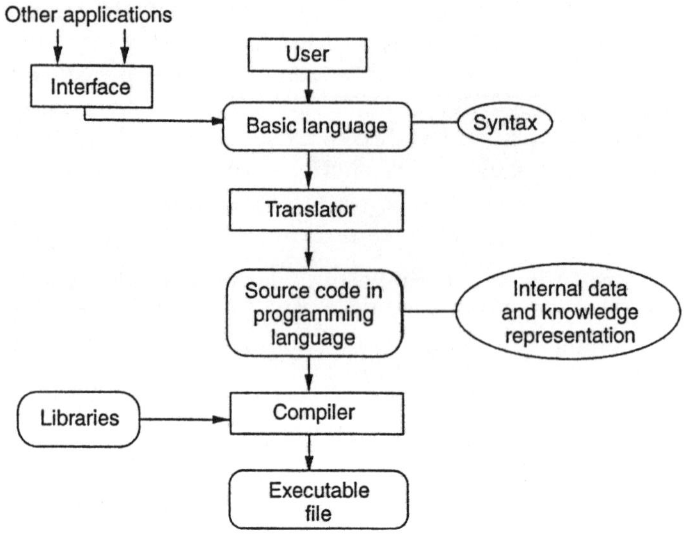

Figure 5.8 Different phases of implementing an application.

The advantages of the object-oriented approach for data modeling are now well known and generally acknowledged. Figure 5.8 shows the different stages of implementing a specific application with the system described. The user can introduce the data manually or import them from a database. The data are structured in objects grouped by classes (depending on the levels of abstraction or aggregation). The interface permits (among other things) the querying of the factory databases. When the application is described in our internal language, the syntax is verified and this internal code is translated to the programming language (in our case, C++). The source code in programming language is then compiled and the external functions are incorporated to produce an executable file.

We believe that this implementation is very interesting, because (among other things):

- the system is hardware independent;
- the architecture is open, and it easily permits us to extend the functionalities of the system;
- the executable code is fast.

More details and first application results of the first version of this system can be found in Artiba *et al.* (1994), Aghezzaf *et al.* (1994) and Artiba (1995).

REFERENCES

Adrion, W.R., Branstad, M.A. and Cherniavsky, J.C. (1982) Validation, verification and testing of computer software. *ACM Computing Surveys*, **14**(2).

Aghezzaf, E.H., Artiba, A. and Yasinovky, S.I. (1994) *AGV-based Systems Dispatching in Manufacturing*. International Dedicated Conference/Agile Manufacturing, October 1994, Aachen, Germany.

Alasuvanto, J., Eloranto, E., Fuyuki, M., Kida, T. and Inoue, I. (1988) Object oriented programming in production management – two pilot systems. *International Journal of Production Research*, **26**(5), 765–76.

Artiba, A., Tahon, C., Soenen, R. and Vandepeutte, G. (1989) *Baxpes: An Expert System for Planning Manufacturing Lines*, SCS International, San Diego, January 1989, pp. 79–83.

Artiba, A. (1990) Contribution à la construction d'un système d'aide à la planification et à l'ordonnancement de lignes parallèles multiproduits. PhD thesis, Université de Valenciennes et du Hainaut-Cambrésis, France.

Artiba, A. (1994) A rule-based planning system for parallel multiproduct manufacturing lines. *Production Planning and Control*, **5**(4), 349–59.

Artiba, A. (1995) *Open System Architecture of a Multi-model Loading and Scheduling System*. Proceedings of IASTED Applied Informatics Conference, Innsbruck, Austria, February 1995, pp. 449–51.

Artiba, A., Bourgeois, S. and Yasinovsky, S. (1994) Technical Report on a Job Shop Scheduling Problem using Multi-Model Approach. Fucam, Industrial Management Division, March 1994.

Artiba, A. and Tahon, C. (1992) Production planning knowledge-based system for pharmaceutical manufacturing lines. *European Journal of Operational Research*, **61**(1–2), 18–29.

Cox, B.J. (1986) *Object Oriented Programming – An Evolutionary Approach*. Addison-Wesley, Reading, UK.

Falster, P. (1987) Planning and controlling production systems combining simulation and expert systems. *Computers in Industry*, **8**, 161–72.

Ganascia, J.G. (1983) MIRLITHO validation des résultats et détection des contradictions dans les systèmes de diagnostic. Thèse de Docteur-ingénieur, Université de Paris-sud, France.

Gupta, J.D. (1988) Two-stage hybrid flowshop scheduling problem. *Journal of Operational Research Society*, **39**(4), 359–64.

Hart, A. (1989) *Knowledge Acquisition for Expert Systems*. Chapman and Hall, London.

Krause, F.L. and Seliger, G. (1987) *Software Structures for Factory Integration*. IFIP WG5.3 Working Conference on Software for Factory Automation, Tokyo, October 1987.

Kusiak, A. (1987) Designing expert systems for scheduling of automated manufacturing. *Industrial Engineering*, **19**(7), 42–6.

Michel, C., Botti, G., Fieschi, M., Joubert, M., Casanova, P. and San Marco, J.L. (1986) *Validation d'une base de connaissances destinée aux médecins généralistes dans l'aide à la thérapeutique du diabète: étude en aveugle*. 6ème Journées Internationales: Les systèmes experts et leurs applications, Avignon, 28–30 April 1986, pp. 139–52.

Muller, H., De Samblanckx, S. and Matthys, D. (1987) The expert system approach and the flexibility–complexity problem in scheduling production systems. *International Journal of Production Research*, **25**(11), 1659–70.

Musier, R.F.H. and Evans, L.B. (1990) Batch process management. *Chemical Engineering Progress*, June, pp. 66–77.
Nassiet, D. (1987) Contribution à la méthodologie de développement des systèmes experts: application au domaine du diagnostic technique. Thèse, Université de Valenciennes et du Hainaut Cambrésis, France.
O'Keefe, R.M. (1985) Expert systems and operational research – mutual benefits. *Journal of the Operational Research Society*, **36**(2), 125–9.
Smith, S.F. (1992) Knowledge-based production management: approaches, results and prospects. *Production Planning and Control*, **3**(4), 350–80.
Van Dam, P., Gaalman, G. and Sierksma, G. (1992) Scheduling of packaging lines in the process industry: Recent trends. Research Memorandum no. 475, May 1992, Institute of Economic Research, Faculty Of Economics, University of Groningen, The Netherlands.

CHAPTER 6

Production control in hybrid flowshops: an example from textile manufacturing

S.E. Elmaghraby and R.E. Karnoub

ABSTRACT

In a real life application, we model the production planning and control function in a textile plant as a problem in planning and scheduling in a flexible flowshop. For practical considerations the total problem is divided into two parts: a 'capacity loading function' and a 'scheduling function', both undertaken over a finite planning horizon of six weeks. The 'loading' was optimized via linear programming. The 'scheduling' was effected via a number of heuristics that accommodated a variety of (oftentimes contradictory) objectives. The input to the combined control scheme was implemented on a spreadsheet. Several reports resulted containing information of interest to shop floor personnel as well as to management. A side benefit to the production planning system was a **marketing** model that helped the design of the firm's marketing strategy.

6.1 PROBLEM CONTEXT

6.1.1 Hybrid flowshops

We are concerned with the production planning of a plant that may be modeled as a **hybrid flowshop** process (sometime also referred to as **flexible flowshop**, or a **series/parallel** process). This type of production is quite prevalent in the process industries such as chemical, metal-lurgical, pharmaceutical, petroleum, food and textile industries. A hybrid flowshop consists of a series of production stages, each of which has several facilities operating in parallel. (Some stages may have only one facility – such as a heat treatment oven – but for the plant to qualify as a hybrid flowshop at least one stage must have several facilities.) The flow of products through the plant is unidirectional. Each product is processed at only one facility in each stage and at one or more stages before it exits the plant. (For more background on the context of the hybrid flowshop problem and the approaches that have been proposed for its solution, see Pinedo (1995), sections 12.2–12.4, and the references cited therein.)

The problem in its more general form is notoriously difficult to resolve optimally under most commonly used criteria (proofs of the 'NP-hardness' of these problems abound in the scientific literature), and resort to approximations and heuristic solutions of 'reasonable', albeit non-provable, performance is quite common.

To give an example of the fundamental difference between the treatment of (standard) flowshops and hybrid flowshops, consider the issue of **permutation schedules** versus **non-permutation schedules**. A permutation schedule is such that the order of performing the jobs is the same in all stages (obeying the so-called **no-overtaking rule**: if job i precedes job j in stage k then job i continues to precede job j in stage $k + 1$, for $k = 1, \ldots, K - 1$, where K is the total number of stages). A non-permutation schedule permits an arbitrary sequence of jobs at each stage. Now it is well known that for a flowshop of no more than three stages, and for all (regular) criteria, there exists an optimal sequence which is a permutation sequence; it simply does not pay to switch the sequence and permit overtaking. Unfortunately, for more than three stages, one cannot guarantee the optimality of permutation sequences, in general. Still, most researchers and shop floor operators accept the limitation to permutation schedules as a 'reasonable' constraint which does not detract greatly from the optimality of the solutions obtained. Such is not the case under hybrid flowshop: it is in the nature of things that the order of performing the jobs shall be different in different stages. Imposing the restriction of permutation schedules is simply unrealistic, and may cause serious degradation in the quality of the solution from optimality.

Treatments of hybrid flowshops in the scientific literature are rather simplistic and abstract in nature; they assume away many of the real life restrictions that must be taken into consideration, and which complicate the problem well beyond the capabilities of the standard methodologies of operations research. In practice, optimization of performance recedes as a realizable goal, though it remains as a desired objective.

Rather than talk in generalities, and in order to render our discussion concrete, we couch the discussion in the language of production in a textile plant for which the development was made. We are certain that you, the reader, if you are not a *bona fide* textile production engineer, shall discover many similarities between the real life issues and problems presented here and issues and problems you face in another manufacturing context. Put differently, it is possible that the 'is-ness' of the problems faced is different, but their fundamental **form** and **nature** remain invariant.

Additionally, to render this chapter more accessible we have relegated the mathematical formulation of the Loading Module (see below) to the Appendix. In the main, the narrative adheres to a description of the problems faced and the approaches that were adopted to resolve them.

6.1.2 The textile plant

The plant to which reference is made throughout this chapter is composed of three 'shops' or **stages** of production: **warping**, **slashing** and **weaving**; each stage is composed of several machines operating in parallel. The overall process is evidently a hybrid flowshop. The input to the plant is yarn, and the output is woven cloth. A **product** is a woven cloth of a particular **cloth style**; the latter is defined by the quality of the yarn used, the number of threads (or 'ends') along the length of the cloth (the warp), the number of threads across (the weft) per inch, and the weight of the yarn used in either. In the sequel we shall use **product** and **item** interchangeably. A warping machine transfers the yarn from the bobbins (in which the yarns are supplied) to drums, which are called **section beams**. A slashing machine applies a chemical coating to the yarn to give it consistency and added strength. The input to the slasher is section beams, and the output is called **loom beams**, which contain a different number of **ends** (threads) on them. A loom weaves the yarn into the finished cloth. The input to a loom is a loom beam (the output of the slasher shop), and the output is a **cloth beam**. The latter is shipped to the customer (who may be an end-user or another dying shop). The process described verbally above is depicted in Fig. 6.1.

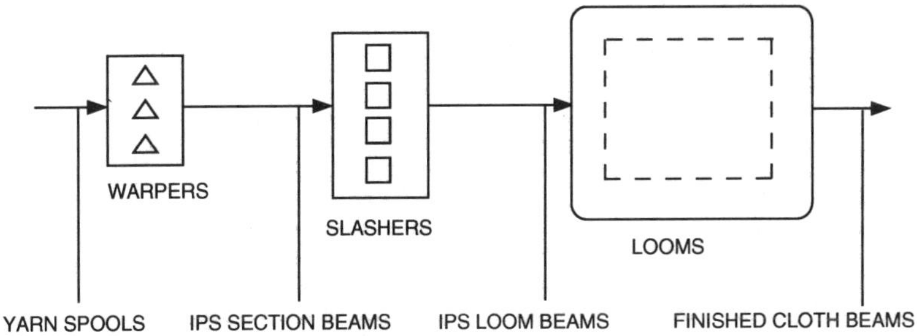

Figure 6.1 Process flow in textile plant.

Typically, the warper and slasher shops contain a small number of machines each (of the order of two to five machines), while the weaver shop contains tens or even hundreds of looms. The machines in the warper and slasher shops are 'in parallel' but are of different capabilities (usually referred to as 'unrelated machines in parallel'); the looms typically may be classified into no more than three or four **classes**, also of different capabilities, each class containing between 50 to 150 looms.

6.1.3 The objectives of the study

The ultimate objective of the study, as expressed by management, is to optimize the performance of the plant. This 'optimization of performance' has several dimensions, of which we enumerate the following four: maximize the 'net value of production' (equal to: (sale price) – (cost of goods sold)); maximize machine and labor utilization (plant efficiency); minimize in-process-stock (IPS); and minimize finished goods inventory. Optimization is to be realized through a computer-based decision support system that takes as its input the projected demand on the various products for a finite planning horizon, and produces as its output the detailed loads and schedules of the various pieces of machinery in all three stages of production.

6.1.4 The approach

The real life problem is complicated by several factors which render a frontal attack through an all-encompassing analytical model (such as a mathematical programming model) computationally infeasible: the end result is a mixed integer linear program of such dimensionality that it defies any existing software within the reach of the producing firm. An attempt to formulate such a model was initially attempted but quickly abandoned. The approach we finally adopted was to 'divide and conquer': we separated the **loading** problem from the **scheduling** problem, and treated each one separately. By loading we mean **capacity loading**, which ignores indivisibilities in the production capacity in the form of individual machines, and indivisibilities in the production lots in the form of individual (section or loom) beams. And by scheduling we mean the determination of the start time and completion time of each operation on each machine in the plant.

We are cognizant of the resulting **sub-optimization**, but we believe that the benefits gained from such approach far outweigh any loss in optimality.

Figure 6.2 depicts the flow of information required for the production control system described here. Many of the terms used shall become evident as the discussion progresses (also see Glossary).

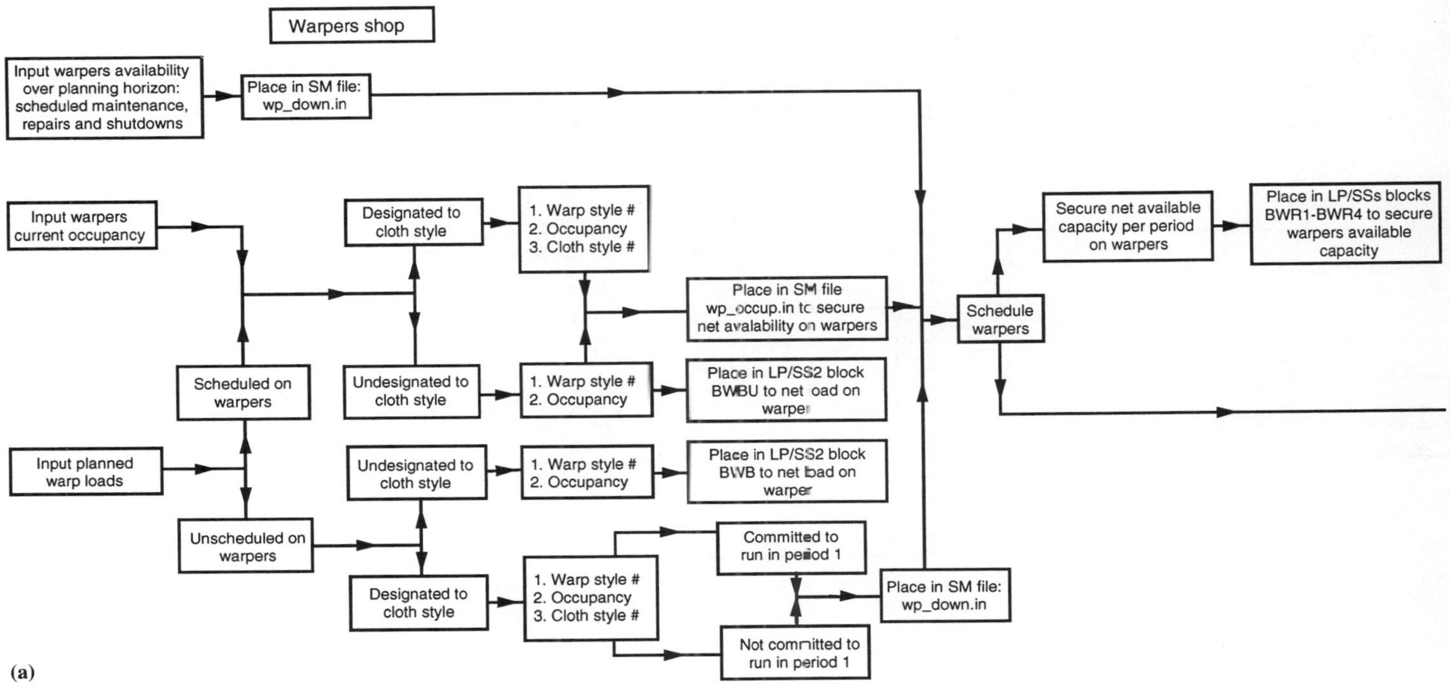

(a)

Figure 6.2(a,b,c) Information flow for control.

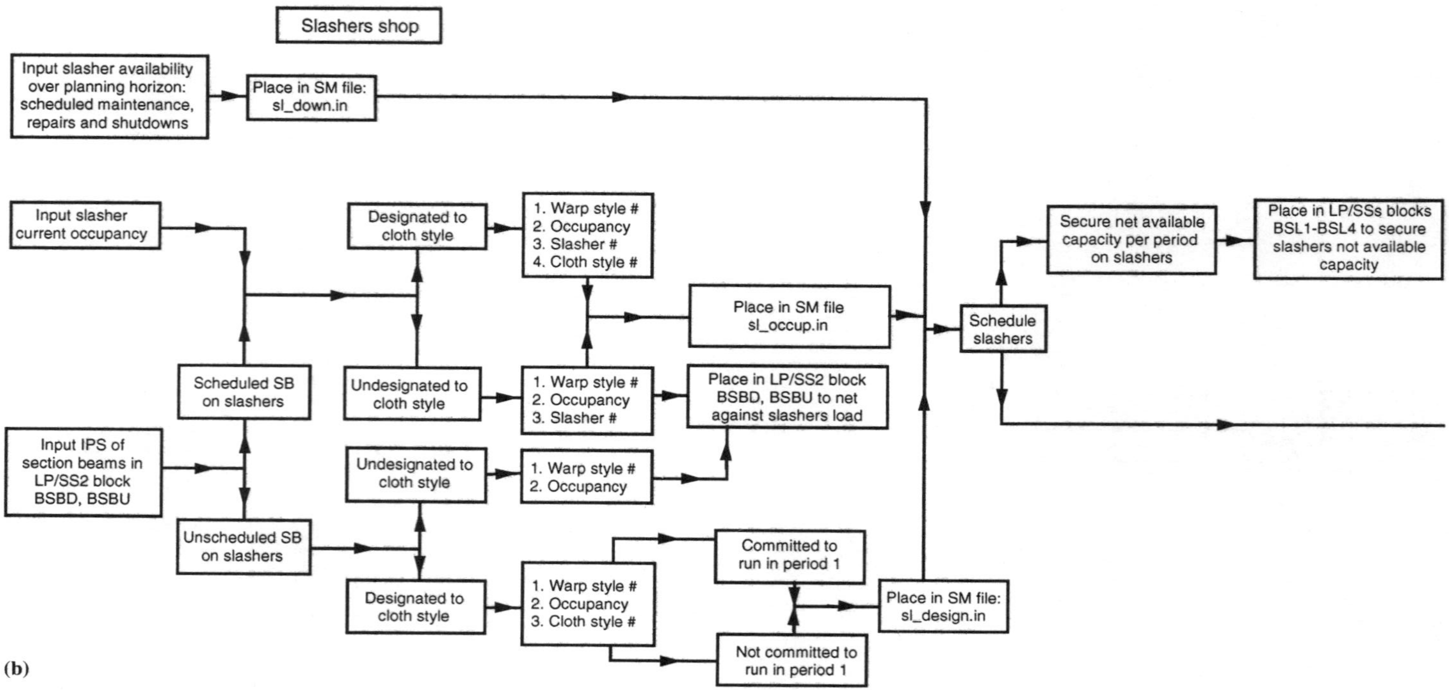
Slashers shop
Input slasher availability over planning horizon: scheduled maintenance, repairs and shutdowns
Place in SM file: sl_down.in
Input slasher current occupancy
Scheduled SB on slashers
Input IPS of section beams in LP/SS2 block BSBD, BSBU
Unscheduled SB on slashers
Designated to cloth style
Undesignated to cloth style
Undesignated to cloth style
Designated to cloth style
1. Warp style #
2. Occupancy
3. Slasher #
4. Cloth style #
1. Warp style #
2. Occupancy
3. Slasher #
1. Warp style #
2. Occupancy
1. Warp style #
2. Occupancy
3. Cloth style #
Place in SM file sl_occup.in
Place in LP/SS2 block BSBD, BSBU to net against slashers load
Committed to run in period 1
Not committed to run in period 1
Place in SM file: sl_design.in
Schedule slashers
Secure net available capacity per period on slashers
Place in LP/SSs blocks BSL1-BSL4 to secure slashers not available capacity

(b)

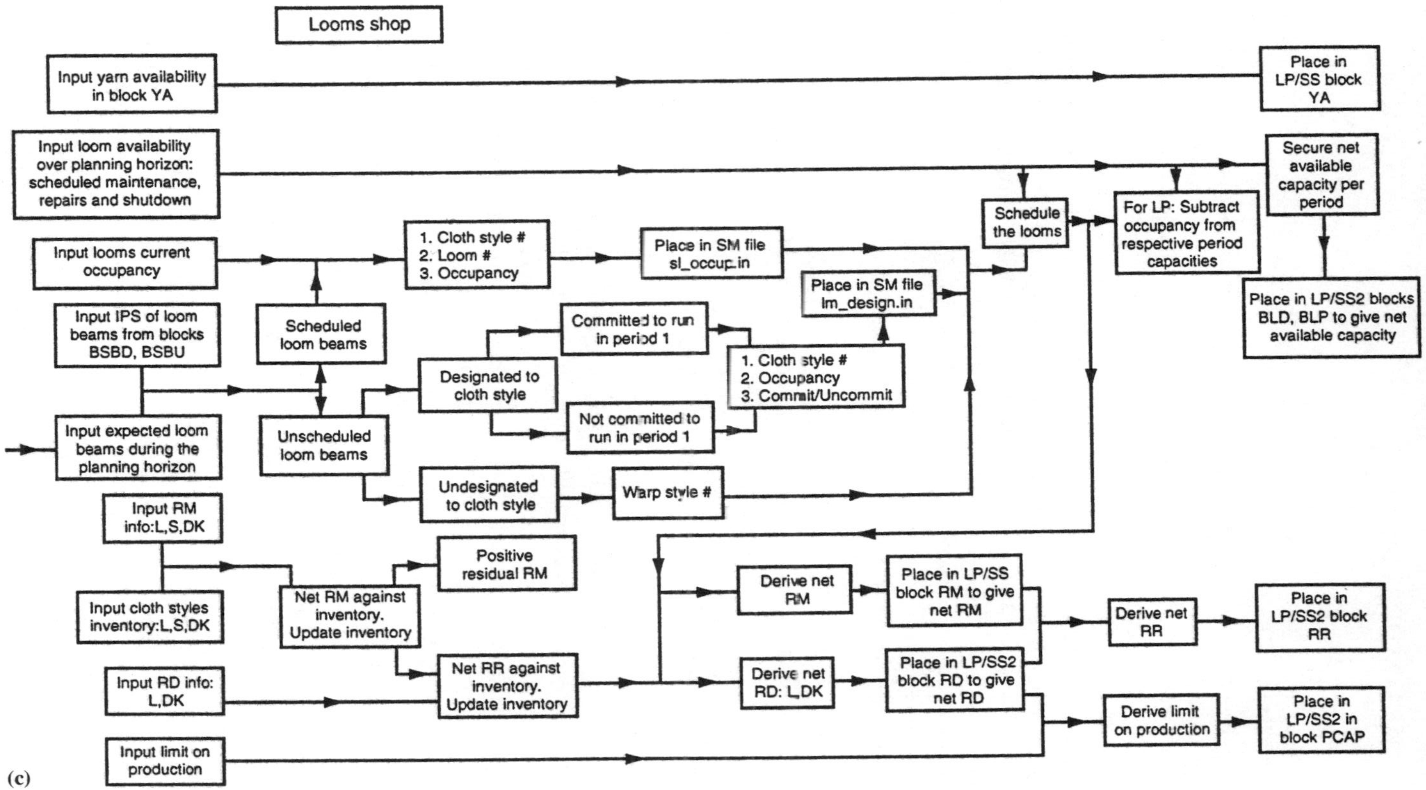
Looms shop
Input yarn availability in block YA
Place in LP/SS block YA
Input loom availability over planning horizon: scheduled maintenance, repairs and shutdown
Secure net available capacity per period
Input looms current occupancy
1. Cloth style #
2. Loom #
3. Occupancy
Place in SM file sl_occup.in
Schedule the looms
For LP: Subtract occupancy from respective period capacities
Place in LP/SS2 blocks BLD, BLP to give net available capacity
Input IPS of loom beams from blocks BSBD, BSBU
Scheduled loom beams
Place in SM file lm_design.in
Committed to run in period 1
1. Cloth style #
2. Occupancy
3. Commit/Uncommit
Designated to cloth style
Not committed to run in period 1
Input expected loom beams during the planning horizon
Unscheduled loom beams
Undesignated to cloth style
Warp style #
Input RM info:L,S,DK
Positive residual RM
Input cloth styles inventory:L,S,DK
Net RM against inventory. Update inventory
Derive net RM
Place in LP/SS block RM to give net RM
Derive net RR
Place in LP/SS2 block RR
Input RD info: L,DK
Net RR against inventory. Update inventory
Derive net RD: L,DK
Place in LP/SS2 block RD to give net RD
Derive limit on production
Place in LP/SS2 in block PCAP
Input limit on production

(c)

6.2 THE LOADING MODULE

The procedures for the optimal loading of the plant are collectively referred to as the **loading module** (LM). The idea underlying its construction is the following.

Demand is classified into two categories (exclusive of **emergencies**, see below): the **must satisfy** market requirements, denoted by RM, and the **may satisfy** market requirements, denoted by RR. Demand under RM represents customer orders that must be satisfied by a particular date, either from available stock or from new production. Demand designated as RR is anticipated demand (either in the form of realized customer orders or the result of market forecasts with no firm orders) that may be backlogged to later periods if capacity is unavailable for its production over the immediate planning horizon.

As far as RM is concerned, the only decisions left are relative to when to produce it and on which machine (in the case of the warper and slasher shops) or on which class of machines (in the case of looms) to produce.

The treatment of RR is different since, in addition to the when and the where questions, there is the question of whether to produce any of it at all. This has to be determined on the basis of profitability within the available capacities in all shops over the planning horizon. The methodology used is **linear programming** (LP).

This represents a radical departure from past practice in the plant, when products were loaded on the basis of their 'loom marginal profitability' as measured by the ratio (variable margin) ÷ (time on loom per 1000 yards). (The 'variable margin' is the shop term for the gain (over cost) per 1000 yards of finished cloth.) The inadequacy of this measure is evident, since it ignores the limited capacities in the two shops (warping and slashing) prior to weaving, and also ignores the desire to balance the load on the weaving shop itself. Its limitations were manifested in the existence of bottlenecks in either the warper or slasher shops at various times, which forced the plant to either purchase already-warped yarns to alleviate the load on the warping shop or delay production of the lower-ranked products in the slashing shop (which caused serious backlogs of demand, to the chagrin of the marketing organization!).

The **planning horizon** was put at six weeks, with two weeks of **lead time**. The planning horizon was divided into three **periods**, each of duration two weeks. Thus the LM may be viewed as loading the plant over three periods, with one additional period of lead time. Note that there is no attempt to 'time-phase' the load in the shops according to the flow of processing; it was sufficient to ensure that the production in one shop in one period shall 'clear' the other shops in one period also. Shop loading is on a **rolling carpet** basis each week: a week is dropped and one added, and the LM re-run with the new **state of the plant** and any new information about demand for the products.

Why a planning horizon of three periods, each of duration two weeks? We have conducted a statistical analysis of the time to weave (which is by far the longest processing time among the three phases of production) and discovered that it varies between ≃ 60 hours (or approximately 0.36 weeks)[1] to ≃ 640 hours (or approximately 3.8 weeks) per beam, with some 67% of the products completing the weaving operation in no more than two weeks. Thus a 'period' of two weeks is sufficient for about two thirds of the load on the looms. In other words, loading in multiples of two-week periods ensures that at least two thirds of the load shall be completed in the first period. The remaining one third shall linger in the weaving shop for at most one more period. But we would know about the loom occupancy through the state of the plant report at the time of running the LM again. A 'forecast horizon' of eight weeks was deemed by management to be a 'reasonable' span of time in which the forecast of demand is relatively accurate. The first period (two weeks) of the forecast horizon is considered lead time: the first week of the lead time typically finds the looms almost fully loaded with continuing work on loom beams already mounted, and the second week is approximately 40% loaded. It was deemed advisable to leave the extra capacity available in the first two weeks of the forecast horizon 'free' to accommodate any variations (random or intentional) from the planned production. In any event, these three parameters (the length of the forecast horizon, the length of the lead time, and the amount of capacity to load during the lead time) were left for future 'fine tuning' after gaining some experience with the actual operating system.

The **objective function** of the LP model is composed of three expressions. The first measures the 'gain' from production. This gain is part real and part fictitious. The real part gives the net profit as measured by the variable margin per 1000 yards. The fictitious part measures a fiducial 'reward' accruing to the plant from utilizing in-process-stock. This encourages the consumption of any IPS whenever available instead of embarking on new production, and thus helps reduce the IPS. The second expression of the objective function subtracts from the amount gained by producing an item the cost of carrying it in inventory from period t to period $t + 1$.[2] The third expression subtracts the cost of **backlogging** demand from period t to

[1] All calculations are based on a 7-day week, 24 hours per day continuous operation, i.e. 168 hours.

[2] In this application the cost of carrying inventory was taken at 20% of the variable margin. Thus any profit made on the product shall be wiped out if it is carried in inventory for 5 periods (= 10 weeks).

period $t + 1$.[3] Note that the last two costs are fiducial, and represent a 'penalty' that is imposed to discourage the activity.

Now, to a brief overview of the **constraints** of the LP. There are six sets of constraints:

1. Demand constraints, which relate production to required demand in RM and forecasted demand in RR, as well as prevent any runaway production of any item (because of its high profitability) by putting a 'cap' on its total production.
2. Loom constraints, which force production not to exceed available capacity on each class of looms available.
3. Slasher constraints, which relate production in the slasher shop to production in the weaver shop, and force the production in any period not to exceed available capacity on each slasher, or set of slashers, in the shop.
4. Warper constraints, which are very similar to the slasher constraints.
5. Whole beam constraints, which simply translate the result of the load from yards to whole beams.
6. Yarn constraints, which limit production in any period not to exceed the anticipated availability of yarn in that period.

The LP is given in the Appendix to this chapter. It has been coded using the Lingo (Industrial version) software package [Lindo Systems, Inc., Chicago, IL 60622]. It runs in approximately 4 minutes on a PC with Pentium chip and 32 MB of RAM.

6.2.1 Complicating factors in the loading module

The following is a recitation of some of the more salient factors which complicated our attempt at a 'straightforward' global optimization model (which, incidentally, was originally attempted) or even a simple-minded LP model, and which could not be ignored in any serious effort at constructing a viable decision support system in the realistic conditions prevailing in the plant.

Problem size

It must be evident that the sheer numbers of looms in the weaver shop defies straightforward application of known schedule optimization methodologies. Fortunately, the numbers were large enough that the total capacity available can be approximated reasonably well by the sum of individual loom capacities in each loom class. Such is not the case in

[3] In this application the cost of backlog was taken at 40% of the variable margin per period. Thus any profit is wiped out if the demand is backlogged for 2.5 periods (= 5 weeks).

the warping and slashing shops, where the small number of machines necessitate individual loading and scheduling of each. Unfortunately, the setup times are sequence-dependent as well as machine-dependent, and the processing times are machine-dependent. Optimization of the sequence of literally hundreds of products under such conditions in these two shops to meet the due dates imposed by the weaving shop proved to defy analytical solutions, and resort to heuristics was necessary, at least for the moment.

Demand classification

As explained above, demand under RM represents customer orders that must be satisfied by a particular date, either from available stock or from new production, while demand under RR is anticipated demand that may be backlogged to later periods if capacity is unavailable for its production over the immediate planning horizon. Naturally, continued delay in satisfying the RR demand shall cause its priority to be upgraded to RM eventually, at which time it will have to be produced by the time it is required. This limits management's freedom to load and schedule the various operations leaving, as it does, very little room for maneuver to optimize anything! It eventually leads to a severely constrained system that is dealing with 'rush orders' almost exclusively, a situation that was prevalent prior to the conduct of the study.

Variable units of production

Production is typically in cloth beams of standard lengths, which range from ≃ 4000 to ≃ 10 000 yards. (A 'beam' is the shop name for a cylindrical drum of about 2 ft in diameter and 8 ft long. The warp is on 'section beams' ahead of slashing, and is on 'loom beams' ahead of weaving. The woven cloth is mounted on 'cloth beams'.) Unfortunately, production over the various stages is not of the same beam, or of the same number of beams throughout the process. For instance, several section beams (the output of the warpers) – the number ranges from 6 to 14 – combine to produce a number of loom beams, the latter ranging from 2 to 6. On the other hand, one loom beam produces one cloth beam of equal length (except for shrinkages). Demand for finished products need not be in integer multiple of the number of loom beams after slashing. For instance, it is possible that the output of the slashing stage is in multiples of 5 loom beams, while the demand for the finished (woven) cloth is for 48 000 yards, which translates into 12 loom beams (assuming 4000 yards per beam). The shop must produce either 10 beams (= 40 000 yards) and be 8000 yards short, or produce 15 beams (= 60 000 yards) and be 12 000 yards over demand, or weave only the required 12 out of the available full complement of 15 beams and keep the remaining 3 loom beams as IPS, which withholds blanks from use in other products (see next concern)! The artful balance of these three alternative modes of operation has kept floor management on its toes for a long time, without much success as evidenced by the huge stocks of finished products accumulated in the warehouse.

Limited availability of blank beams

There are relatively few blank (i.e. empty) beams, and the IPS between stages as well as loom production should be planned to accommodate their availabilities. Once a blank is utilized as a section beam it remains as such until it is transformed (at the slashing operation) into a loom beam. This latter occupies the blank until it produces a cloth beam, which is produced on a new blank. The occupancy of the blanks either as section beams, loom beams, or cloth beams cannot exceed their individual availabilities. The re-circulation of these blanks is an important feature of successful operation.

Difficulties with JIT philosophy

The objective of minimizing the IPS calls for adopting the 'just-in-time' (JIT) philosophy of production: the output of the warper shop should feed directly into the slasher shop, which output, in turn, should feed directly into the weaver shop. This philosophy is hampered by two practical considerations. First, there is the issue of **time for change-over**[4]: it would be unwise to incur the required time for the change-over from one style to another on any machine in any of the three shops, which is appreciable, if the same product, or one close to it, shall be produced soon. Grouping 'like products' for the sake of minimizing the change-over time (which also induces better machine utilization) results in the building up of IPS, which plays havoc with the concept of JIT! Second, there is the issue of the requisite number of section beams to produce a loom beam: if 14 section beams are required to 'assemble' the warped yarn into loom beams, there is no escape from waiting for all 14 section beams to be produced in the warper shop before production is initiated in the slasher shop. The first section beam produced must wait for the other 13 section beams to be produced!

Designated vs undesignated IPS

A section beam carries a **warp style** (= yarn type). But the same warp style may be used in several **cloth styles** (= finished product). This fact

[4] The name 'change-over time' is typically reserved for the time required between two different products, while the name 'setup time' is typically reserved for the time required between two lots of the same product. Evidently, the former is much longer than the latter.

adds considerable flexibility to the management of production through greater flexibility in meeting customer demand. It permits the delay in the final decision as to which cloth style to produce to the last minute immediately before the weaving operation, thus carrying the concept of JIT one step further. However, this very flexibility introduces a slight, albeit significant, complication, due to the following consideration. The in-process-stock itself is divided into two sub-categories: in-process-stock that is **designated** to a subset of products, and in-process-stock that is **undesignated**. The reasons for such division are beyond the scope of this treatise, but suffice it here to say that management wishes to retain the prerogative of allocating a portion of the IPS to a subset of the products to guarantee their immediate loading/scheduling on the looms, irrespective of issues of profitability. Both designated and undesignated in-process-stocks ahead of weaving may contain anticipated arrivals from the slashing and warping shops. Consequently, the available undesignated IPS of that warp style should be counted against the load of a **set of products**, not just an individual product. As a result, we now must group together products that require the same warp style and constrain the total of the group to the available IPS, not individual products.

Emergencies

Consideration of emergencies are really a subset of the concept of designated IPS described above. The scenario is as follows. Suppose that, due to emergency demand (the nature of such emergency is of no concern to us here) it is desired to 'run through' a product from start to finish with minimal delay. Then, from the outset, the end product is always in sight, and the IPS is said to be 'designated' for that product. The loading of this product on the three shops is outside the regular production control scheme, albeit its impact on the plant operations must be taken into account (since the operations in the various stages of production occupy capacity). Some of the emergency demands are not only 'designated' but also **allocated**, in the sense of having their machines (warper, slasher or loom) on which they are to be processed specified. In such cases the decision of allocating the product to machines is also outside the production control scheme, except for the specification of the time slot in which the operation is to be performed (the schedule).

Alternate routings

A practical consideration that had to be taken into account is the presence of **alternate routings** for the majority of the products in almost all stages of production. For instance, a yarn may be warped on warpers

#2 or #3, but not on #1 or #4; another yarn may exhibit the opposite possibilities! The same is true for the slashing shop: a few warp styles may be produced by only one slasher; the overwhelming majority of styles may be produced on two, three or even all slashers. (The two or three slashers on which the item may be produced vary from product to product.)

Already warped yarns

Another practical consideration is related to the purchase of **already warped** yarns. Briefly, items which require such yarns do not impose any load on the warping stage, albeit they do impose a load on the slashing and weaving stages. Therefore, careful account must be taken of the inventories of such yarns, which imposes additional constraints on the quantities produced of these items in those two stages.

Netting against IPS

A significant, practical consideration is the issue of netting demand against IPS. Now, netting demand against finished goods inventory is a time-honored standard procedure that is routinely performed by almost all producing firms. But netting demand against **loaded but yet to be completed** items is peculiar only to plants with long lead times, which is the case here. As can be easily imagined, a demand for 80 000 yards of a particular cloth style in period 2 (that is, five to six weeks hence) should impose no additional load at all on the weaving stage if more than this amount is already loaded on some looms, (this amount of cloth production may occupy between 8 and 20 looms), or are being processed in the slashing shop and are slated to be processed in the weaving shop immediately thereafter, so that they shall be ready in three to four weeks! The load of the existing products on the various stages is taken care of in the estimation of the **net available capacity** in that stage, and should not be counted twice. Therefore, the validity of the LM is crucially dependent on the availability of accurate information on the status of each machine and each product in the plant. In the absence of such precise information the load specified by the optimal solution of the LP may be the optimum for the wrong problem!

6.3 THE SCHEDULING MODULE (SM)

Recall that the objective of the analysis is to develop a computer-based procedure that loads **and schedules** the three stages of production in the plant to achieve the objectives stated in section 6.1. By 'scheduling' we mean the specification of the time at which each operation is to start and finish in each of the three stages: the warping, slashing and weaving stages.

6.3.1 Module design

We have developed a **pull system** to approach, as much as possible, the JIT philosophy. The controlling element is the weaving shop, naturally enough, since it is the last stage of production, and the time to weave a beam dominates the times in the two previous stages by some two orders of magnitude. Controlling the weaving shop has always been recognized by plant personnel as the key to controlling production in the plant.

The specified load on the looms is scheduled first. This is accomplished knowing three elements: the status of each loom (the product actually on the loom, if any, and its **residual processing time**), the status of any designated and/or scheduled IPS, and the load on each loom class in each period as specified by the LP loading module. Figure 6.3 is a flow diagram that gives a global view of the logic of the SM.

The SM has two primary objectives to accomplish which are often conflicting: satisfy the demand on time, and minimize the cost of change-over.[5] The large number of looms in each class (in the hundreds) prevents any optimization of the resulting schedule, even if we were to limit the analysis to only one objective. (See below for 'complications' that render such optimization more impractical!) Therefore, we had to resort to heuristics. The basic logic of the SM runs as follows (see Figure 6.3):

- Products are ranked first by their due time (the period in which they are supposed to complete the weaving stage) and second by their **priority**. This latter is a number between 2 and 100, with priority 1 reserved for the products that are already on the looms (no pre-emption); the lower the number the higher the priority.

  There are several ways in which priority is assigned to a product. For instance, it may be based on the 'shadow price' of the product's 'cap on production' constraint in the LP solution, which correctly measures the contribution to profit from a unit relaxation of the constraint; or it may be based on the product's variable margin; or it may be assigned arbitrarily by the shop floor manager. Currently, the priorities are based on the product's variable margin, by default.

[5] Note that the number of 'setups', and their time occupancy on the same loom, which represent 'tie-in of ends' of the same item, is fixed by the number of cloth beams that must be produced of a particular item, say j. However, the number of 'change-overs', which represent changing the product, is a control variable that depends on how many looms are used to produce the cloth beams of the item and how frequently other items are interspersed between two beams of item j.

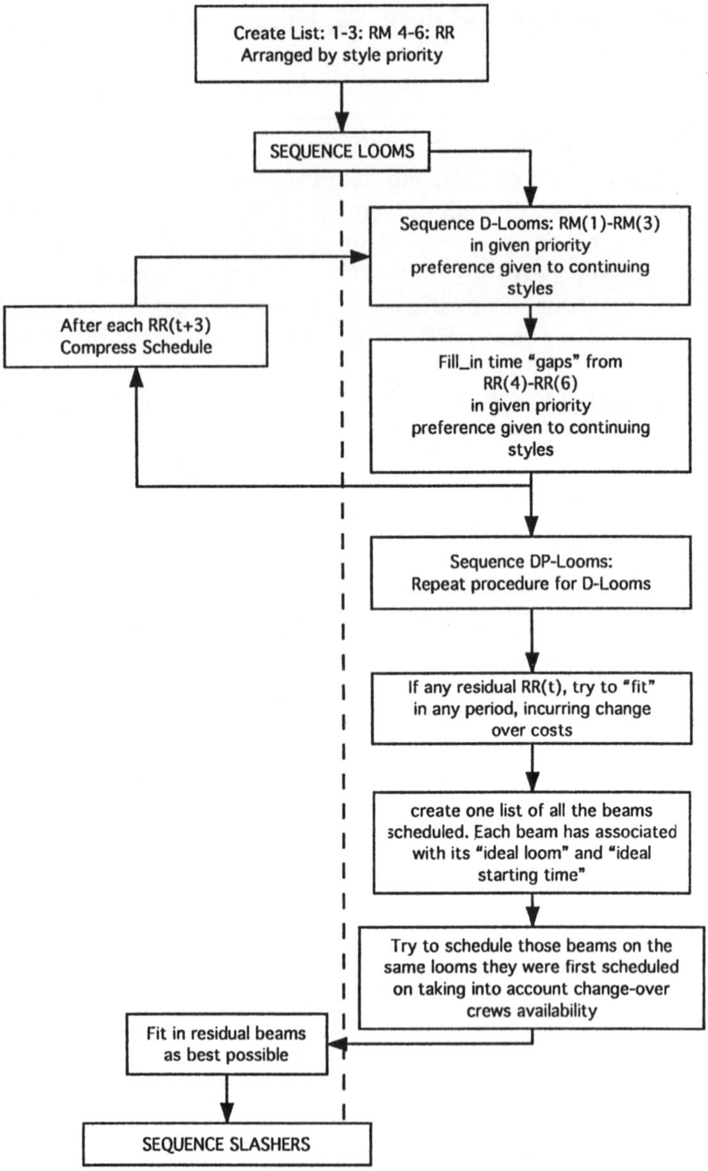

Figure 6.3 Logic of SM.

- The looms are scanned for the same 'current occupancy' as the LM-loaded products. Whenever a match is discovered the beams are assigned to the loom(s) sequentially, with proper allowance for setup time between beams, as long as the due date is respected. If none is present, or should the sequential allocation of beams to looms cause the due date to be exceeded, the product is scheduled on an empty loom if any is available, or on a loom with a 'similar' product (i.e. a product which requires minimal change-over time).[6] The process is continued until the LM-load of the product is satisfied.

- The products are selected sequentially in order of priority. We are assured, by the LM, of adequate capacity for at least **initiating** each product in its required period, and completing most of it in the specified periods.
- After the weaver shop has been scheduled for the planning horizon (with possibly some 'overhang' on the looms – i.e. occupancy beyond the 8-week planning horizon), scheduling the slasher shop is initiated. The guiding principle in scheduling the slashing shop is to complete the processing on the slashers in time to transport the loom beams to the weaver shop and load the beams on the looms.

  A word of explanation is necessary relative to this 'overhang occupancy' on the looms at the end of the planning horizon. Consider the following abbreviated (and drastically simplified) example of the real life problem. Suppose we have only three looms that are free. Their capacity in any period is $3 \times 168 \times 2 = 1008$ hours. Suppose that the looms are loaded to capacity as follows: 2 loom beams of product #1 of 380 hours processing time plus 5 hours of setup time each, and one loom beam of product #2 of 230 hours processing time plus 6 hours of setup time. The load on the looms is $= 2 \times 380 + 10 + 230 + 6 = 1006$, which is only 2 hours short of full capacity in each period. Close enough. Consider the two looms of product #1: they will actually consume only 336 hours each in period 1: 5 hours of setup time for each loom beam plus 331 hours of processing. This leaves 49 $(= 380 - 331)$ hours to be completed in period 2 on each loom. The finished cloth beams shall be ready at time 385, which is 49 hours into period 2. Now two new (loom) beams shall be loaded on the same two looms, and they shall complete at time 770, which is 98 $(= 770 - 2 \times 336)$ hours into period 2. Finally, two more beams shall be loaded in period 3 on the same two looms, and they shall complete at time 1155 which is 147 $(= 1155 - 3 \times 336)$ hours into period 4. This is the 'overhang occupancy' on those two looms at the end of the planning horizon![7]
- After scheduling the slashers, the SM schedules the warpers. Again, the guiding principle is to complete the processing in the slasher in time sufficient to transport the section beams to the slasher shop and load the beams on the slashers.

[6] In the vernacular of the 'traveling salesman problem' this is the 'closest city heuristic.

[7] You are urged to plot the Gantt chart for product #2, to discover that all three beams shall complete by time 708, thus leaving 300 hours of **idle time** on the loom. This idle time is also identified in the SM report.

6.3.2 Complicating factors in the SM

In the weaving phase

1. Scheduling the products on the looms is somewhat complicated by the need to take into account the availability of the **crews** required to effect the setup/change-over. Thus, in reality, the SM considers two types of resources: (a) the looms, and (b) the personnel. Typically, two operators are required to perform the setup/change-over of product loom beams, and there is a small number of operators relative to the number of looms. (The ratio varies between 1:15 to 1:50, depending on the reliability of the looms and the relative cost of downtime of a loom to the idle time of an operator.) If a setup/change-over is called for and a crew is not 'free' to attend to it the SM shall postpone the initiation of production until the crew is available. If, in the meantime, one (or several) setup/change-over is called for, a queue is formed which is serviced on a first-come-first-served basis by the crews as they become 'free'. This is also reflected in the SM.
2. It is also complicated by consideration of IPS. Recall that the LM loaded the looms in such a way as to consume any IPS (in the form of loom beams) before any new production is generated in the slasher shop. In particular, the slashers are not scheduled to produce any new loom beams of a particular warp style before the existing IPS is consumed.
3. Finally, the SM must also consider any unplanned machine stoppages, as reflected in the status of the shop at the time of running the SM. Planned stoppages for regular maintenance of the looms are typically done over plant vacations and therefore do not interfere with the regular operations of the plant. (This remark is equally applicable to the slashing and warping shops.)

In the slashing phase

The slasher shop scheduling process, which is straightforward in principle, is complicated by three factors.

1. The first is the presence of IPS. Recall, again, that the LM loads the slashers in such a way as to consume any IPS (in the form of section beams) before any new production is generated on the warpers. This fact has to be reflected in the manner the SM loads the various slashers.
2. The second is the need to produce in fixed multiples of section beams and loom beams. The issue of slashing in 'lots' of fixed number of section beams and loom beams increases the complexity of the SM to a considerable degree. To see this, consider the following example.

Suppose the production plan of a product (as dictated by the LM) called for the following number of loom beams to be produced:

	Period 1	*Period 2*	*Period 3*
RM	2	3	1
RR	5	6	8

Suppose that a 'lot' is 8 loom beams, which require 12 section beams to produce. The first question is: should RM be satisfied first over the planning horizon (all three periods), after which attention may be paid to RR, or should the demand in each period (for RM and RR) be completed before proceeding with the schedule for the following period?

Under the first rule the SM shall be faced with the requirement to produce 16 loom beams in period 1 (to guarantee the satisfaction of the 6 loom beams for RM through period 3, and the 5 loom beams for RR in period 1), 8 loom beams in period 2 (to satisfy the RR demand for the one remaining loom beam in that period) and 8 loom beams in period 3 (to satisfy the RR demand for the one loom beam remaining in that period). Under the second rule the SM shall be faced with the requirement to produce 8 loom beams in period 1, 8 loom beams in period 2, and 16 loom beams in period 3. The first rule demands more capacity in the first period but guarantees the satisfaction of **all RM** over the planning horizon. The second rule delays the heavy demand on capacity to the third period. Which will it be? Furthermore, the IPS shall be quite different over the three periods under the two rules: the first shall have 9, 8, 7 and the second shall have 1, 0, 7. (Evidently, the price paid for securing the availability of RM is higher IPS, as to be expected.)

We have opted for the second rule for three good reasons. First, it presents a more balanced load to the SM in the near horizon. Second, the LM would not have programmed any RR unless it is profitable, therefore it should be produced. (Unfortunately, some RM is in fact produced at a loss!) Third, if the need be, the loom beams produced to meet RR in the first period may be used to satisfy RM in the second, or later, periods if capacity is 'tight' in these periods. The commitment for RM shall still be met.

The second question is: should we plan on a reduced lot size in period 3, in the spirit of 'wait and see'? After all, the requirements in the first two periods would be completely satisfied by exactly two lots. The third period is concerned with forecasted demand in weeks 7 and 8 from 'now', and changes may take place in the meantime that would modify the figures considerably. In this case, the SM would schedule eight loom beams in each period, leaving one loom beam short over the planning horizon. The advantage of such a strategy would be to conserve capacity in the third period, and provide an underestimate of the load on the slashers, hence remain within the feasible region of the LM. The disadvantage would be a logically more complex SM because of its need for the added logic. (Note that if the demand in the last period for RR were four instead of eight, then the RM would schedule five loom beams only in the last period. This reduced lot size may be modified in later runs of the program.)

We have opted for such curtailment of production in the last period.

3. The third is the desire to group like 'warp styles' and run them sequentially to minimize the setup time on the slashers. This is an ever-present desire to minimize change-over time, which often contradicts the JIT philosophy. The artful balance between these two desiderata is embedded in the computer logic.

6.4 THE OUTPUTS

The main output of the LM is in the form of three values corresponding to the total production plan for each product in each of the three periods:

FACSIMILE OF LP OUTPUT

CLOTH STYLE ==>		160/60 1	360/38 2	364/52 3	365/52 4	630/50 5	650/50 6
XLDPOUT (Yrds)	PER. 1	0	40	94	25	0	78
	PER. 2	0	0	4	0	0	0
	PER. 3	25	15	0	0	25	0
TLDPOUT (Yrds)		25	55	98	25	25	78
TLRMOUT (Yrds)		25	55	86	23	25	78
TLRROUT (Yrds)		0	0	12	2	0	0
Y/OUT (Yrds)	PER. 1	0	0	0	0	0	0
	PER. 2	0	0	0	0	0	0
	PER. 3	0	0	0	0	0	0
Z/OUT (Yrds)	PER. 1	0	0	0	0	0	0
	PER. 2	0	0	0	0	0	0
	PER. 3	0	0	0	0	0	0
Whole Beams : XLRMI/OUT	PER. 1	1	10	21	6	1	9
	PER. 2	1	1	1	1	1	1
	PER. 3	6	4	1	1	3	1
Whole Beams : XLRRI/OUT	PER. 1	1	1	2	1	1	1
	PER. 2	1	1	1	1	1	1
	PER. 3	1	1	1	1	1	1

Figure 6.4 Sample output of LM.

$T(j,t)$, $T(\mathrm{RM},j,t)$, and $T(\mathrm{RR},j,t)$. It is presented in spreadsheet format (Fig. 6.4).

A variety of other auxiliary output reports are also produced according to management requests, which vary over time, such as planned machine occupancy, shadow prices, planned finished goods inventory, and planned backlog. (The word 'planned' refers to the fact that it is planned by the loading module.)

The main output of the SM is a listing of each machine in the plant and the specification of its occupancy by the various products throughout the planning horizon. A facsimile of such output for the looms is presented in Fig. 6.5.

```
                       DETAILED SCHEDULE PER LOOM

THE SCHEDULE OF PLAIN LOOM 101 IS AS FOLLOWS

Style no    160/60   Start-Time    0   Fin-Time     200     Priority        1
Style no    160/60   Start-Time  513   Fin-Time     761     Priority      100
Style no    160/60   Start-Time  778   Fin-Time    1026     Priority      100

IDLE TIME ON LOOM 101 IS 324 HOURS
OVERRUN ON LOOM 101 IS 18 HOURS

THE SCHEDULE OF PLAIN LOOM 102 IS AS FOLLOWS

Style no  5075/36G   Start-Time 1203   Fin-Time    1341     Priority      100

IDLE TIME ON LOOM 102 IS 1200 HOURS
OVERRUN ON LOOM 102 IS 333 HOURS

THE SCHEDULE OF PLAIN LOOM 103 IS AS FOLLOWS

Style no    360/38   Start-Time    0   Fin-Time     500     Priority        1
Style no    360/38   Start-Time  781   Fin-Time     938     Priority      100
Style no    360/38   Start-Time  941   Fin-Time    1098     Priority      100

IDLE TIME ON LOOM 103 IS 278 HOURS
OVERRUN ON LOOM 103 IS 90 HOURS

THE SCHEDULE OF PLAIN LOOM 104 IS AS FOLLOWS

Style no  50000/50   Start-time    0   Fin-Time       0     Priority        1
Style no   8760/38   Start-time  803   Fin-Time    1126     Priority      100

IDLE TIME ON LOOM 104 IS 798 HOURS
OVERRUN ON LOOM 104 IS 118 HOURS

THE SCHEDULE OF PLAIN LOOM 105 IS AS FOLLOWS

Style no    360/38   Start-Time   21   Fin-Time    178      Priority      100
Style no    360/38   Start-Time  492   Fin-Time    649      Priority      100
Style no    360/38   Start-Time  652   Fin-Time    809      Priority      100
Style no    360/38   Start-Time  812   Fin-Time    969      Priority      100

IDLE TIME ON LOOM 105 IS 329 HOURS
TIME REMAINING ON LOOM 105 IS 39 HOURS
```

Figure 6.5 Sample output of SM.

Again, a variety of summary statistics are produced according to management request, such as: the planned number of looms occupied by the same product in each week (which may be up to 30 in some instances); the planned time for setups/change-overs; the average flowtime; the number of beams started per week per cloth style; the total 'unutilized' shop capacity due to idle time (time waiting for the change-over crew); the total overrun time (time of loom occupancy beyond the six weeks planning horizon); the total setup time; and the total crew utilization.

6.5 A SIDE BENEFIT: A MARKETING APPLICATION

As is the practice in many large enterprises, Marketing is divorced from Production, not only organizationally and administratively but also physically: the Plant, which is one of several plants belonging to the same parent company, is located in one region to be closer to the raw material and labor, while Marketing is located closer to the financial centers of New York City. More importantly, the marketing organization normally strives to promote the styles with the highest profit margins which, on the surface, seems to be the 'logical' thing to do, and commits the plant to a set of deliveries. The plant, however, experiences severe imbalances in its production loads on the various shops. Excessive delays in responding to demand has traditionally necessitated renegotiating contracts, to the consternation of the marketing organization and the chagrin of the customer.

A study of the discrepancy between the commitments made by Marketing and the response of the plant to the demands imposed on it revealed that dependence on the products' profit margin as the sole parameter that guides the selection of the styles to promote by Marketing is flawed because it does not take into account the limited capacity of the various shops in the plant. It is elementary to observe that a style's 'variable margin' is a correct measure of its profitability if it is the only one produced or, equivalently, under the assumption of infinite capacity of the producing Plant. The moment one is concerned with two or more products which compete for a limited capacity – even a single one – the variable margin is no longer a valid measure of profitability! Rather, in the case of a single resource, one must seek the **marginal profit per unit of the limited resource**. When more than one resource are limiting because of the multiplicity of shops required to process each style, as is in our case, simple marginal analysis fails and one must resort to mathematical programming to accommodate all the constraints simultaneously. Having constructed such an LP for the purpose of optimal loading of the shops, it was a simple step to use it also for marketing purposes.

The question posed was the following: can the marketing organization be directed to promote styles that are more in harmony with the plant's capacities?

6.5.1 The approach

Our approach is iterative in nature and runs as follows. At the outset we consider all the styles and utilize the LP model with the objective to maximize net profit from production, assuming no 'cap' on production. The result is a plan which produces a subset of the styles whose production would 'saturate' the capacities of the shops in the plant as much as possible and yield the maximal total net profit. We shall refer to this set of styles as Class 1. Note that by the very nature of LP one cannot 'rank' the styles in this class: together they constitute an optimal set, and together they are the most profitable and should be the prime target for promotion by the Marketing organization.

As to be expected, the quantities produced of these woven cloth styles were excessively large – in the order of millions of yards of each style per week! – which are far in excess of what the market could possibly absorb. At this point Marketing was called into the picture to give their best estimates of the maximum that can be sold of each of these styles over the next planning horizon. Typically, this is a small fraction of the quantities originally specified. We use this information to either place a 'cap' on the quantities that may be produced of each of these products, or fix their production at the values provided by Marketing, deduce their load from the available capacities that would be required to produce these maxima, and remove them from the list of products under consideration. We repeat the LP optimization procedure with the reduced set of styles and reduced shop capacities, which results in another set of preferred styles, now called Class 2. Again, the resulting plan usually calls for the production of excessive amounts of each product, and again Marketing is called upon to give estimates of the maximum that can be sold of each; and the cycle repeats.

Let M denote the number of cycles needed to exhaust the plant capacities. Typically, M is no more than 15. At termination we have in hand the classification of the styles into no more than $M + 1$ classes: the M selected by the LP, and the residual styles which were never selected. The directive to Marketing is complete and unambiguous: promote the styles in the order of Class 1 through Class M.

6.5.2 Implementation

The longest phase of the iterative scheme was in securing the feedback from Marketing concerning the limits on sales of the styles selected by the solution. The input data to the LP model was in the form of a Lotus

1-2-3 spreadsheet, which greatly facilitated the reduction in available capacities according to the 'fixing' of the production of the various styles at successive iterations. The maximal number of iterations to capacity saturation was eight. The model has been in use for the past year, run on a monthly basis. Satisfaction by management is high.

6.5.3 Remarks

It is common practice to use aggregate measures of production that are based on some fictitious unit such as 'yards of cloth' in a textile firm, or 'tons of steel' in a steel mill, etc., in all issues ranging from marketing to production. Such measures are even used in comparing manufacturing plants. In the overwhelming majority of cases these aggregate measures are devoid of meaning, and cannot account for variations in capacity utilization among different products. Consequently, they result in a distorted picture of what is feasible to produce or what is profitable. When combined with the separation of Marketing from Production they usually spell trouble, if not disaster. We believe that simple LP models, such as the one presented here, provide a reasonable alternative that is both economically feasible (since they can be implemented on a PC) and practically useful (since they do not ignore the capacity limitations of individual shops in the plant).

Two side benefits of such a model are worth emphasis. First, **sensitivity analysis** of the LP model can guide Marketing in deciding on any contemplated changes in the **selling price** of a style, either to promote it or to discourage its demand. Second, the LP model can be used by Production to evaluate any alternate technology that may change the 'cost of goods produced' which, in turn, changes the 'net gain' for the styles affected, and therefore may change their ranking.

6.6 LESSONS OF IMPLEMENTATION

Apart from the Marketing model and its application, which is a 'freebie', since its realization did not involve any major developmental effort, we can safely state that three lessons were learned from the attempt to implement the mathematical and decision models that were generated in the process of our research/development.

The most important lesson is that the demands of exact mathematical models and precise algorithms have forced the plant to re-evaluate its data processing system and to redesign it to provide more accurate and, more importantly, more timely data. For instance, the 'state' of the plant, as given by the occupancy of each machine and the status of every order, had to be acquired and reported expeditiously enough to permit the results of the LM and the SM to be timely and meaningful. It was quickly learned that a state that is one or two days old is worthless! The construction of the spreadsheet which contains the (more-or-less) fixed parameters of the plant (such as the time standard of setup and processing on the different machines) consumed over 75% of the development time!

Under less tight control mechanisms, the parameters used in the planning of production (such as estimates of machine breakdowns, time standards, cost standards) had to have a number of allowances to absorb the imprecision in the loading and scheduling functions. (The parameters had considerable 'fat' in them.) The development of a more timely information system acted as an incentive for management to consider the updating of these parameters in shorter intervals of time, and to use more refined estimates. For instance, there is no longer need to use the same processing time standard that is used for payment (which may contain a 20% incentive) in the loading or the scheduling modules that are used for production planning. The delineation of functions elicited a rethinking of the very database on which decisions are made.

The plant personnel were introduced, perhaps for the first time, to several notions that play a dominant role in the decision to produce one product in preference to another. An excellent example is the concept of the capacity constraints of the various stages of production. The modeling process of these constraints was a new experience that seems to 'make sense' to shop personnel.

Management was introduced to the concept of shadow prices of facilities (such as a slashing machine) and of products (such as the 'cap on production' of a style, or the 'must-satisfy requirement' (RM)). Management now realizes that it has the capability to measure, albeit approximately, the 'worth' of adding (or subtracting) one more looms or one more warpers. This 'worth' has nothing in common with the accounting measures of purchase price or depreciation. On the other hand, it came as a surprise to management to be told that certain cloth styles were produced at a loss; and to be apprised of the exact amount lost per 1000 yards. This was more revealing since the loss (as measured by the shadow price) had little in common with either the selling price, or the marginal value of the style, or even the standard cost of producing it! Management now realizes that it has a tool with which it can investigate different 'what if' scenarios (what if a warper or slasher is out of service for one or two weeks? What if a particular yarn is not available?), though they have not used the facility placed in their hands in this manner yet.

A great deal has been written on 'resistance to change', the reasons for its presence, how it is manifested, and the approaches suggested for 'winning the "resistor" to your side'. We encountered a singular lack of hostility to new ideas. The shop floor personnel are naturally inquisitive if not downright cooperative, though somewhat skeptical (who isn't?).

But we discovered, through bitter experience, that the most important aspect of smoothing the road to acceptance is the presentation of information in a format that is understandable to the person using it. At the beginning of the development the output of the loading module was presented in the language of the LP, such as: $XLRM(I,J) = 24\,850$ for the quantity to be produced to meet the 'must-satisfy requirements' of style I in period J! We soon learned that we are not getting anywhere! A simple change to a spreadsheet format with the style number written in plain English rendered the result acceptable.

6.7 BIBLIOGRAPHICAL REMARKS

The linear programs used here are of the usual garden variety that may be found in any introductory book on linear programming or operations research, such as Winston (1994). Sequencing and scheduling flowshops and hybrid flowshops is a 'hot' topic that has occupied, and continues to occupy, researchers and practitioners alike. A recent book on the subject is Pinedo (1995). A more elementary discussion of the basic principles of the theory may be found in the book by French (1990). The LP software used in the application is Lingo (Industrial Version), available from Lindo Systems, Inc., P.O.B. 148231, Chicago, IL 60614. The scheduling software was developed in-house and tailored to the specific requirements of the sponsoring firm.

6.8 GLOSSARY

$b(LB,j)$: available loom beams of product j
$b(LB,w)$: available loom beams of warp style w
$b(S,t)$: total available capacity in the slasher shop in period t
$b(SB,j)$: available section beams of product j
$b(W,t)$: total available capacity in the warper shop in period t
$b(\zeta,t)$: available capacity of subset ζ of loom classes
$b(\theta,t)$: available capacity of subset θ of slashers
$b(\varrho,t)$: available capacity of subset ϱ of warpers
$b(LB,D,w)$: available 'designated' loom beams of warp style w
$b(LB,U,w)$: available 'undesignated' loom beams of warp style w
$b(YN,r,t)$: cumulative available yarn r in period t
$B(k,j)$: quantity backlogged from period k to period $k+1$ of product j
$BSIZE(j)$: yards in one loom beam of product j (beam size)
$HC(S,t)$: load in hours on the slasher shop in period t
$HC(W,t)$: load in hours on the warper shop in period t
$HR(S,t)$: residual capacity, in hours, in the slasher shop after LP load

HR(*W*,*t*): residual capacity, in hours, in the warper shop after LP load
I(*k*,*j*): quantity carried in inventory from period k to period $k + 1$ of product j
m(*L*,*w*,*j*): yards required of warp style w by 1000 yards of product j produced on the looms
m(*S*,*w*,*j*): yards required of warp style w by 1000 yards of product j produced on the slashers
$m(\zeta,j)$: hours occupancy per 1000 yards on any loom in class ζ
$m(\theta,j)$: hours occupancy per 1000 yards on any slasher in subset θ
$m(\varrho,j)$: hours occupancy per 1000 yards on any warper in subset ϱ
m(*YN*,*r*,*j*): yards required of yarn r per 1000 yards of product j
RM(*t*,*j*): 'must satisfy' requirements in period t of product j
RR(*t*,*j*): 'may satisfy' (or residual) requirements in period t of product j
T(*L*,*j*): total quantity produced of product j on the looms to satisfy both RR and RM over the planning horizon
T(*L*,*RM*,*j*): total quantity produced of product j on the looms to satisfy RM over the planning horizon
T(*L*,*RR*,*j*): total quantity produced of product j on the looms to satisfy *RR* over the planning horizon
T(*L*,*N*,*j*): total quantity produced of product j on the looms from new production
T(*L*,*V*,*j*): total quantity produced of product j on the looms from loom beams inventory
T(*L*,*VD*,*j*): total quantity produced of product j on the looms from loom 'designated' beams inventory
T(*L*,*VU*,*j*): total quantity produced of product j on looms from loom 'undesignated' beams inventory
T(*S*,*N*,*j*): total quantity produced of product j on slashers from new production
T(*S*,*V*,*j*): total quantity produced from section beams inventory of product j on the slashers
VM(*j*): 'variable margin' of product j
X(*L*,*t*,*j*): total quantity produced of product j on the looms to satisfy both RR and RM in period t
X(*L*,*RM*,*t*,*j*): quantity produced on the looms to satisfy RM in period t of product j, yards
XI(*L*,*RM*,*t*,*j*): quantity produced on looms to satisfy RM in period t of product j, integer loom beams
X(*L*,*RR*,*t*,*j*): quantity produced on looms to satisfy RR in period t of product j, yards
XI(*L*,*RR*,*t*,*j*): quantity produced on looms to satisfy RR in period t of product j, integer loom beams
YD(*RM*,*t*,*j*): yards planned for production to satisfy RM of product j in period t

YD(*RR*,*t*,*j*): yards planned for production to satisfy RR of product *j* in period *t*
YN(*r*,*t*): cumulative yarn *r* required up to period *t*

6.9 APPENDIX: THE LOADING MODULE

For the definition of the terms used in this LP, see the Glossary above.

6.9.1 Demand constraints

- Produce the **must-satisfy (RM)** requirements:

$$\sum_{t \leqslant k} [X(L,RM,t,j) - RM(t,j)\ ] \geqslant 0,\ \forall k,j \tag{6.1}$$

- Production may exceed or fall short of **residual requirements (RR)**:

$$\sum_{t \leqslant k} [X(L,RR,t,j) - RR(t,j)\ ] - I(k,j) + B(k,j) = 0;\ \forall k,j \tag{6.2}$$

  Note that if the cumulative production exceeds the cumulative RR (sometimes referred to as the **may satisfy** demand) then inventory shall be carried over from period k into period $k + 1$. Otherwise, part of the demand shall be backlogged for that period. Note also that both RM and RR must have been 'netted' against available finished goods inventory and/or in-process production that is expected to be completed in each period. This pre-processing of the demand parameters is achieved prior to running the LP.
- Do not exceed the permissible 'cap' on production (PCAP) for each individual product over the planning horizon. To formulate this aggregate constraint we must identify the production of each item designated for RM as well as RR, then secure the total production over the planning horizon (which is used in the objective function), and finally impose the desired constraint. We have,

$$X(L,t,j) = X(L,RM,t,j) + X(L,RR,t,j);\ \forall t,j \tag{6.3}$$

$$T(L,j) = \sum_{t=1}^{H} X(L,t,j);\ \forall j \tag{6.4}$$

$$T(L,j) \leqslant PCAP(j);\ \forall j \tag{6.5}$$

- Respect any specified production imposed by management, either from available loom beams (LB) or section beams (SB) to be processed in slashing:

$$T(L,j) \leqslant b(LB,j);\ \forall j \tag{6.6}$$

$$T(L,j) \leqslant b(SB,j);\ \forall j \tag{6.7}$$

6.9.2 Loom constraints

- Respect loom capacity on each class of looms, individually as well as in combination (to accommodate alternate routings of a product); that is, several classes of looms are eligible to produce the product. To represent this flexibility in the model, let $L(j)$ denote the set of classes of looms which are eligible to produce item j. (For instance, loom classes 1 and 3 are eligible to produce item 1, loom classes 2 and 3 are eligible to produce item 2, etc.) A constraint must be written to reflect the desire to respect the total capacity of each subset of classes. Let the loom classes be denoted by l_k for $k = 1, \ldots, K$; i.e. there are K different classes of looms. Let Q_{lm} denote the mth subset of the set of classes, and $U[Q_{lm}]$ denote the **universal set** (of subsets) of looms **less the null set**. The cardinality of $U[Q_{lm}]$ is denoted by n_l, where $n_l = |U[Q_{lm}]| = 2^{|U[Q_{lm}]|} - 1$, and $|\gamma|$ is the cardinality of set γ. Therefore, the sets $\{Q_{lm}\}$ shall be indexed by the numbers $i = 1, \ldots, n_l$, written as $Q_{l1}, \ldots, Q_{1n_l}$. We then have the constraints

$$\sum_{\zeta \in Q_{li}} \sum_{j} m(\zeta,j) * X(L,t,j) \leq \sum_{\zeta \in Q_{li}} b(\zeta,t);\ \forall t, Q_{li} \tag{6.8}$$

For instance, suppose that a subset of the products, $J = \{j_1, \ldots, j_{10}\}$ may be woven on any of three classes of looms, symbolically denoted by l_1, l_2, l_3 of capacities $b(l_k,t)$ for $k = 1, 2, 3$ and $\forall t$. Then the set

$$U = \{\ (1), (2), (3), (1, 2), (1, 3), (2, 3), (1, 2, 3)\ \}$$

i.e. there are **seven** sets of capacity constraints corresponding to the $2^3 - 1$ alternative possibilities of loading, as follows (here, the coefficient $m(l_k,j) = 0$ whenever loom $k \notin L(j)$):

$$\sum_{j \in J} m(l_1,j) * X(L,t,j) \leq b(L_1,t);\ \forall t \tag{6.9}$$

$$\sum_{j \in J} m(l_2,j) * X(L,t,j) \leq b(L_2,t);\ \forall t \tag{6.10}$$

$$\sum_{j \in J} m(l_3,j) * X(L,t,j) \leq b(L_3,t);\ \forall t \tag{6.11}$$

$$\sum_{j \in J} m(l_1,j) * X(L,t,j) + \sum_{j \in J} m(l_2,j) * X(L,t,j) + \sum_{j \in J} m(l_1 l_2,j) * X(L,t,j) \leq b(l_1,t) + b(l_2,t);\ \forall t \tag{6.12}$$

$$\sum_{j \in J} m(l_1,j) * X(L,t,j) + \sum_{j \in J} m(l_3,j) * X(L,t,j) + \sum_{j \in J} m(l_1 l_3,j) * X(L,t,j) \leq b(l_1,t) + b(l_3,t);\ \forall t \tag{6.13}$$

$$\sum_{j \in J} m(l_1,j)*X(L,t,j) + \sum_{j \in J} m(l_3,j)*X(L,t,j) +$$
$$\sum_{j \in J} m(l_2l_3,j)*X(L,t,j) \leqslant b(l_2,t) + b(l_3,t);\ \forall t \tag{6.14}$$

$$\sum_{j \in J} m(l_1,j)*X(L,t,j) + \sum_{j \in J} m(l_2,j)*X(L,t,j) +$$
$$\sum_{j \in J} m(l_3,j)*X(L,t,j) + \sum_{j \in J} m(l_1l_2,j)*X(L,t,j) +$$
$$\sum_{j \in J} m(l_1l_3,j)*X(L,t,j) + \sum_{j \in J} m(l_2l_3,j)*X(L,t,j) +$$
$$\sum_{j \in J} m(l_1l_2l_3,j)*X(L,t,j) \leqslant b(l_1,t) + b(l_2,t) + b(l_3,t);\ \forall t \tag{6.15}$$

The first three constraints (6.9)–(6.11) specify that whatever is produced of the items that can be woven only on a specific loom class cannot exceed the capacity of that class. Note that the definition of $m(l_k,j)$ as $= 0$ if loom class k is **not** the only element in the set $L(j)$ causes the summation to add those products that are specified to be processed on loom k only. The next three constraints (6.12)–(6.14) are for products that have an alternate route on two classes of looms. Consider, for example, constraint (6.13). It accounts for the capacity of the subset $Q_{l_5} = \{1, 3\}$. If product j is eligible to be produced on any loom in class l_1 and l_3 then $m(l_1l_3,j)$ shall be >0, otherwise, it is $= 0$. The total load on the looms in the two classes l_1 and l_3 is the sum of the load resulting from items that may be processed only on class l_1, plus items that may be processed only on class l_3, plus items that may be processed on either class l_1 or l_3. Hence the constraint given. Finally, the last constraint (6.15) accounts for the capacity of the subset $Q_{l_7} = \{1, 2, 3\}$, which represents the items that may be produced on any of the three classes of looms.

- Production on the looms is either from available IPS – in the form of loom beams inventory ahead of the weaving shop – or from **new** production. (Recall that 'raw' demand has already been 'netted' by the finished goods inventory and the 'on looms' load.) The distinction is necessary in order not to overestimate the capacity required from the slashing shop (the loom beams available as in-process-stock are already slashed).

  In order to identify the production from each category we need to first identify the total production over the planning horizon for each item in both the 'must satisfy' (RM) category and 'may satisfy' (RR) category, then identify the total loom production over the planning horizon for each item as the sum of these two entities, and, finally, divide this total between **in-process-stock** and **new**. We shall identify production from in-process-stock by the letter V (for 'inventory'), and new production by the letter N. The desired constraints would read as follows,

$$T(L,RM,j) = \sum_t X(L,RM,t,j);\ \forall j \tag{6.16}$$

which gives the total production to satisfy RM over the planning horizon,

$$T(L,RR,j) = \sum_t X(L,RR,t,j);\ \forall j \tag{6.17}$$

which gives the total production relative RR over the planning horizon,

$$T(L,j) = T(L,RM,j) + T(L,RR,j);\ \forall j \tag{6.18}$$

which gives the sum of production of item j over the planning horizon. Finally, this total is now divided between IPS and new,

$$T(L,j) = T(L,V,j) + T(L,N,j);\ \forall j \tag{6.19}$$

Note that we had already defined $T(L,j)$ in terms of the individual period production quantities $X(L,t,j)$, see (6.4). Evidently, equation (6.19) ties the total production of item j for RM and RR with the individual period production.

Having defined the total loom production of item j from the IPS, we must now limit it to the available quantity,

$$\sum_j m(L,w,j)*T(L,V,j) \leq b(LB,w);\ \forall w \tag{6.20}$$

in which 'LB' stands for loom beams present as in-process-stock.

The complication due to the dichotomy between the designated and the undesignated IPS can be taken care of by the following device. Note that both designated and undesignated IPS may contain anticipated arrivals from the slasher and warper shops. Since the processing times in these two shops are some two orders of magnitude smaller than the length of the loading period (which is two weeks, or 336 hours) we assume that the 'arrivals' are available during the first period. We also assume that the designated in-process-stocks are destined to be loaded in the first interval.

Let the in-process-stocks $T(L,V,j)$ now be divided between VD (for designated) and VU (for undesignated).

$$T(L,V,j) = T(L,VD,j) + T(L,VU,j);\ \forall j \tag{6.21}$$

The capacity constraints become,

$$\sum_{j \in N(w)} m(L,w,j)*T(L,VD,j) \leq b(LB,D,w);\ \forall w,\ N(w) \tag{6.22}$$

$$\sum_{j \in N(w)} m(L,w,j)*T(L,VU,j) \leqslant b(LB,U,w);\ \forall w,\ N(w) \qquad (6.23)$$

in which $b(LB,y,w)$ is the quantity available of warp style w in category $y,y = D$ or $y = U$, and the sets $N(w)$ define the products that utilize loom beams of warp style w.

Note that under certain circumstances the designated load $T(L,VD,j)$ may be **required production**, in which case there is no decision to make and this quantity should be loaded external to the LP. (This subtle point usually escapes the attention of field operators.) Its effect would be felt in the reduction of the available capacity in the loom classes on which it has been loaded via its occupancy of these looms, and in the netting of demand of individual products, rather than groups of products, by the anticipated available production.

There still remains the quantity $T(L,VU,j)$ (which is now equated to $T(L,V,j)$, since there is no decision to be made relative to the quantity $T(L,VD,j)$), to be determined by the LP proper according to constraints (6.23). The formulation of this set of constraints is slightly more complicated. The difficulty arises from the very interpretation of an 'undesignated' section beam, as described above.

Since the processing times in these two shops are some two orders of magnitude smaller than the length of the loading period (which is two weeks, or 336 hours) we assume that the 'arrivals' are available during the first period (in fact, early in the first period). We also assume that the **designated** in-process-stocks are destined to be loaded in the first interval.

To this end, let $S_r = \{j$: product j requires warp style $r\}$. The constraint on production from available *IPS* of section beams may now be expressed as,

$$\sum_{j \in S_r} m(w,j)*T(L,VU,j) \leqslant b(w);\ \forall w \qquad (6.24)$$

6.9.3 Slashing constraints

- Respect slasher capacities for all subsets of alternate routings. In a manner similar to the looms, let $S(j)$ denote the set of slashers eligible to produce item j. Let the slashers be denoted by s_k for $k = 1, \ldots, K$; i.e. there are K different slashers. Let Q_{sm} denote the mth subset of the set of classes, and $U[Q_{sm}]$ denote the *universal set* (of subsets) of slashers *less the null set*. The cardinality of $U[Q_{sm}]$ is denoted by n_s, where $n_s = |U[Q_{sm}]| = 2^{|U[Q_{sm}]|} - 1$, and $|\gamma|$ is the cardinality of set γ. Therefore, the sets $\{Q_{sm}\}$ shall be indexed by the numbers $i = 1, \ldots, n_s$, written as $Q_1, \ldots, Q_{n_s}$. We then have the constraints

$$\sum_{\theta \in Q_{s_i}} \sum_{j} m(\theta,j)*X(L,t,j) \leq \sum_{\theta \in Q_{si}} b(\theta,t); \ \forall t, Q_{si} \qquad (6.25)$$

- Identify the load on the slasher shop (for control purposes),

$$HC(S,t) = \sum_{\theta \in Q_{si}} \sum_{j} m(\theta,j)*X(L,t,j); \ \forall t, Q_{s_i} \qquad (6.26)$$

- Identify 'residual capacity' in the slasher shop (for control purposes),

$$HR(S,t) = \sum_{k \in S} b(S,t) - HC(S,t); \ \forall t \qquad (6.27)$$

- Respect the availability of in-process-stock in the form of **section beams** ahead of the slasher shop. Production on the slashers is either from available IPS – in the form of section beams inventory – or from **new** production. Again, the distinction is necessary in order not to overestimate the capacity required from the warp shop (the section beams available as in-process-stock are already warped, either in the warper shop or purchased as such). The first set of constraints divide total production between in-process-stock and new,

$$T(L,j) = T(S,V,j) + T(S,N,j); \ \forall j \qquad (6.28)$$

The next set of constraints limit the production from in-process-stock to availabilities,

$$\sum_{j} m(S,w,j)*T(S,V,j) \leq b(S,w); \ \forall w \qquad (6.29)$$

Note that some of the section beams present as IPS ahead of the slasher shop may be yarn that is purchased already warped. Constraints (6.29) do not make distinction between the two classes of section beams available, albeit such distinction should be made relative to each warp style w. Note also that some of that IPS may be **anticipated** arrivals from the warper shop. No distinction is made here either between these two possible kinds of IPS.

Similar considerations are given to 'designated' vs. 'undesignated' in-process-stock as under section 6.9.2, which shall not be repeated here.

6.9.4 Warping constraints

The constraints repeat those for slashing except for considerations of section beams, since the input to the warping shop is the raw yarn. In place of constraints (6.29) we now have constraints on yarn availabilities over the planning horizon.

- Respect warper capacities for all subsets of alternate routings. The definitions of the variables is similar to both slashing and weaving shops, except for substituting W (for warping) in place of S.

$$\sum_{\varrho \in Q_{wi}} \sum_{j} m(\varrho,j)*X(L,t,j) \leq \sum_{\varrho \in Q_{wi}} b(\varrho,t);\ \forall t,Q_{wi} \tag{6.30}$$

- Identify the load on the warper shop (for control purposes),

$$HC(W,t) = \sum_{\varrho \in Q_{wi}} \sum_{j} m(\varrho,j)*X(L,t,j);\ \forall t,Q_{wi} \tag{6.31}$$

- Identify 'residual capacity' in the warper shop (for control purposes),

$$HR(W,t) = \sum_{k \in W} b(W,t) - HC(W,t);\ \forall t \tag{6.32}$$

6.9.5 Whole beam constraints

The 'integerization' of the LP output in whole beams is achieved by a pair of inequalities, written for the RM and RR demand categories. Note that this is determined **after the fact**, in the sense that the *XI*s are determined from the (continuous variables) Xs, and are not specified as integer variables **a priori**.

$$XI(L,RM,t,j) \geq X(L,RM,t,j)/BSIZE(j) - 0.50;\ \forall t,j \tag{6.33}$$

$$XI(L,RM,t,j) \geq X(L,RM,t,j)/BSIZE(j) + 0.50;\ \forall t,j \tag{6.34}$$

$$XI(L,RR,t,j) \geq X(L,RR,t,j)/BSIZE(j) - 0.50;\ \forall t,j \tag{6.35}$$

$$XI(L,RR,t,j) \geq X(L,RR,t,j)/BSIZE(j) + 0.50;\ \forall t,j \tag{6.36}$$

One must then impose the integrality constraints:

$$XI(L,RM,t,j),\ XI(L,RR,t,j) \geq 0 \text{ and integer}$$

Should the running time of the *LP* prove to be unacceptable long due to the stipulation of integer values for the *XI*s,[8] it is suggested that this integerization process be carried out externally to the LP proper.

If it is desired to determine the exact number of yards planned for production corresponding to the number of whole (cloth) beams just determined, one simply evaluates these from

$$YD(RM,t,j) = BSIZE(j)*XI(L,RM,t,j);\ \forall t,j \tag{6.37}$$

$$YD(RR,t,j) = BSIZE(j)*XI(L,RR,t,j);\ \forall t,j \tag{6.38}$$

[8] Unfortunately, the Lingo software takes the specification of **any** variable(s) as integer at its face value and assumes it is solving an integer (or a mixed-integer) linear program, which demands a procedure that is quite different from the solution of (continuous variables) LP.

6.9.6 Yarn constraints

- Respect the availability of yarn in storage. Here, one must take into account **anticipated deliveries** over the planning horizon in order not to curb production too severely if limited to the on-hand stock. The desired constraints are based on the determination of the cumulative production for each item in each period of the planning horizon, which is determined from the following equation,

$$X(L,t,j) = \sum_{\tau \leq t} X(L,\tau,j); \; \forall t,j \tag{6.39}$$

The cumulative yarn required for each product in the first t-period of the planning horizon is next determined from,

$$YN(r,t) = \sum_{j} m(YN,j) * X(L,t,j); \; \forall r,t \tag{6.40}$$

Finally, the desired availability constraints are given by,

$$YN(r,t) \leq b(YN,r,t); \; \forall r,t \tag{6.41}$$

6.9.7 The objective function

It is desired to maximize the total **gain** from production as given by the **variable margin** of the items produced, while recognizing the four factors that play a dominant role in such determination. First, it is desired to account for the gain accruing from the production of an item, as represented by its **variable margin** per 1000 yards. Second, it is desired to consume any available IPS before initiating any new production. This is taken care of by according a fictitious gain equal to about 5% of the variable margin of the product variable margin to production from IPS (the variables $T(L,V,j)$ and $T(S,V,j)$). Third, any inventory accumulation should be penalized at a rate high enough to prevent excessive build up of finished goods. This was accomplished by simply pricing the inventory (variable $I(k,j)$) at about 20% of the variable margin of the product per period. Finally, any backlogging of RR demand, though permitted (see constraint (6.2)), should be penalized high enough to make it undesirable to backlog demand for too long, lest most RR demand would become RM and thus obliterate any flexibility management may have in managing the plant! This was accommodated by costing the backlog (variable $B(k,j)$) at about 40% of the variable margin of the product per period. The objective function is as follows,

$$\text{Total gain} = \sum_{j} VM(j) * [T(L,j) + 0.05 * (T(L,V,j) + T(S,V,j))] - \sum_{j} C_I(j) * \sum_{t} I(t,j) - \sum_{j} C_B(j) * \sum_{t} B(t,j) \tag{6.42}$$

ACKNOWLEDGEMENTS

We are grateful to the splendid cooperation of the personnel of Glen Raven Mills, Burnsville, NC, USA, without which this developmental study would have never seen the light. In particular, we acknowledge the vision and foresight of Mr Donald Burhoe who initiated the project, and the continued support and encouragement of Mr Scott Gillen, Mr Danny Hardin, and Mr Tony Allen, and the untiring help of Ms Renee Styles and Ms Janet B. Daniels.

REFERENCES

French, S. (1990) *Sequencing and Scheduling: An Introduction to the Mathematics of the Job-Shop*, Ellis Harwood, New York, NY.

Pinedo, M. (1995) *Scheduling: Theory, Algorithms, and Systems*, Prentice-Hall, Englewood Cliffs, NJ 07632.

Winston, W.L. (1994) *Operations Research: Applications and Algorithms*, 3rd edn, Duxbury Press, Belmont, CA.

CHAPTER 7

The design and operation of flexible manufacturing shops

J. Slomp

ABSTRACT

Many firms recognize the need to make their manufacturing departments more flexible. This chapter explains the possibilities and limitations of flexible manufacturing systems (FMSs). Furthermore, the chapter elaborates on three important conditions which determine the success of an FMS for a firm. The first condition concerns the integration of the FMS in the goods flow of a company. The second condition focuses on the design of the FMS. The third condition, finally, refers to the operational control of the FMS.

7.1 INTRODUCTION

Rapid advances in computers and microelectronics have been major driving forces behind the development of new manufacturing technology in recent years. While tracing the evolution of computer-controlled manufacturing, three partly overlapping phases can be identified (Economic Commission for Europe, 1986). The first phase, which can be labelled **stand-alone machine automation**, commenced in the early 1950s with the advent of numerically controlled machine tools (NC-machines). The main characteristic of this phase was that more and more machines were equipped with digital control units of a rapidly increasing degree of sophistication. Important advantages of stand-alone machine automation include the higher efficiency and quality of the manufacturing processes compared with conventional manufacturing. The second phase, which may be called **systems integration**, started in the late 1960s with direct numerical control (DNC) by which several individual machine tools were controlled by a central computer. In the early 1970s, the concept of flexible manufacturing systems (FMSs) started to evolve.

An FMS is a computer-controlled system consisting of automated work stations linked by a material handling system and capable of processing different jobs simultaneously. Developments in the phase of systems integration enable firms to realize shorter manufacturing lead times and a higher flexibility (i.e. the ease with which the machines in a system can be reset for processing other parts). The third and final phase may be termed the **computer-integrated factory**. In this concept, all functions in manufacturing as well as other functions in a factory (such as design, engineering, process planning, and production planning and control) are computer-controlled, integrated and, to various degrees, self-optimizing. The promise of the concept includes a faster, more efficient, and effective introduction of new products. This concept may be more fully realized in the next century.

During the 1980s, there seemed to be a general agreement that the use of FMSs would spread widely. The general idea is that FMSs enable firms to achieve the efficiency of automated, high-volume mass production while retaining the flexibility of low-volume job-shop production. This is of utmost importance for firms which are being confronted with considerable market pressures concerning price, variety of products, and delivery performance. These market pressures can be found almost everywhere in current industry. The manufacturing departments of firms have to be efficient and flexible.

Despite the potential advantages of FMSs, the use of FMSs is spreading somewhat slowly. To a considerable extent, this is because flexible manufacturing systems have not been as profitable to users as some other innovations (Mansfield, 1993). Technical, organizational, and environmental (changes in the market place) problems have also had a negative impact on the profitability of FMSs. Some of these problems will be dealt with later in this chapter (section 7.2.4). The less-than-expected profitability of FMSs has had a negative impact on the imitation rate in industry. But, despite this, Mansfield (1993) found that the majority of firms in the automobile, electric equipment, machinery and aerospace industries in Japan, the United States and Western Europe will have installed flexible manufacturing systems by the year 2000. Jaikumar (1986), describing the application of FMSs in several countries, found that industry worldwide seems to prefer small FMSs with large buffer storages for pallets/products and simple routings of workpieces. This chapter will therefore be focused on the impact of small FMSs on the realization of flexible manufacturing shops.

The integration of the system in the goods flow of the manufacturing firm is important for the success of (small) FMSs. Section 7.2 will concentrate on this issue. Failures of FMSs are frequently caused by a lack of understanding what this aspect of integration means. Section 7.2

also presents an overview of experiences of firms with FMSs. A perfect integration of an FMS in the goods flow of the firm can be frustrated by a poor design of the system. Section 7.3 is devoted to the FMS design problems. The success of an FMS also depends on the solution of the operational control problems of the system. Rachamadugu and Stecke (1994) categorize FMSs from a scheduling point of view, and discern basically two types of FMSs: flexible flow systems (FFS) and general flexible machining systems (GFMS). An FFS can either be a flexible assembly system (FAS) or a flexible transfer line (FTL). A GFMS can be subdivided into dedicated flexible manufacturing systems (DFMS) and nondedicated flexible manufacturing systems (NFMS). These types of FMSs can be described in terms of their system, environmental and operational characteristics (Table 7.1). Each type has its own specific operational control problems. Furthermore, even within one class of FMSs, the specific characteristics of the FMSs and their environment may differ significantly. Therefore, it is likely that each particular FMS application requires its own operational (or production) control hierarchy and accompanying algorithms and/or heuristics. In order to deal with the complexities of an FMS, many authors propose a hierarchical framework for the operational control of FMSs (see e.g. Stecke, 1985, and Van Looveren *et al.*, 1986). Section 7.4 is devoted towards the design of an hierarchical production control system for an FMS. Such a system is of utmost importance for gaining the advantages of an FMS. Section 7.5 of this chapter presents some important research issues, derived from theory and practice. These issues relate to the integration, the design and the operational control of FMSs.

7.2 THE INTEGRATION OF AN FMS IN A FIRM

An FMS can offer a firm important advantages in terms of more efficient production, shorter throughput times, lower stocks, and higher quality of work. The realization of these advantages, however, is not obvious, as will be made clear here. Section 7.2.1 presents a small FMS, a so-called flexible manufacturing cell. This small system does have most of the characteristics of larger FMSs and can be seen as a reference for the remaining part of the chapter. Section 7.2.2 focuses on the performance characteristics of an FMS. Understanding these characteristics is of utmost importance for getting insight in the advantages and limitations of an FMS in a particular situation. Section 7.2.3 shows how the advantages of an FMS depend on the integration of the system in the goods flow of the company. Section 7.2.4 reviews the experiences of firms with FMSs.

Table 7.1 Types of FMSs and their characteristics

	System characteristics	*Environmental characteristics*	*Operational characteristics*
Flexible flow system (FFS)	• Is dedicated to produce a specific set of (few) part types (FTL) or assembled products (FAS) • Material flow is usually unidirectional	• Part types are asked in relatively large volumes	• **Performance indicators**: System outputs and machine utilization • **Main operating problems**: Determination of the production ratio (only if not defined externally) and determination of the part of input sequence
General flexible machining system (GFMS)	• Is able to produce a wide variety of part types simultaneously • Part routings can be different, even for parts of the same type		

Dedicated flexible machining system (DFMS)		• Few part types are asked in particular ratios	• **Performance indicators**: System output and machine utilization
			• **Main operating problems**:
			  • Determination of the input ratio
			  • Determination of the part input sequence
			  • How to keep the bottleneck machine busy

Nondedicated flexible machining system (DFMS)	• Part types and production requirements change frequently	• **Performance indicators**: Throughput time, machine utilization and meeting due dates • **Main operating problems**: • Dealing with routing flexibility • Dealing with large number of orders with different due dates • Dealing with limitations concerning fixtures, cutting tools and tool magazine capacities

7.2.1 An example of an FMS

Several authors have recognized a trend in industry towards small FMSs with large buffer storages for pallets and simple routings of workpieces (see Jaikumar, 1986; O'Grady, 1989). These small FMSs, also called Flexible Manufacturing Cells, can be seen as a stage towards larger systems. Werner und Kolb GmbH, a German machine tool manufacturer, distinguishes four realization stages of flexible manufacturing systems (Werner und Kolb Werkzeugmaschinen GmbH, 1987):

- **system-compatible machine tool**: NC machining centres with standardized workpiece, tool and control interfaces;
- **flexible manufacturing cell**: automatic machining of workpieces on several complementary machines of the same kind by means of flexible material-flow interlinkage, tool supply system and integrated control;
- **flexible manufacturing island**: complete, automatic machining of workpieces with a complement of supplementary machines by means of flexible material-flow interlinkage, tool supply system and integrated computer control;
- **interlinked flexible manufacturing system**: manufacturing islands, manufacturing cells and stand-alone machines with production control system, interlinked to form a computer-integrated manufacturing system.

Many firms regard a flexible manufacturing cell as a good alternative to a flexible manufacturing island, without the many technical, cost and justification problems associated with larger systems. These firms do not study the possibility of extending an FMC to a complete manufacturing island.

This section will give an example of a flexible manufacturing cell, since this type of FMS plays a central role in industry and in the development of flexible manufacturing. This example can be seen as a reference for the subsequent sections of this chapter.

The FMS in Fig. 7.1 consists of two basically identical machining centres linked together by a pallet transport vehicle and an integrated pallet buffer system, having a capacity of 15 pallets. Each of the machines can hold a limited number of tools in the tool magazine. The changing of tools in the tool magazines is performed automatically. The connection between the central tool storage and the tool magazines is realized by means of a tool robot. An FMS-computer takes care of the coordination of all activities within the FMS. The clamping and unclamping of parts on pallets/fixtures (a fixture being a device to position and hold a workpiece during the machining operation; it either may be built from modular components or it may be a special construction; fixtures are located on pallets which circulate in the FMS) is done manually on one or more integrated clamp/unclamp stations.

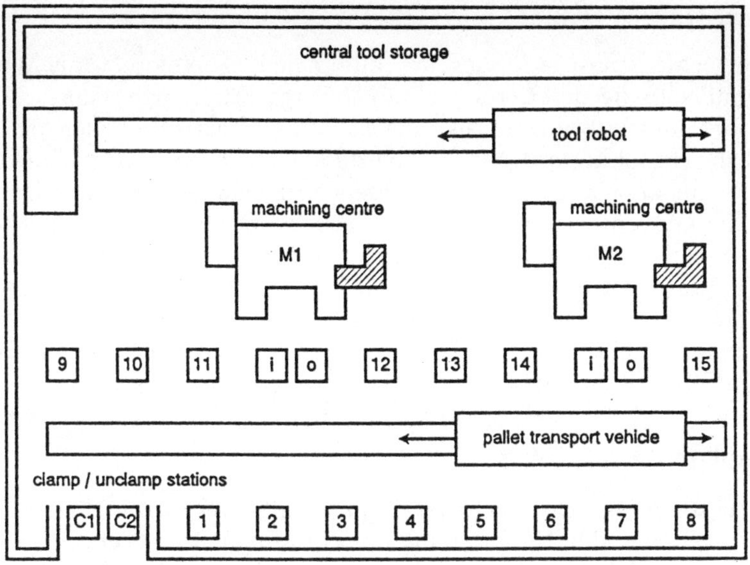

Figure 7.1 Schematic representation of a small FMS.

Before manufacturing of an order can start on the FMS, an operator has to (a) prepare the required pallet(s)/fixture(s) for the specific order, (b) build up and pre-set the required cutting tools,[1] (c) load the central tool storage with these tools, and (d) instruct the FMS-computer with respect to the release of the order. Next, one or more items of the order can be clamped on a pallet/fixture. This is done at a clamp/unclamp station (C1, C2). The pallet transport vehicle transports the palletized items either to the pallet pool (1–15) or directly to a machining centre (M1, M2). After machining, the pallet transport vehicle transports the palletized items either to the pallet pool or directly to a clamp/unclamp station. On this clamp/unclamp station the items are replaced by new items. The refilled pallet/fixture repeats the cycle through the system. The (unclamped) items on which an operation has been performed may wait for their next operation.

7.2.2 Performance characteristics of an FMS

The advantages of an FMS are largely determined by the degree of **efficiency**, **flexibility** and **multifunctionality** of the system.

[1] A cutting tool is built up of a tool holder, one or more adaptors, a tool body, and one or more inserts. Presetting of a cutting tool is needed in order to compensate the difference between a reference value and the actual dimension of the cutting tool.

Efficiency

In an FMS, most activities are automated. Therefore, only a few operators are needed to run a relatively large FMS. Furthermore, all the manual preparatory work, like the clamping of parts on pallets/fixtures and the presetting of tools, can be done during machining time. This results in an almost complete absence of changeover times on the machining centres. An important aspect of efficiency furthermore concerns the capability of an FMS to operate during an unmanned night shift. The efficiency of an FMS, however, is limited. Often, the buffer storage of pallets in the FMS is not large enough to support a fully unmanned night shift. Sufficient workload on each machining centre of an FMS is also an important condition for an efficient operation.

Flexibility

Due to the presence of all essential equipment in the system, like pallets, fixtures, tools and NC-programs, and through the absence of large changeover times, an FMS is able to produce different part types simultaneously. An FMS can easily change from the processing of one mix of orders to another mix. Rush orders can be performed without a loss of efficiency. However, the flexibility of the system is limited by the availability of equipment (e.g. a unique cutting tool cannot be used simultaneously on different machining centres). Furthermore, the introduction of new part types may cause significant preparation times (NC-programs, clamping plans) and loss of production time (caused by testing activities).

Multifunctionality

An FMS is capable of performing several processing steps of an order which traditionally were executed on subsequent production units. This simplifies the materials flow control in the firm and may reduce throughput times of orders. The multifunctionality can be realized in two different ways: (a) through the use of multifunctional machining centres or (b) through the presence of various types of work stations in one system. The multifunctionality of an FMS, however, is limited. Certain finishing processes, such as tempering and painting, are seldom integrated in an FMS.

The advantages of an FMS for a manufacturing shop are not simply the result of the hardware and software of the system. An optimal use of an FMS demands a carefully considered choice of part types, or operations to be performed on the FMS. This will be explained in section 7.2.3. Also the presence of a well-designed production control system is a key condition for an effective use of the FMS. Section 7.3 is devoted to this issue. Furthermore, several organizational measures may be needed in order to support the FMS. Section 7.2.4 will mention some of these measures, based upon experiences of firms.

7.2.3 The relative advantage of an FMS

For the most part, the FMS has been considered as an independent system in past literature. However, in order to realize all potential advantages from an FMS, it is necessary to consider the overall logistics context within which the FMS operates. In particular, it is advantageous to consider the flow of materials through the manufacturing department, the way in which production orders are generated, the objectives of the firm, and the potential role of the FMS in meeting these objectives efficiently and effectively.

An FMS is usually only responsible for a part of the processing steps (operations) that have to be performed on the part types assigned to the system. The production orders for this part of the processing steps can be generated either by means of forecast/inventory stock or directly by customer demand. This depends, to a large extent, on the position of the so-called **customer order decoupling point** (CODP) in the firm. The potential advantages of an FMS depends heavily on the part of the goods flow for which the FMS is responsible and on the position of the FMS with respect to the CODP. This will be explained in this section. Firstly, the concept of 'customer order decoupling point' will be clarified and, secondly, a small case will be discussed which shows the importance of the logistic context for the successful operation of an FMS.

Figure 7.2 illustrates the concept of CODP. Hoekstra and Romme (1992) define the CODP as follows:

> The decoupling point separates the customer-order part of the activities (to the right of the CODP) from the activities that are based on forecast and planning (to the left of the CODP). The customer order penetrates as far as the decoupling point, and from there the goods ordered are supplied to the customer. In general, the decoupling point will coincide with a main stock point.
>
> *(Hoekstra and Romme, 1992, p. 6)*

The CODP at firms can be found before the installation of the final product, before the shipment of the final product, before assembly, before manufacturing, or before the purchase of raw material. These different positions of the CODP categorize firms in five basic logistic structures:

(i) make and ship to stock;
(ii) make to stock (central stock);
(iii) assemble to order (assembly for one specific customer);

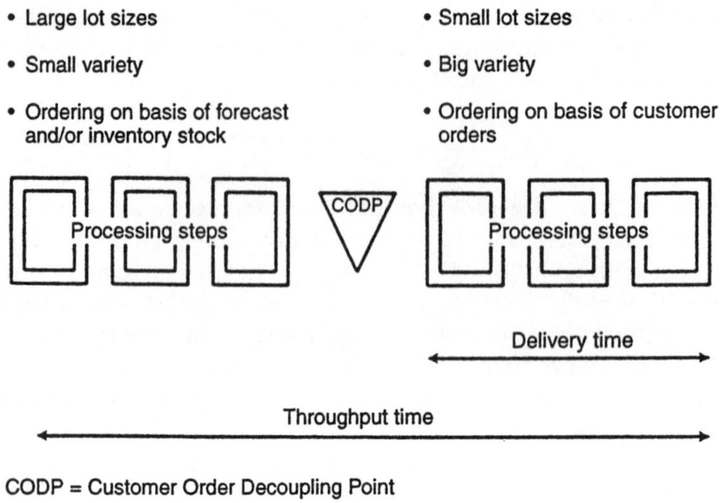

Figure 7.2 The concept of the customer order decoupling point (CODP).

(iv) make to order; and
(v) purchase and make to order.

These structures are ideal types and reality may be more complex. In a manufacturing firm, for instance, the CODP for each part type may have a different position. Figure 7.2 gives an example of a part type with six processing steps. Each processing step may increase the number of variants of the part type that will be produced by the manufacturing firm (e.g. a drilling machine may have to realize different hole patterns depending on each particular production order). In Fig. 7.2, the CODP is positioned in the middle. The choice for this position can be seen as a trade-off between the delivery time of a part type and the stock level (or the number of variants of the part type) at the CODP.

Figure 7.3 contains information of a small case study which will illustrate the importance of the logistic context of an FMS. The firm in the case study manufactures two part types (A and B) which need to be assembled in the assembly department. Within the manufacturing department, part type A has to undergo seven processing steps. Production orders for part type A are generated on the basis of forecasts and/or inventory stock. The number of variants of part type A remains the same after each processing step (= V_{Astart}). This means that all the production orders for part type A undergo the same machining operations at each processing step. Part type B requires four processing steps in the manufacturing department. The ordering of two of these steps are based on forecasts and/or inventory stock (in CODP), the other processing steps are activated by customer orders. Each processing step increases the number of possible variants of part type B (V_{Bstart}, V_{B1}, V_{B2}, V_{B3}, V_{B4}). V_{B2} is the number of variants of part type B that is on stock in the CODP. Suppose that the production control system of the firm reserves the same time period for each of the processing steps.

The application of an FMS in the situation of Fig. 7.3 may offer important advantages. These advantages, however, depend on the specific assignment of processing steps, or operations, to an FMS. Suppose, in this particular case, five assignment possibilities are conceivable, see Fig. 7.4. An FMS may perform processing steps 6 and 7 of part type A, with the CODP before or after the FMS manufacturing step (situations (i) and (ii)). Another possibility is the assignment of processing steps 2 and 3 of part type B to an FMS, with the CODP before or after the FMS step (situations (iii) and (iv)). The last possibility concerns the assignment of processing steps 3 and 4 of part type B to an FMS (situation (v)). In all the five assignment possibilities, the multifunctionality of the FMS enables the integration of two processing steps.

In order to analyse the specific advantages of the five assignment possibilities, the attention will first be focused on the impact of the multifunctionality of an FMS on the performance of the firm. After this, the impact of the efficiency and flexibility characteristics of the system will be dealt with briefly.

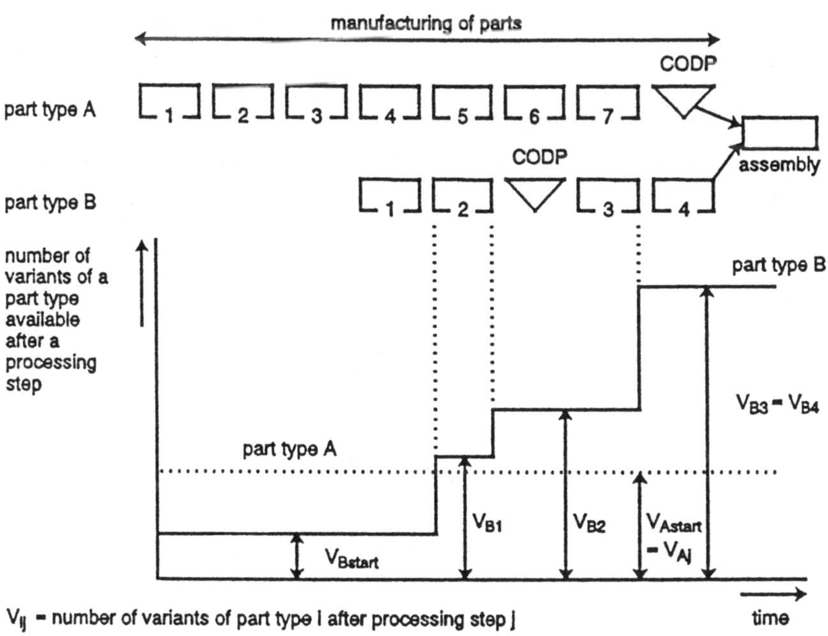

Figure 7.3 The manufacturing of part types A and B.

part type		application of FMS	1. shorter forecast/ planning horizont	2. shorter delivery times	3. less variety in CODP
A	i	CODP / FMS 6.7	+	0	0
A	ii	FMS 6.7 / CODP	+	0	0
B	iii	FMS 2.3 / CODP	0	+	-
B	iv	CODP / FMS 2.3	0	0	+
B	v	CODP / FMS 3.4	0	+	0

- = disadvantage 0 = neutral + = advantage

Figure 7.4 Impact of the multifunctionality of an FMS on the performance of a firm, depending on the assignment of processing steps to the system.

Figure 7.4 shows the impact of the five assignment possibilities on some important performance indicators: (1) the forecast/planning horizon, (2) the delivery times, and (3) the variety in the CODP. The scores of the assignment possibilities will be explained briefly.

1. **The forecast/planning horizon**. The multifunctionality of the FMS decreases the number of processing steps and consequently the reserved manufacturing lead time of the particular part type. As can be seen in Fig. 7.3, the total manufacturing lead time of the final product is determined by part type A. Therefore, only the assignment possibilities (i) and (ii) have a positive effect on the required forecast/ planning horizon. The minimally required forecast/planning horizon corresponds with the so-called 'frozen period', that is the period during which the planning of final products theoretically cannot be altered due to the required lead time. The shorter the frozen period, or forecast/planning horizon, the better manufacturing will correspond with actual demand.
2. **The delivery times**. The delivery time of an order is determined by the part type with the longest lead time, to the right of the CODP. In the case of Fig. 7.3, part type B determines the delivery time. This explains the +s at the assignment possibilities (iii) and (v), since the multifunctionality of the FMS reduces in these cases the number of

processing steps after the CODP of part type B. In all the other cases, the FMS does not impact the delivery times of orders.

3. **The variety in the CODP**. The number of variants in the CODP only changes if the position of the CODP alters with respect to the processing steps needed for part type B. As can be seen in Fig. 7.3, the number of variants of part type A is constant over all the processing steps. With assignment possibility (iii), the number of variants in the CODP will increase from V_{B2} to V_{B3}. Assignment possibility (iv) leads to a decrease of the number of variants in the CODP of V_{B2} to V_{B1}. The number of variants in the CODP determines the stock holding costs, to certain extent.

As can be seen in Fig. 7.4, the advantages of multifunctionality of an FMS depend to a large extent on the specific assignment of operations to the FMS. The impact of the efficiency and flexibility characteristics of an FMS require additional attention. For many firms, the possibility of unattended production was the most important reason to install an FMS. These firms are focused on gaining efficiency. The flexibility of an FMS may offer a firm important additional advantages, such as the possibility of decreasing the lot sizes and, consequently, the manufacturing lead times of production orders. It also offers a firm the possibility of dealing with rush orders.

This section has shown that the selection of operations, or processing steps, for an FMS can be seen as an important element in the design and operation of a flexible manufacturing shop. In particular, we stress the importance of the position of the CODP.

7.2.4 Experiences with FMSs

An almost endless list of reasons can be given as to why many firms did not realize the original objectives with their FMS. Based upon seven case studies, Boer *et al.* (1990) distinguish three categories of reasons:

1. **Technical problems**. These problems refer to engineering faults in, and problems with, standardization and integration of both hardware and software. It is likely that the technical problems will be less dominant in the future due to technical developments.
2. **Changes in the market place**. Most FMSs are dedicated towards a family of products. The demand for these products, however, is usually not completely predictable and may change with respect to mix and volume. This will complicate the balancing of the production orders on (the various machines and fixtures/pallets of) the FMS and may therefore cause inefficiencies.
3. **Insufficient knowledge of and attention to the organizational prerequisites for the effective operation of FMSs**. The implementation of an FMS is frequently seen as a technical problem to be solved

> by process engineers. This, however, is a too narrow view on the application of FMSs. Operators have to be educated in order to optimize the systems performance. A TQC-type system is needed to ensure that the FMS will get the correct raw material, cutting tools and NC-programs in time; the number of failures have to be minimized. Preventive maintenance is needed to reduce the number of breakdowns. And if a breakdown occurs, the maintenance department has to be adapted to solve the problem quickly. The production control system has to be adapted to the abilities of the FMS in order to realize shorter throughput times, smaller lot sizes, etc. In many cases, manufacturing, production planning, maintenance and quality control were insufficiently involved in the innovation process.

Boer *et al.* (1990) furthermore argue that an FMS can only contribute to the company's business success if the system is used for the reduction of a strategic bottleneck. An FMS, for instance, will not reduce manufacturing lead time if other parts, which are not manufactured on the FMS, determine this lead time (see section 7.2.3). Also the impact of an FMS on other, perhaps strategic, functions within a firm should be dealt with carefully. Boer *et al.* (1990) report a case in which the reduced manufacturing lead time was fully absorbed by increased process planning lead time. Finally, it may be possible that other departments in a firm, for instance design or process planning, are of more strategic importance than manufacturing and that an FMS, consequently, deserves a low priority.

Some specific experiences with FMSs (Slomp, 1993) may furthermore illustrate the operational issues with which firms have to deal when optimizing the use of their systems.

1. Tool management is sometimes a problem in FMSs. Idle time occurs if more than one machining centre requires the same unique cutting tool simultaneously. Also the limited availability of fixtures may cause operational problems; more than one order may require the same unique fixture simultaneously.
2. The number of serious breakdowns of an FMS is usually less than with conventional machines. The duration of a serious breakdown is usually longer. Therefore, the weekly output of an FMS may show a more erratic pattern than the output of conventional machines. Furthermore, due to the complexity of an FMS, the cause of a breakdown is sometimes difficult to discover.
3. An FMS with identical machining centres theoretically offers the advantage of some routing flexibility of orders. However, in practice, it appears that the accuracy of the identical machining centres frequently differs. Furthermore, in the case of various types of raw material (e.g. aluminium and steel), it is more or less necessary to

dedicate the machining centres in order to avoid a mixture of chips.

4. Within an FMS, several production orders usually run in a mix on each of the machining centres. This is done in order to avoid idle time caused by the transport, the unloading, and the loading of a (unique) pallet/fixture. This situation causes a lack of transparency with respect to the cutting tools needed on each machining centre, the moment at which the processing of an order can start, and the throughput times of orders.
5. The flexibility of an FMS offers the possibility to decrease lot sizes of orders. However, the inflexibility of pre- or post-manufacturing processes may reduce this advantage significantly.
6. In many practical cases, an MRP system is used for the firms' production control. In most cases, it is difficult to adapt the MRP system towards the operational characteristics of the FMS. An MRP system, for instance, is not able to deal with the fact that production orders run in a mix on the machining centres (see experience 4 above) with an unpredictable throughput time. Furthermore, an MRP system may not support a more JIT like production.
7. Some firms had planned to produce to customer order with their FMS (e.g. situation (iv), Fig. 7.4). However, due to a varying demand pattern of customers, these firms were forced to produce to forecast/planning (e.g. situation (iii), Fig. 7.4) in order to balance the workload over time.

The material in this section can be seen as a warning for future FMS users. It shows that a careful introduction of an FMS is required in order to gain advantages with the system. Most of the experiences mentioned above concern production control issues. The next section will concentrate on the development of a production control system for an FMS which fits in the overall production control of the firm.

7.3 THE DESIGN OF AN FMS

The integration of an FMS in the goods flow of the firm can be seen as a major aspect which has to be considered at the beginning of the planning process for an FMS. Perfect integration, however, can be frustrated by a poor design of the system. This section will briefly review the design problems and the issues which have to be dealt with during the design of an FMS.

Kouvelis (1992) discerns five FMS design problems:

1. Optimal system configuration (i.e. determination of number and types of machines, level of work-in-process in the system);
2. Specification of the FMS layout;

3. Selection of a storage system (size of local buffers and/or central storage;
4. Specification of the type and capacity of the material-handling system (MHS);
5. Determination of other important system resources (i.e. number of pallets, number and types of fixtures, number and types of tools).

The general objective in the design phase is to minimize the cost of operation, subject to the constraint of achieving a minimum required system throughput. There are several conceivable approaches for dealing with the design problems, such as simulation, queuing theory (e.g. CAN-Q or Mean Value Analysis) and mathematical programming. In practice, however, simple calculations (e.g. the bottleneck model described by Solberg, 1981) may lead to results which are not necessarily inferior to the solutions gained by more advanced methods. The gap between design theory and practice is due to several factors (Kouvelis, 1992). In many firms, the input data for the models may be absent or inaccurate. It may be difficult to forecast the product mix and the required volume because of the dynamic environment of a firm. The theoretical models, furthermore, may be not able to deal with real size problems. Also the focus of many theoretical approaches is not appropriate. In the FMS literature the sub-problems are merely addressed independently. The design of an integrated system, however, requires an integrated solution.

While designing, or selecting an FMS, it is of utmost importance to assess the impact of all the limitations of the system (see section 7.2.2) carefully and systematically. These limitations determine the applicability of the system and the complexity of the operational control of the system. An important consideration in the design may be the possibility of relaxing some limitations, if necessary, in the future; many firms have decided to purchase modular FMSs which can be expanded on a step-by-step basis, using more or less standard equipment.

The decision for a specific FMS will depend on the potential benefits and costs. Arbel and Seidmann (1984) suggest the application of the 'analytic hierarchy process' (AHP) approach (Saaty, 1980) within an evaluation methodology for selecting a flexible manufacturing system. By doing this, they are able to deal with many concerns, of which some are difficult to quantify. Typical concerns relate to machine capabilities, cost reduction, lead time, vendor's reputation, service lead time, and so on.

7.4 PRODUCTION CONTROL FOR FMSS

As described in section 7.2.2, an FMS is an efficient, flexible and multifunctional system. These performance characteristics, however, are limited by the capacity of the pallet pool, the number of unique cutting tools, the number of unique fixtures, the size of the tool magazines, etc. The production control system of the FMS has to deal with these limitations.

A production control system of an FMS can be defined as the system which is responsible for the planning, scheduling and control of activities that have to be performed by the FMS in order to realize certain goals. A production control system of an FMS usually has a hierarchical character (see e.g. Stecke, 1985, and Van Looveren *et al.*, 1986). Several arguments can be given to justify a hierarchical approach to production control problems:

- **Reduction of complexity**. The production control problem of an FMS can be characterized by the presence of multiple, sometimes contradictory, objectives and a number of complicating and to a certain extent conflicting constraints. A hierarchical approach offers the possibility of splitting up complex interrelated production control problems into several small solvable parts.
- **Separation of short-, medium-, and long-term aspects**. Production control problems on a long-term level are generally more strategic in nature than medium-term and short-term problems and therefore demand different solution methods.
- **Improving stability and controllability**. Production control problems may arise at regular and/or irregular intervals. Without a hierarchical decision structure all the (interrelated) production control problems are affected by any disturbance. A hierarchical approach offers the possibility of solving problems on one level without the need to replan on higher levels. This improves the stability of the production control decisions and severly reduces the amount of information required.

The above-mentioned arguments for a hierarchical approach to production control problems of FMSs are not different from considerations in the absence of FMSs. However, the arguments gain special importance in the case of FMSs since these systems face complex limitations (see sections 7.2.2–7.2.4), they have to fulfil important strategical as well as more tactical and operational objectives, and they are vulnerable to disturbances (see section 7.2.4, experience 2).

The main components of a production control system of an FMS, as it is viewed in this chapter, are:

- a decision hierarchy (which describes the various decisions and their connections),
- an organization hierarchy (which describes the positions of the human decision makers), and
- an information system and decision support tools.

Section 7.4.1 categorizes the design problems of a production control system for an FMS. Such a categorization is useful in order to get a complete overview of the problems which have to be dealt with in a particular situation. Section 7.4.2 concerns the design of the decision and organization hierarchy of the production control system. Section 7.4.3 briefly reviews some approaches for the design of decision support tools. A case study will illustrate the concepts mentioned in this section (Slomp and Gaalman, 1993).

7.4.1 Design problems for production control

Figure 7.5 presents a schematic view of the components of the production control of an FMS and the problem categories which can be distinguished with respect to the design of a production control system for an FMS.

The meaning of the arrows in Fig. 7.5 can be explained as follows: given the limitations of the FMS (= input), design decisions (= measures) may be taken such that the production control system (= system) will improve with respect to the realization of its goals (= output). Following this systemic view, there are essentially four interrelated categories of problems to be dealt with in the design: (a) goal problems, (b) limitation problems, (c) scope problems, and (d) design capability problems. The

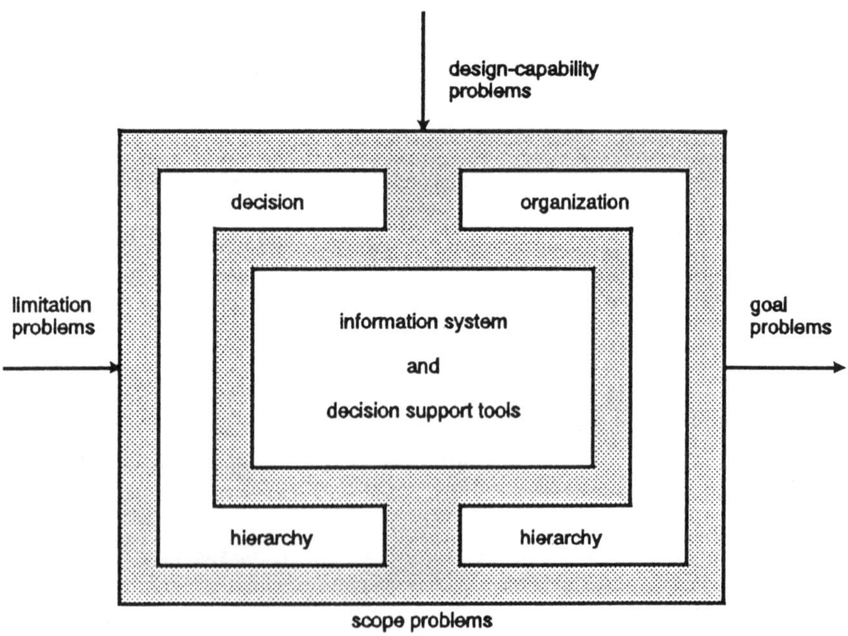

Figure 7.5 Components of the production control of an FMS and the problem categories.

goal problems concern the desirable effect of the production control system. The system, for instance, should lead to short throughput times of orders and a high delivery performance. The limitation problems refer to those characteristics of the FMS with which the production control system has to deal. Important are the limitations of the FMS with respect to the available capacity, pallets/fixtures, cutting tools, the pallet storage capacity, etc. Scope problems arise due to the fact that in most cases the production control system can only be partly redesigned. Several parts of the system may be unchangeable; for instance, the MRP system or some organizational arrangements. The design capability problems refer to the capability of the firm to design and implement changes in the production control. In the design process of a production control system for an FMS, one has to deal with all four categories of problems together.

Illustrative Case 1

The FMS of Enraf Nonius BV, The Netherlands, consists of two identical machining centres, linked together by a pallet transport vehicle and an integrated pallet buffer system (see Fig. 7.1). Orders have to be finished before their due dates (= goal problem). One of the production control problems in this case arises due to the fact that a unique cutting tool cannot be used simultaneously on both machining centres (= limitation problem). Another production control problem concerns the fact that a unique fixture cannot be used for more than one order at the same time (= limitation problem). The presence of an MRP system and the priority logic (FIFO) of the FMS-computer, which controls the transport vehicle, limits the design space for the production control (= scope problem). An external consultant was hired by the firm in order to design a production control system for the FMS (= design-capability problem).

7.4.2 Production control hierarchy

The design of a production control hierarchy for an FMS is very important because here the decision tasks and the responsibilities of the employees involved will be determined. Within the production control hierarchy two aspects can be discerned: a decision hierarchy and an organization hierarchy.

Decision hierarchy

Based upon literature and the organizational practice in many companies, a generic description can be given of the decision hierarchy of the production control for an FMS (Stecke, 1985; Dhar, 1991; Slomp, 1993). The hierarchy consists of three decision levels (Fig. 7.6).

On the assignment level orders are distributed among the various production units, including the FMS. The assignment level is responsible for generating realizable throughput times and a realistic workload of

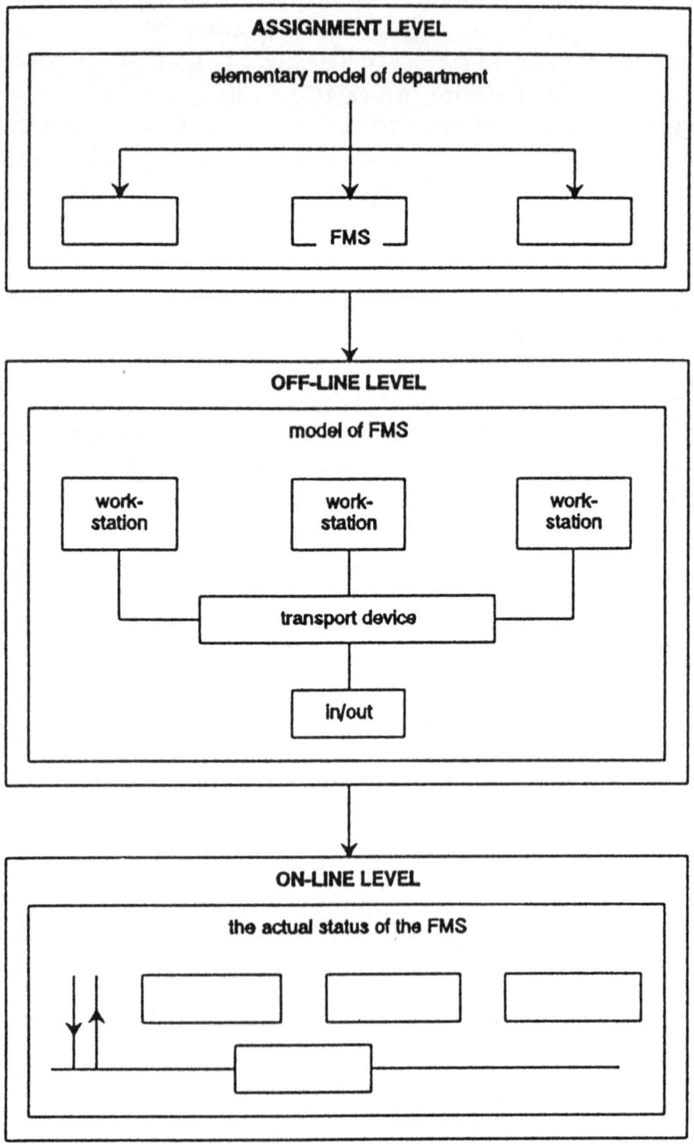

Figure 7.6 Decision levels in the production control for an FMS.

the FMS. In many practical situations, the activities on the assignment level are executed periodically, for instance monthly, weekly or daily. Often an MRP system is used at the assignment level. Within the assignment level it might be difficult to take account of the specific characteristics (or limitations) of the FMS. In an MRP system, for instance, an FMS is just seen as a capacity resource. There are few attempts to integrate FMS production planning problems into a closed-loop MRP system (see, for instance, Mazzola *et al.*, 1989).

The off-line level receives orders from the assignment level. The off-line level is responsible for getting a good fit between the received orders and the characteristics (limitations) of the FMS. Off-line decisions are based upon a model of the FMS, in which the most important characteristics (limitations) of the FMS are incorporated. Most off-line decisions are performed periodically, for instance monthly, weekly or daily. The off-line decisions may concern the batching of orders to be produced in the same period and the loading (assignment) of operations and tools to the various workstations. Sometimes, scheduling is also done at the off-line level.

On-line activities are based upon the information from the off-line level **and** upon the actual status of the FMS. On-line decisions are activated by real-time information from the FMS and can be taken at any moment. The decisions concern the release of orders to the FMS and the sequence in which competitive activities have to be performed (dispatching). The extent of detail at the off-line level determines the degree of freedom at the on-line level.

An alternative for batching, loading, and release is the so-called 'flexible approach' (Stecke and Kim, 1988), in which decisions about the actual order mix are based primarily on the actual status of the system with the objective of optimizing certain performance measures such as machine utilization and due-date performance. Basically, the flexible approach can be seen as an intelligent form of releasing.

The three-level hierarchy can be recognized in most production control systems. Differences are often due to the location of certain subproblems on another level (for instance on-line loading) or even the complete absence of the off-line level.

Organization hierarchy

The decision levels of Fig. 7.6 are determined by the information flow required to and from the FMS (aggregated versus real-time information) and the knowledge needed about the specific characteristics of the FMS (global versus detailed). These aspects are also of importance for the assignment of decision tasks to the organizational levels of the firm. Fig. 7.7 illustrates how, in many practical cases, the levels of the decision hierarchy correspond with the levels of the organization hierarchy. The

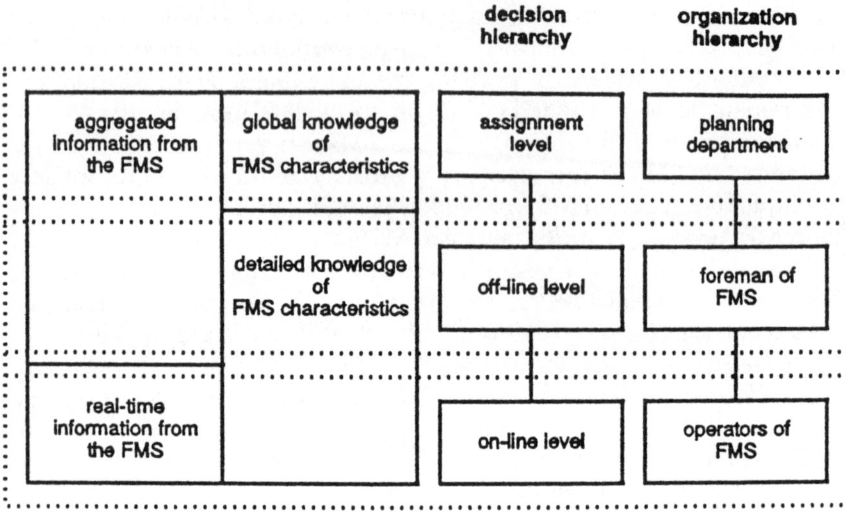

Figure 7.7 Correspondence between decision hierarchy and organization hierarchy.

planning department has a global knowledge of the FMS and bases its assignment decisions on aggregated information of the FMS. The foreman has detailed knowledge of the FMS, but his production control decisions are usually based on aggregated information of the FMS. The main reason for this is that a foreman of the FMS usually works on a day shift while the FMS operates in two or three shifts. The operators of the FMS have detailed knowledge of the FMS and their production control decisions are based upon real-time information of the FMS.

The assignment of production control tasks to the levels of the organization hierarchy plays an important role in solving production control problems. The assignment, for instance, determines the quickness of reaction on breakdowns and/or rush orders. Figure 7.7 should be seen as just one example of assigning production control tasks to the levels of the organization.

7.4.3 Decision support tools and the information system

The production control hierarchy gives information about the assignment of decision tasks and responsibilities to individuals or groups of people. This assignment has to be seen as the starting point for the design of decision support tools and an information system.

The need for decision support tools depends on the complexity of the problems at each level of the production control hierarchy. There are a number of articles proposing solutions for a variety of production control problems. The solution methods used are, among other things, simulation, integer programming and heuristic algorithms. Overviews of methods and solutions are given in Buzacott and Yao (1986), O'Grady and Menon (1986), Van Looveren *et al.* (1986), Kouvelis (1992), Shanker and Agrawal (1991), Dhar (1991), Stecke and Kim (1988) and Rachamadugu and Stecke (1994). Dhar (1991) states that only a few examples exist of actual applications of DSS/ES (decision support systems/expert systems) in FMS.

An important aspect of a decision support tool concerns the interaction between human and computer. Ammons *et al.* (1986) report that the role of people is only rarely considered when designing decision support tools for FMSs. In their opinion, this has led to a frequent bad fit between human and automated production control functions of an FMS. They subsequently argue that the tasks of human decision makers should be designed and described as precisely as the algorithms applied in the computer software. An example of such an approach is given in an article of Sharit and Elhence (1989). They discuss the computerization of tool-replacement decision making in FMSs from a human–systems perspective and conclude that a human–computer cooperative arrangement has the potential to realize the best balance between the more direct **local** consequences and less direct **global** consequences of tool-replacement decisions on economic and throughput objectives. An essential aspect of a human–computer arrangement (or decision support tool) is the design of the interface, e.g. screen layout and communication (software) ergonomy.

Illustrative Case 2

Figure 7.8 shows the production hierarchy which is designed and implemented for the FMS of Enraf Nonius BV. The hierarchy basically consists of four levels:

- **Level I**. At the beginning of each week, an MRP print-out is generated which contains all future orders of the FMS. The planning department selects a subset of these orders for immediate release to the FMS. This selection is based upon the due dates generated by the MRP system and the workload of the FMS. The total amount of work-in-process is limited to 1.5 weeks. The FMS should manufacture each order within at most two weeks. In this way the FMS has a slack of 0.5 weeks to produce all orders in time.
- **Level II**. The foreman allocates the released orders to each machining centre. Several aspects complicate this allocation activity. As far as possible, order allocation should be done such that each fixture and each unique cutting tool is needed

Illustrative Case 2 *contd*

on just one machining centre. By doing this, the machining centres can be considered independently of each other during the remainder of the procedure.

- **Level III**. The third level in the production control hierarchy deals with the scheduling of orders on each machining centre. The human-scheduler (= foreman) has to deal with several aspects, such as the limited capacity of the tool-magazine and limited availability of pallets/fixtures. Orders which need the same pallet/fixture should not be scheduled sequentially since this would cause idle time. Finally, the throughput time of an order should not exceed two weeks. In this particular case, an interactive scheduling tool is developed to support the foreman's decision making, see Slomp and Gupta (1992).
- **Level IV**. The operators are responsible for the monitoring and control of the FMS.

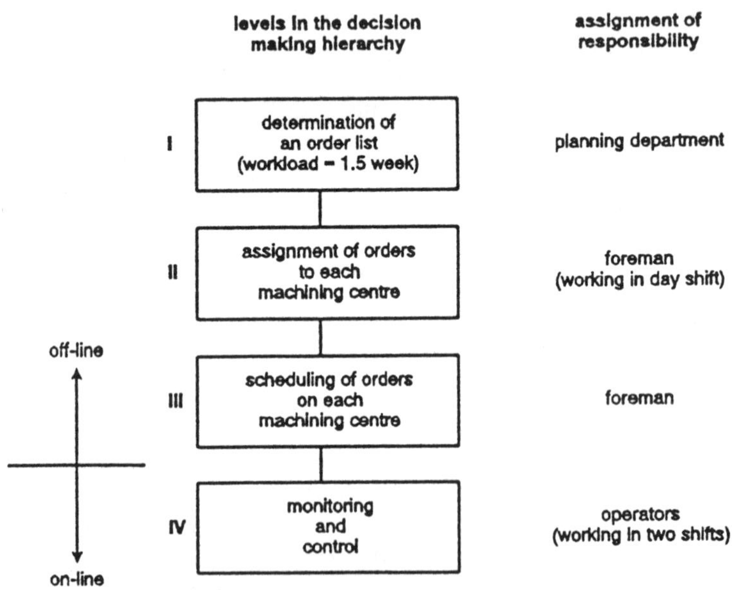

Figure 7.8 The hierarchical structure of the production control system.

The successful use of any production control system will heavily depend upon the availability of information on each of the levels of the decision and organization hierarchy. Therefore, the design of a well-working and reliable information system, either automated or not, is of utmost importance.

7.5 RESEARCH ISSUES

The research issues described in this section are arranged in conformity with the previous sections and concern issues of integration, design and production control.

7.5.1 Integration

Section 7.2.3 has showed the importance of the so-called customer order decoupling point. Many firms had planned to apply their FMS in such a way that more processing steps would have been performed in the customer-order part of the goods flow (e.g. assignment possibility (iv), Fig. 7.4). However, many of these firms did not succeed because of (a) the dynamic nature of the demand which would cause an irregular load (in volume and mix) of the FMS and (b) the inefficiency of the FMS when producing small lot sizes. The latter inefficiency is caused by the time needed to prepare a (sometimes modular) workholding system. Research is needed with respect to the (optimal) application area of an FMS in the goods flow of a firms. It should be noted that the integration of an FMS in the goods flow of a firm is not a problem which has to be solved only once; the assignment of part types (processing steps) to the FMS is an ongoing activity since, in most cases, the products change frequently.

7.5.2 The design of the FMS

Flexibility is one of the most important performance characteristics of an FMS (see section 7.2.2). Suppliers of FMSs may suggest the acceptability of very small lot sizes (lot size one) of parts. In most applications, however, this flexibility is not available because of the substantial amount of time needed to prepare a workholding system (fixture). This activity can be seen as the least automated part of the FMS operation. O'Grady (1989) emphasizes the desirability of fixtures which are sturdy, precise, universal, able to handle a wide product variety and which can be made at relatively low cost (and in little time). It is a challenge for engineers to develop flexible fixturing possibilities.

The design or selection of an FMS is a complex problem because of the interdependence of the various design problems (see section 7.3) and the existence of various other considerations. There is a need for integrated design methods.

7.5.3 Production control of the FMS

In literature much attention is paid to the design of sophisticated solutions for specific production control problems of FMSs. Less attention is paid to the question of whether or not the proposed solutions really address the problems encountered in practice. There is a need for application-oriented research in the field of production control for FMS. This will have impact on at least two aspects of production control research. As mentioned by Rachamadugu and Stecke (1994), most FMS scheduling research to date concentrates on criteria related to machine utilization and the avoidance of bottlenecks. Application-oriented research will be directed towards developing scheduling procedures, with the primary objective of meeting due dates. Furthermore, application-oriented research will stress the necessity of short response times of on-line scheduling methods.

Literature rarely deals with organizational aspects while presenting new methods for production control problems. However, in practical situations the human organization plays an important role in the production control function. The limited working hours of a foreman, for instance, may play an important role in the design of the decision hierarchy. More knowledge is needed about the interdependence of the decision and organization hierarchy of the production control. Another important organizational aspect which has an impact on the production control of FMSs concerns the fact that different tasks may have to be done by the operators during the various shifts. No human task at all can be done in unmanned periods. In other periods the operators may have assistance from other employees. The impact of the 'temporal division of labor' on the production control of complex systems needs to be studied further.

7.6 SUMMARY

An FMS may be an important tool in the design of a flexible manufacturing shop. This chapter has focused on three conditions for a successful application of an FMS. First, the integration of the FMS in the goods flow of the firm requires sufficient attention. This also includes understanding of the necessary changes in other parts of the organization, such as the maintenance department and the quality control of raw material. Secondly, the FMS itself has to be designed carefully; in particular, the impact of the limitations of the FMS requires sufficient attention. Finally, the production control system of the FMS needs to be designed towards the specific characteristics of the FMS.

REFERENCES

Ammons, J.C., Govandaray, J. and Mitchell, C.M. (1986) *Human Aided Scheduling for FMS: A paradigm for human-computer interaction in real time scheduling and control*. Proceedings of the Second ORSA/TIMS Conference on Flexible Manufacturing Systems: Operations Research Models and Applications, Elsevier Publishers BV, Amsterdam, pp. 443–54.

Arbel, A. and Seidman, A. (1984) Performance evaluation of flexible manufacturing systems. *IEEE Transactions on Systems, Man, and Cybernetics*, **14**(4), 606–17.

Boer, H., Hill, M.R. and Krabbendam, J.J. (1990) FMS implementation management: promise and performance. *International Journal of Operations and Production Management*, **10**(1), 5–20.

Bolwijn, P.T., Boorsma, J., van Breukelen, Q.H., Brinkman, S. and Kumpe, T. (1986) *Flexible Manufacturing. Integrating Technological and Social Innovation*. Elsevier, Amsterdam.

Buzacott, J.A. and Yao, D.D. (1986) Flexible manufacturing systems: a review of analytical models. *Management Science*, **32**(7).

Dhar, U.R. (1991) Overview of models and DSS in planning and scheduling of FMS. *International Journal of Production Economics*, **25**, 121–7.

Economic Commission for Europe (1986) *Recent Trends in Flexible Manufacturing*. United Nations, New York.

Hoekstra, S. and Romme, J. (1992) *Integral Logistic Structures – Developing Customer-oriented Goods Flow*. McGraw-Hill Book Company.

Jaikumar, R. (1986) Postindustrial manufacturing. *Harvard Business Review*, **64**(6), 69–76.

Kouvelis, P. (1992) Design and planning problems in flexible manufacturing systems: a critical review. *Journal of Intelligent Manufacturing*, **3**, 75–99.

Mansfield, E. (1993) The diffusion of flexible manufacturing systems in Japan, Europe and the United States. *Management Science*, **39**(2), 149–59.

Mazzola, J.B., Neebe, A.W. and Dunn, C.V.R. (1989) Production planning of a flexible manufacturing system in a material requirements planning environment. *The International Journal of Flexible Manufacturing Systems*, **1**, 115–42.

O'Grady, P. (1989) Flexible manufacturing systems: present development and trends. *Computers in Industry*, **12**, 241–51.

O'Grady, P.J. and Menon, U. (1986) A concise review of flexible manufacturing systems and FMS literature. *Computers in Industry*, **7**(2), 155–67.

Rachamadugu, R. and Stecke, K.E. (1994) Classification and review of FMS scheduling procedures. *Production Planning and Control*, **5**(1), 2–20.

Saaty, T.L. (1980) *The Analytic Hierarchy Process: Planning, Priority Setting Resource Allocation*. McGraw-Hill, New York.

Shanker, K. and Agrawal, A.K. (1991) Loading problem and resource considerations in FMS: a review. *International Journal of Production Economics*, **25**, 111–19.

Sharit, J. and Elhence, S. (1989) Computerization of tool-replacement decision making in flexible manufacturing systems: a human-systems perspective. *International Journal of Production Research*, **27**, 2027–39.

Slomp, J. (1993) *Production Control for Flexible Manufacturing Systems – An Application-Oriented Approach*. PhD Thesis, Department of Mechanical Engineering, University of Twente, Enschede.

Slomp, J. and Gupta, J.N.D. (1992) Interactive tool for scheduling jobs in a flexible manufacturing environment. *Computer-Integrated Manufacturing Systems*, **5**, 291–9.

Slomp, J. and Gaalman, G.J.C. (1993) A production control system for a flexible manufacturing system – description and evaluation. *International Journal of Flexible Automation and Integrated Manufacturing*, **1**(2), 93–103.

Solberg, J.J. (1981) Capacity planning with a stochastic workflow model. *AIIE Transactions*, **3**(2).

Stecke, K.E. (1985) Design, planning, scheduling and control problems of flexible manufacturing systems. *Annals of Operations Research*, **3**(1), 3–12.

Stecke, K.E. and Kim, I. (1988) A study of FMS part type selection approaches for short-term production planning. *International Journal of Flexible Manufacturing Systems*, **1**, 7–29.

Van Looveren, A.J., Gelders, L.F. and van Wassenhove, L.N. (1986) A review of FMS planning models, in *Modelling and Design of Flexible Manufacturing Systems* (ed. Andrew Kusiak), Elsevier Science Publishers BV, Amsterdam, pp. 3–31.

Werner und Kolb Werkzeugmaschinen GmbH (1987) *Guidepost to the Solution of Individual Manufacturing Problems. From a Stand-alone Machine to a Flexible Manufacturing System*, Berlin.

CHAPTER 8

Constraint logic and its applications in production: an implementation using the Galileo4 language and system[1]

D. Bahler and J. Bowen

ABSTRACT

The flexible and efficient support of the process of product design and production requires combined expertise from many aspects of the product life-cycle, and the management and use of this knowledge in turn increasingly require sophisticated computer-based advice systems. In this chapter we present a constraint-based approach to making the full first-order predicate calculus available for knowledge representation purposes in intelligent networked colocation advice systems for concurrent engineering. Domain knowledge is represented as a theory written in a first-order logical language. The set of models of the language under which the theory is satisfied is the set of solutions of a constraint network. Solutions to the network are computed by a hierarchical and opportunistic constraint inference method. Galileo4, an implemented language and companion run-time system based on these ideas, is discussed and its efficacy is demonstrated by presenting applications in which high problem-solving versatility is provided by concise and readable programs.

[1] This work was partially supported by the National Science Foundation under Grants DDM-8914200 and DDM-9215755 and by IBM Corp.

8.1 INTRODUCTION

Concurrent engineering (CE) is an approach to the organization and management of design, production, marketing, etc. in which, in order to avoid costly downstream redesign, aspects of the product life cycle such as manufacturability, testability, repairability and marketing are taken into account as early in the design process as possible. Since most life-cycle costs are dictated in the design phase, it is important that life-cycle information be available to the designer so that it can influence product design. Once a product has completed the design stage, it is too late to make significant changes in life-cycle costs. The US National Science Foundation, for example, has estimated that 70% or more of a product's manufacturing cost is dictated by design decisions.

Currently many companies attempt to implement concurrent engineering at the detailed design phase by maintaining checklists or rulebooks that a designer must adhere to and then scheduling face-to-face meetings with the appropriate experts at critical project points. In our interaction with companies we have found several problems with this approach. First, the rulebooks or checklists are getting so large and complex that even the experts, let alone the designer, cannot help letting design deficiencies through to manufacture. Second, the need for physical meetings reduces time-to-market. One way to overcome these difficulties is to use **networked colocation**. In this approach, the team members supplement face-to-face meetings with electronic communication over a computer network.

The concept of networked colocation can be realized with varying degrees of sophistication. At its simplest, it may amount to no more than some combination of electronic mail and shared access to a CAD database. However, something much more sophisticated is needed if we are to succeed in CE. Product development teams often find themselves overwhelmed by the volume and variety of information that arises in CE. As the design evolves, the number of concerns grows quickly beyond what even a team of persons can successfully manage. We need software which can relieve some of this cognitive overload. What is needed is an **intelligent networked colocation advisor** (INCA) which not only relieves the logistic and scheduling difficulties but also reduces the problem complexity perceived by team members. These design advice tools are intended for use by and for multiple clients, including both computer systems and human users.

In our experience [3, 4, 5] constraint-based systems constitute a good basis for such tools; much of our effort has emphasized expressive competence in constraint-based languages in order to provide a convenient and sufficient representation of the kind of real-world problems encountered in concurrent engineering. Moreover, we believe that an approach to constraint processing which is strongly grounded in a formal logical model offers the greatest chance to gain insight and make progress. Accordingly, we view a network of interlocking constraints as a collection of sentences in **first-order predicate calculus** (FOPC). Constraint processing then becomes exactly the process of semantic modeling, that is, finding an interpretation for the symbols appearing in the theory that renders the constraint sentences true according to some well defined measure of truth. In this way the notions of constraint logic and constraint network are mutually intertwined.

The remainder of this chapter is organized as follows. In section 8.2, we briefly present an overview of Galileo4, the latest of a set of constraint programming languages and companion run-time systems we have developed for constructing INCAs, and illustrate the utility of this system for supporting a production task application in design for testability of printed wiring boards. In section 8.3, we further illustrate the practical utility of the constraint-based approach by discussing another application, this time in the realm of selecting lowest-cost insertion robot equipment while fulfilling a set of user-supplied specifications. In section 8.4, we compare our approach with other constraint-based techniques.

Beginning in section 8.5, we elaborate in a more technical vein on the relationship between constraint networks and semantic models in logic. In this view the consistency of a constraint network corresponds to the satisfiability of a theory in logic, and we argue that a suitable computational approach to the full FOPC can be based on generalizing the consistency algorithms [21] that have been presented in the literature on **constraint satisfaction problems** (CSPs). In section 8.6, we look at Galileo4 from a more formal perspective, using a progression of sample programs to illustrate the correspondence between constraint processing and semantic modeling. In section 8.7, we discuss human-machine interaction in Galileo4 in terms of a model of machine-interlocutor interaction.

Finally, in section 8.8, we make some concluding remarks.

8.2 THE UTILITY OF CONSTRAINT PROCESSING FOR PRODUCTION

Galileo4 is a language and run-time system we have developed for the construction and use of INCAs. In this section, we will consider a Galileo4 program called KLAUS2, which is an INCA for printed wiring board design. In this section, we will consider only a few extracts from KLAUS2 (Fig. 8.1). For a more complete presentation of KLAUS2, see [4].

A program in Galileo4 specifies a frame-based constraint network and comprises a set of declarations and definitions. The declaration statements include declarations of the parameters that exist in the network and the constraints that exist between these parameters. The definition statements include definitions for: partitions which divide the network into different regions that may be seen by different classes of user; application-specific domains, functions, and relations that are used in constraint declarations; and icons for application-specific domains, if it is intended to present certain regions of the network to users graphically rather than textually, a textual presentation being the default.

In a Galileo4 program, statements can be written in any order, although good software engineering practice dictates that some agreed standard ordering be used, in order to facilitate program maintenance. In Fig. 8.1, we have taken advantage of this free-ordering by listing the statements in an order which facilitates discussion in this chapter; in KLAUS2 itself, however, the statements are ordered differently. Although the statements are numbered in Fig. 8.1, this is purely for the purposes of the current exposition; the numbers are not part of the program text. The discussion that follows will be very dense, because of the need to illustrate many features of a richly expressive language in a very short space. Reviewers wanting a more leisurely overview of the language are encouraged to read [4]. Galileo4 is a research prototype which is available on an as-is basis. The current version runs on IBM PS/2s and compatible hardware.

8.2.1 Frame-based product–process–organization modeling

In a frame-based constraint network, frames can be used to represent a model of product, process and organization (PPO). Consider, for example, statement 1 of Fig. 8.1 which declares the existence of a parameter called `the_test_facility` and specifies that it is of type, or domain, `test_facility`. The string delimited by apostrophes is a long synonym which specifies the name by which the parameter will be referenced in all output given to users of the program. From this, it can be seen that this parameter represents `the facility where the board will be tested`.

This parameter is a frame, as can be seen in statement 2, which defines the application-specific domain `test_facility` referenced in statement 1. The statement specifies that a test facility is represented by a frame containing three slots. For each slot, we specify its name, its long synonym and its domain. In output given by the system to users, the long synonyms of the slots in a frame are concatenated with the long synonyms of the frame-valued parameters.

The constituent slots of a `test_facility` frame, as specified in statement 2, have scalar application-specific domains. These are defined in statements 3, 4 and 5. Notice that, while scalar domains may be defined extensionally by listing each possible value (statement 3), they may also be defined intensionally by specifying an appropriate logical formula (statements 4 and 5). Notice the use of long synonyms in statement 3; internally, the system will use values such as `erdsys`, but at run-time the user will see names like `Erdsys TX`.)

Frame-based inheritance is illustrated in statements 6, 7, 8, 9, 10 and 11. Statement 6 defines the frame domain `component`; for simplicity, we give only three slots, although in KLAUS2 this frame has many more slots. Notice, from statement 7, that there can be several different types of component. By using inheritance, frames for these different types of component can be defined in terms of the generic component frame. In statement 8, for example, it is specified that an `osc_crystal` frame, which is used to represent a crystal oscillator, is a `component` frame which has the value `crystal` in its `type` slot and which also has an additional slot `freq` that represents the frequency at which the crystal oscillates. In statement 9, a `divider` frame, which is used to represent a frequency divider, is specified to be a `component` frame in which the `type` slot has the value `divider` and which has several other slots – these additional slots are not specified here, however. A `cpu` frame is defined similarly in statement 10, and a `resistor frame` in statement 11.

8.2.2 Using constraints to represent life-cycle interdependencies

In a frame-based constraint network, constraints can be used to represent the life-cycle interdependencies that exist between parts of the PPO model. However, because of the large variety of life-cycle interdependencies that can exist in different product domains, the constraint language must be richly expressive. Knowledge representation research in artificial intelligence indicates that 'for . . . versatile [systems], the language in which declarative knowledge is represented must be at least as expressive as first-order predicate calculus' [26]. We consider the discovery that constraint networks could provide the full expressive power of the predicate calculus to be an important conceptual breakthrough to arise from our work. Thus, in Galileo4, any sentence in first-order predicate calculus is a well-formed constraint. (The language is not restricted to classical logic, as we will see below.)

Consider, for example, statement 12 in Fig. 8.1. This is a universally quantified constraint which specifies that the number of test points required by each component on the board (which, for our purposes, we take to be equal to the number of leads on the component) must be recorded in a list. This list is referenced in statement 13, which states that the total number of test points in the circuit being designed is the

```
1   'the facility where the board will be tested' (the-test-facility) : test-facility.
2   domain test-facility =::=
         ('the equipment at' (name) : tester,
          'the maximum clock frequency testable at' (maxfreq) : frequency,
          'the maximum number of test points testable at' (maxtpoints) : counter).
3   domain tester =::= {'Dynamax Tester Mark I' (dynatest1)
                        {'Dynamax Tester Mark II' (dynatest2), 'Erdsys TX' (erdsys)}.
4   domain frequency =::= {X : number (X) and 3 = < X = < 40}.
5   domain counter =::= {X : integer(X) and X > = 0}.
6   domain component =::= {'the type of' (type) : component-type,
      'the number of leads on' (nleads) : counter, 'the power consumed by' (power) : positive number).
7   domain component-type =::= {processor,dsp,comm-controller,divider,
                                crystal,resistor,capacitor,diode}.
8   domain 'crystal oscillator' (osc-crystal) =::=
      {X : component (X) and X.type=crystal and
           exists (X.'the oscillation frequency of' (freq) : positive number)}.
9   domain 'frequency divider' (divider) =::= {X : component (X) and X.type=divider and ...}.
10  domain cpu =::= {X : component (X) and X.type=processor and ...}.
11  domain resistor =::= {X : component (X) and X.type=resistor and ...}.
12  all X : component (X) implies X.nleads in !1st-tpoints.
13  num-tpoints = sum(1st-tpoints).
14  num-tpoints = < the-test-facility.maxtpoints.
15  'the numbers of test points required by components' (1st-tpoints) : bag of counter.
16  'the total number of test points on the board' (num-tpoints) : counter.
17  not (exists X : tester (X) and cost (X) < cost (the-test-facility.name) and satisfactory(X)).
```

```
18  function 'the total cost per hour of' (cost) =::= {X -> Y : 1.8 * price(X)}.
19  function 'the direct cost per hour of usage of' (price) =::= datafile (qdb,TSTRCSTS).
20  relation 'capable of testing the board' (satisfactory) =::=
      {X :  num-tpoints =< max-tpoints(X) and (all Y : osc-crystal(Y) implies Y.freq =< max-clock(X))}.
21  function 'the maximum number of nails usable by' (max-nails) =::= datafile(qdb,TSTRNALS).
22  function 'the clock limit for' (max-clock) =::= datafile(qdb,TSTRCLKS).
23  all X : component (X) and X.type in {processor,comm-controller,dsp}
    implies exists (X.'the oscillator for' (crystal) : osc-crystal) and
            exists (X.'the pullup resistor for' (pullup) : resistor).
24  all X : osc-crystal(X)
    implies X.freq = < the-test-facility.maxfreq else
            exists (X.ancillary-circuit : divider).
25  'the power consumptions of the components' (1st-powers) : bag of positive number.
26  'the total power consumed by the board' (total-power) : positive number.
27  all X  : component (X) implies X.power in !1st-powers.
28  total-power = sum(1st-powers).
29  'the goal power consumption' (power-goal) : positive number.
30  'the maximum acceptable power consumption' (max-power) : positive number.
31  'the power consumption discrimination increment' (power-inc) : positive number.
32  total-power = < (power-goal to max-power step power-inc).
33  no-tpoints < 200.
34  hardness (no-tpoints < 200 = 0.9.
35  field 'the circuit designers perspective' (configuration) =::= {X : component (X)}.
36  field 'the perspective taken by test engineers' (testability) =::= {the-test-facility}.
37  permission({X:component(X)},configuration).
```

Figure 8.1 Extracts from KLAUS2, a Galileo4 program.

sum of the numbers of test points required by the individual components. Statement 14 then expresses an important life-cycle interdependency between this total number of test-points and the intended test facility. It is interesting to note that, because of the natural language synonyms given in statements 1, 2 and 16, the constraint in statement 14 would be expressed to KLAUS2 users (in justifications, for example) as follows: `the total number of test points on the board must be less than or equal to the maximum number of test points testable at the facility where the board will be tested.`

The two parameters `lst_tpoints` and `num_tpoints` referenced in statements 12 and 13 are declared in statements 15 and 16. Statement 15 illustrates another feature of the language – its ability to handle set-valued parameters. Indeed, the constraint uses a special kind of set, a **bag** or multiset, that is, a set in which multiple copies of the same member are treated as distinct. This is needed in order to deal with the fact that different components can have the same number of leads.

While the life-cycle interdependency expressed in statement 12 required universal quantification, statement 17 shows a constraint which requires the other form of quantification in logic, that is, existential quantification. Statement 17 specifies that the cheapest tester available should be used, by specifying that there should not be any tester available which is capable of testing the board and which costs less than the chosen tester. This constraint is worth further discussion, because it illustrates usage of other features of Galileo4 that are available only because the language provides the full expressive power of the predicate calculus. This statement shows that while Galileo4 provides generic predicates such as < and generic functions such as *, it also allows application-specific functions and predicates to be used.

If we are to use application-specific functions and predicates, their meanings must be defined. In Galileo4, this can be done using either of the notions of extensional or intensional definition from set theory. Consider, for example, statement 18, which defines the meaning of the function `cost`. The meaning of a function is a set of mappings from inputs to outputs. In statement 18, the meaning of cost is defined intensionally; the total cost of using a tester is 1.8 times the direct cost of using the tester, which is denoted by the function symbol `price`. There is a finite number of possible testers, so the meaning of the `price` function is a finite set which can be defined extensionally.

Galileo4 allows extensional set definitions to be given either in the program text or in an external database file. Statement 19, for example shows that the meaning of the function `price` is defined by specifying that the set of pairs of values is in the database file TSTRCSTS. (We have found in our application experiments that tying function and predicate definitions to database files is a very natural way of linking INCA systems to corporate databases.)

As with a function, the meaning of a predicate is also a set, which can also be defined either extensionally or intensionally. In statement 20, we define the meaning of the predicate `satisfactory` by specifying an intensional formula which uses universal quantification and two application-specific function symbols whose meanings are defined in statements 21 and 22.

8.2.3 Non-parametric design

In parametric design, the overall architecture of the product and its life-cycle have already been determined and the task is merely one of establishing appropriate values for the parameters of this architecture. Concurrent engineering is not this simple. Parametric design must be accompanied by what is sometimes called componential design, in which the structure of the product and/or its life-cycle environment themselves are determined.

The belief is still widespread that constraint networks are incapable of addressing non-parametric design. However, another contribution of this research was the discovery that this limitation could be overcome by incorporating into constraint processing theory the notion of conditional existence from free logic [19]. This enables a constraint processing inference engine to deduce that, when certain conditions are true, additional parameters must be introduced into a constraint network. This was a fundamental discovery, since it enables a constraint-based CE system to reason about when to introduce new elements into a product or life-cycle architecture. For further information on the scientific basis for using free logic in constraint networks, see [2].

Consider, for example, statement 23. This is a universally quantified constraint which also uses the notion of conditional existence from free logic to specify that if a parameter of domain `component` is used to represent a CPU, a communications controller or a digital signal processor device, then the parameter must have an extra slot to represent the oscillator which drives the device and must have a further slot to represent a pullup resistor. The `exists` tokens in this constraint are free logic existence specifiers, not existential quantifiers.

A more interesting usage of free logic appears in statement 24, which also uses modal logic. The `else` connective in this constraint comes from modal logic. The constraint specifies that, ideally, every crystal should oscillate at a frequency which does not exceed the maximum clock speed that is testable by the test facility. However, it then goes on to say that if this is not possible, then any crystal which oscillates at a faster frequency must have an ancillary divider circuit. Here, we see the constraint network extending itself by introducing a new parameter when a certain condition arises. This new parameter represents a new component, the necessity of whose existence has been inferred by the system.

8.2.4 Optimization

Statement 17 used existential quantification to require that the cheapest possible tester be used. Since there is a finite number of possible testers, this statement illustrated optimization of a parameter which ranged over a discrete domain. Galileo4 also supports optimization of parameters which range over infinite domains, provided these domains are discretized into finite numbers of equivalence sets.

Consider, for example, statement 32. This specifies that the total power consumed by the board (see statements 25 through 28) should, ideally, not exceed the goal power consumption (statement 29), that it should certainly not exceed the maximum acceptable power consumption (statement 30), and that between those two values the optimization tolerance is equal to a value called the power consumption discrimination increment (statement 31).

8.2.5 Prioritization

By default, all constraints in a Galileo4 program are treated as being equally important, and are treated as hard constraints, that is, as constraints that must be satisifed. However, we can also specify that some constraints in a program are soft. The meaning of a constraint being soft is that the constraint should, if possible, be satisfied but, if there ever arises a situation in which violating the constraint would relieve an over-constrained situation, then it may be ignored. We can have as many soft constraints as we want in a Galileo4 program and can assign them different levels of hardness or priority. Constraint hardness is a number in [0.1], with 1 being the default and 0 being the hardness of a constraint that has been disabled completely. Statement 34 is a second-order constraint which specifies that the hardness of the first-order constraint in statement 33 is 0.9.

8.2.6 Multiple perspectives and interfaces

Galileo4 enables constraint networks to be divided into (possibly overlapping) regions called `fields of view`. A field of view is that region within a constraint network that is currently important to a user interacting with the network. A field of view can be either global or local. The global field of view consists of the entire constraint network. A local field of view contains only a subnetwork. Each field of view contains all the parameters that are of interest to the user, as well as all constraints which reference these parameters.

We define a field of view by specifying the set of parameters which it contains. In Fig. 8.1, for example, we define two of the fields of view that are provided by the KLAUS2 application. Statement 35 defines a `configuration` field of view, which will be seen by a circuit designer, and specifies that it contains the set of all parameters of domain `component`. Statement 36 defines a `testability` field of view and specifies that its set of parameters contains just one parameter, `the_test_facility`

Different fields of view can be presented to their users through different styles of interface. Although the specification of these different types of interface is a simple matter in Galileo4, detailed discussion is beyond the scope of this necessarily brief presentation.

8.2.7 Specifications and decision

Galileo4 programs are interactive. A user can augment the set of constraints in the initial network that is specified in a program, by inputting additional constraints to represent his design decisions. Thus, for example, if a test engineer decides to use an Erdysys TX tester, he can indicate this decision by inputting the following equational constraint: `the_test_facility.name = erdsys`. (Note that the test engineer would not have to type this constraint – the desired decision can be input by using a mouse to select appropriate options in a series of pull-up menus. Furthermore, because of the system's use of long synonyms, the engineer would think that he was entering the following decision: `the equipment at the facility where the board will be tested = Erdsys TX`. The test engineer need never know about such 'unfriendly' tokens as `the_test_facility.name` or `erdsys`.)

Figure 8.2 shows the interface presented by KLAUS2 to the test engineer after he has selected the test equipment to be used for the project, and the system has inferred two of the attributes of this equipment via relational information represented as constraints. The largest window in this screen is a single-column spreadsheet, or 'scrollsheet', in which each cell occupies one or more lines. Various pull-down menus, as well as overlay windows for constraint violation detection and advice generation, also appear when appropriate.

Decisions like the above selection of a tester are parametric design decisions. Componential design decisions can be expressed by adding new parameters to the initial network that is defined by the program. Thus, for example, a circuit designer interacting with the KLAUS2 application can introduce new parameters to represent various parts of his evolving circuit. To introduce a CPU, for example, he can either

```
{}Help      {}File       {}New         {}Utilities {}Search
{}Up        {}Down       {}Focus                   {}Toggle

{} The equipment at the facility where the              Erdsys TX
   board will be tested
{} the maximum clock frequency testable at the          9.8
   facility where the board will be tested
{} the maximum number of test points testable           200
   at the facility where the board will be tested

>>>

KLAUS2 - a PWB Design Advisor (Testability)
```

Figure 8.2 The Galileo4 scrollsheet interface.

introduce a parameter of domain `component` and specify that the `type` slot of this parameter has the value `processor`, or he can achieve exactly the same result, through frame-based inheritance, by introducing a parameter of domain `cpu`.

In Galileo4, we can specify which users of an application that supports multiple fields of view are allowed to introduce new parameters and what classes of parameters they are allowed to enter. Statement 37 of Fig. 8.1, for example, specifies that users of the `configuration` field of view are allowed to introduce parameters of domain `component` or of any domain (such as `osc_crystal`, `divider`, `resistor` or `cpu`) that is a sub-class of the `component` domain.

One further point should be made on the representation of design decisions in Galileo4. Any syntactically well-formed Galileo4 constraint may be used to represent a design decision. We are not restricted to equations of the form seen above in the test engineer's selection of test equipment. Any sentence, atomic, compound or quantified, in first-order predicate calculus, including modal and free logic as well as classical logic, can be used to represent a design agent's decision.

8.2.8 Explanation

As well as specifying information by introducing new parameters and new constraints, the user of a Galileo4 program can ask for information. He can, for example, ask for the range of allowable values for any of the parameters in a network. He can also ask for justification for these ranges – whenever the range of allowable values for a parameter is reduced by a constraint, the rationale for this reduction is noted by the run-time system as a dependency record which can be accessed later for explanation purposes.

Consider, for example, a scenario in which the circuit designer specifies that he wants the crystal to oscillate at 25 MHz. Before doing so, he could have asked KLAUS2 for the range of possible values. If he had done so, he would have been told that the frequency should ideally be in the range [3,9.8] but that, as a last resort, any frequency in the range [3,40] could be used. If he had asked for a justification, the explanation would have referred to the fact that, in general, the only frequencies allowed are those in the range [3,40] (see statement 4 in Fig. 8.1) but that the test engineer's previous choice of an Erdsys TX tester and KLAUS2's preference for avoiding the introduction of ancillary dividers (see the above discussion on statement 24 in section 8.2.3) means that the preferred range is [3,9.8] because the maximum frequency testable by the Erdsys TX is 9.8 MHz.

8.2.9 'What if' design reasoning

A user can always withdraw any constraint or parameter that he has added. Thus, by introducing and withdrawing constraints and parameters, the user can investigate 'what if' scenarios.

8.3 A SECOND APPLICATION: INSERTION ROBOT SELECTION

To demonstrate the practical utility of this constraint-based approach to logic, we present in Fig. 8.3 a Galileo4 program for a type of application

```
domain robot =::= datafile(qdb,$ROBOTS.DBE$).
function price(robot) - > positive =::= datafile(qdb,$PRICES.DBE$).
function mag_capacity(robot) - > pos_int =::= datafile(qdb,$MAGCPTY.DBE$).
function fixtures(robot) - > pos_int =::= datafile(qdb,$FIXTURES.DBE$).
function throughput(robot) - > positive =::= datafile(qdb,$THRUPUT.DBE$).
constant tax_rate =::= 0.06.
positive (reqd_throughput).
pos_int(reqd_mag_capacity).
pos_int(reqd_fixtures).
robot(chosen_model).
positive(cost).
cost = price(chosen_model) * (1 + tax_rate).
fixtures(chosen_model) >= reqd_fixtures.
throughput(chosen_model) >= reqd_throughput.
mag_capacity(chosen_model) >= reqd_mag_capacity,
not(exists X : fixtures(X) >= reqd_fixtures and throughput (X) >= reqd_
  throughput and mag_capacity(X) >= reqd_mag_capacity and price (X) <
  price(chosen_model)).
```

Figure 8.3 A Galileo4 insertion robot selection application.

(manufacturing equipment selection) which is commonly implemented using declarative rule-based programming. In this case, the application involves selecting an insertion robot to provide required functionality at a minimum price. This practical application is a short program in Galileo4 and illustrates well the advantage of constraint-based programming: an equivalent declarative rule-based program would be longer and less perspicuous.

The problem of insertion robot selection is to select a robot model from among a set of available models, each with varying prices, insertion magazine capacities, number of fixtures accommodated, and throughput rates. A suitable robot should have throughput, fixture, and magazine capacity at least as great as those specified by the user of the Galileo4 program. The problem of optimal-cost robot selection is to select the suitable robot of lowest cost, factoring in the tax rate.

All of these aspects of the problem are compactly encoded as constraints in the Galileo4 program listed in Fig. 8.3. The first five statements in this program illustrate an advantage of our approach that was not exercised by the printed wiring board example, namely the natural way in which programs based on this approach can be linked to relational databases. Application-specific domains, functions and predicates can be defined either extensionally or intensionally within a constraint-based program. Alternatively, as in this program, extensional definitions can be tied to the contents of external relational database files.

model_9520
model_8581
model_8582
model_7001
model_7002
model_6001

(a) ROBOTS.DBE

model_9520	4499.99
model_8581	3100.99
model_8582	3500.99
model_7001	2859.99
model_7002	2959.99
model_6001	1999.99

(b) PRICES.DBE

model_9520	628
model_8581	128
model_8582	140
model_7001	88
model_7002	115
model_6001	60

(c) MAGCPTY.DBE

model_9520	3
model_8581	6
model_8582	5
model_7001	6
model_7002	5
model_6001	1

(d) FIXTURES.DBE

model_9520	20
model_8581	25
model_8582	25
model_7001	20
model_7002	20
model_6001	16

(e) THRUPUT.DBE

Figure 8.4 Contents of relational database files.

The first statement in this program declares that the extensional definition of the domain *robot* is the set of all robot models present in the relational file ROBOTS.DBE. Similarly, the next four statements declare that the extensional definitions of the four functions *price, mag_capacity, fixtures* and *throughput* are in the files PRICES.DBE, MAGCPTY.DBE, FIXTURES.DBE and THRUPUT.DBE, respectively. The sixth statement declares that *tax_rate* $\mapsto$ `0.06`.

Suppose that the contents of the files ROBOTS.DBE, PRICES.DBE, MAGCPTY.DBE, FIXTURES.DBE and THRUPUT.DBE are as shown in Fig. 8.4.

There are five values required of the user in order to instantiate all the parameters in the program: *reqd_throughput, reqd_mag_capacity, reqd_fixtures, chosen_model* and *cost*. These correspond, respectively, to the throughput (in parts per minute) that the user requires, the insertion magazine capacity needed, the number of fixtures required, the robot model most appropriate for his needs and the amount he will have to spend. There are constraints in the program to specify type restrictions for these symbols; another constraint specifies that *cost* is the price of the *chosen model* plus 6% tax; still other constraints specify that the *chosen_model* must provide the required functionality, and the final constraint specifies that there should not be anything cheaper than the *chosen_model* which offers this functionality.

8.3.1 Problem-solving versatility: an example scenario

To show the versatility of this program and, in a broader context, to illustrate the problem solving power provided by using constraint networks the way we do, we next present an example scenario which samples the wide variety of problems that the program can be used to solve.

Even this simple a program can be used to solve a wide variety of problems. The user is free to invoke forward-chaining by asserting (and later retracting) additional arbitrary constraints, the only restriction being that each such sentence must reference at least one parameter from the set {*reqd_throughput, reqd_mag_capacity, reqd_fixtures, cost, chosen_model*}. Alternatively, the user can invoke backward-chaining by asking the system to determine the value of any of the parameters in this set. The following example scenario illustrates some of this versatility.

Suppose there is a manufacturing operation, with a limited budget, which hopes to buy a robot which will satisfy two needs: a lot of magazine capacity and sufficient fixtures.

Immediately after invoking this program, the user specifies what he estimates to be the amount of magazine capacity needed, by interactively asserting *reqd_mag_capacity* = 200. The system reports that *chosen_model* $\mapsto$ `model-9520` and that *cost* $\mapsto$ `4769.9894`. If the user asks for a justification, the system will explain that these conclusions were caused by the requirements that

mag_capacity(chosen_model) $\geqslant$ *reqd_mag_capacity*

and that

cost = *price(chosen_model)* * (1 + *tax_rate*).

But suppose that this is more than management wants to spend. The user retracts *reqd_mag_capacity* = 200 and asserts *cost* $\leqslant$ 4000. If the user asks to be shown the current set of possible values for *reqd_mag_capacity*, he will be told that it is any positive integer up to 140, which is the magazine capacity of the `model_8582`, the robot with the largest magazine capacity whose cost does not exceed 4000.

Suppose that the user now decides to stop volunteering information and tells the system to determine the appropriate robot, asking whatever questions it sees fit along the way. Then, triggered by the specification *fixtures(chosen_model)* $\geqslant$ *reqd_fixtures*, the system asks the user to specify a value for *reqd_fixtures* and, in response, suppose the user asserts that *reqd_fixtures* = 6. The system then asks for the *reqd_throughput*; the user has no specific speed in mind but, choosing a ball-park minimum, he asserts that *reqd_throughput* $\geqslant$ 18. The system then asks for the *reqd_mag_capacity*; the user, remembering the previous information about the set of possible values for this parameter, responds by asserting that *reqd_mag_capacity* = 140.

However, this causes a contradiction, because the maximum magazine capacity offered by any robot which can support six fixtures is 128. The system suggests that the user should retract one of the two assertions, *reqd_fixtures* = 6 or *reqd_mag_capacity* = 140.

However, suppose the user wants to know why he cannot have the `model_8582`. Therefore, before adopting any of the above suggestions, he asserts *chosen_model* = *model_8582*. In response, he is told that this contradicts the requirements that *fixtures(chosen_model)* $\geqslant$ *reqd_fixtures* and *reqd_fixtures* = 6.

Recognizing that he cannot get a robot which satisfies all his needs, the user decides that, for now, he will just buy a robot with a smaller insertion magazine. So, he retracts *chosen_model* = *model_8582* and *reqd_mag_capacity* = 140, and asserts that *reqd_mag_capacity* = 10.

The set of possible values for *chosen_model* which is allowed by the user's stated functionality requirements contains two robots, the `model_8581` and the `model_7001`. The system cannot decide for certain, however, that either of these robots is suitable, because the user has not been specific enough about *reqd_throughput*.

Suppose, however, that the user now chooses a ball-park maximum for *reqd_throughput* and that *reqd.throughput* < 29 Both of the robots just mentioned provide the required functionality but, because of the constraint

> not(exists X : fixtures(X) > = reqd_fixtures and throughput (X) > = reqd_throughput and mag_capacity(X) > = reqd_mag_capacity and price(X) < price(chosen_model)).

the system can choose the cheaper of the two robots, reporting that *chosen_model* ↦ `model-7001`.

8.4 COMPARATIVE DISCUSSION

Although it is less widely used than rule-based programming, constraint-based programming is not a new idea; the first constraint-based programming system was developed almost 30 years ago [29]. The major factor inhibiting a widespread application of constraint-based programming has been the highly specialized nature of the constraint-based programming systems available. For example, the first constraint-based system, Sketchpad [29], was oriented towards graphics while Thinglab [6] was developed for simulation applications. An early language which offered more generality was Constraints [28] but several important notions, including inequality, were not available in the language. Magritte [14] was restricted to algebraic relationships and did not support the use of arbitrary application-specific predicates, functions and domains.

More recently, there has been a surge of interest in the relationship between constraints and logic programming. Several languages based on Prolog have been developed, most notably CLP ($\Re$) [17]; Prolog III [8]; and CHIP [11]. In these languages, which are generically known as the CLP languages, unification in the Herbrand universe is supplemented with constraint processing of linear equations or inequalities over the numbers. CHIP also uses arc-consistency to process atomic constraints involving application-specific predicates defined over finite domains.

Galileo4 provides richer expressive power than any other constraint-based programming language because it allows the theory used to specify a constraint network to contain arbitrary FOPC sentences (atomic, compound or quantified) about a many-sorted universe of discourse which includes, besides the real numbers $\Re$, any arbitrary application-specific sorts. Nearly all other constraint languages restrict the theory to ground sentences. For example, in the CLP languages, the theory used to specify a constraint network is the set of conjuncts in a compound goal presented to the top-level interpreter; although logic variables may appear in such a goal, these are implicitly subject to existential quantification, which means that, essentially, they are uninterpreted constant symbols, making the constraints ground sentences. One language which does seem to support quantified constraints is CONSUL [7], but it restricts the universe of discourse to the integers.

Expressiveness in knowledge representation, however, is only one side of the coin. Constraint-based languages also differ in the inferential competence and efficiency offered by their run-time systems. One of the earliest, and most commonly used, constraint propagation mechanisms is local propagation of known states, which is efficient but incomplete. The simplex-like algorithm used in the CLP languages is more complete (that is, can draw more inferences given the same premises) than local propagation of known states and considerable effort has been devoted to developing efficient and fast implementations. Until recently, there was little cross-fertilization between the literature on finite domain CSPs and that on constraint-based programming languages. The first attempt to amalgamate ideas from these two bodies of work seems to have been the CLP language CHIP [11], in which the application of simplex to constraints on the rationals was supplemented by the application of arc consistency to constraints on application-specific finite domains.

The CCP algorithm used in Galileo4 takes the idea of borrowing concepts from the finite domain CSP literature a step further, by integrating local propagation of known states with versions of arc and path consistency that have been generalized to constraints of arbitrary arity on infinite domains. The inferential competence of the resulting algorithm is better in some respects, but worse in others, than the algorithms used in the CLP languages. For example, there are several classes of problem involving sets of simultaneous non-linear equations that can be handled by CCP [27], while the CLP languages can only handle non-linear constraints if they become linear during the propagation process. However, the CCP algorithm is limited to handling systems of simultaneous equations in which each equation involves only two unknowns or is reduced to a binary equation during propagation. Although there has been considerable research into constraint propagation algorithms for finite domain networks, and despite the fact that constraint processing has been around a long time, constraint propagation for infinite domain networks is still a relatively undeveloped field. We view CCP as being capable of further improvement; for example, an obvious topic to investigate is the amalgamation of simplex analysis with the battery of techniques currently used in CCP.

The query interface to CLP languages is basically the same as that provided by the early Prologs. That this interface is inadequate for constraint processing was recognized in [22] where a new interface was proposed. It is interesting to note that the 'answer manipulation commands' which they propose would allow the same kind of non-monotonic editing of goals as the Galileo4 user can achieve by asserting/ retracting assumptions.

8.5 THE RELATIONSHIP BETWEEN CONSTRAINT NETWORKS AND CONSTRAINT LOGIC

This chapter presents an attempt to make the full FOPC available in an INCA-construction language known as Galileo4. In the run-time system for this language, inference is based on a treatment of semantic entailment as constraint propagation. Very simply, constraint propagation is the communication of parameter values through the network to all constraints which refer to those parameters. This propagation process occurs omnidirectionally each time a parameter acquires either a new or revised value or a restriction of its domain of possible values. These new values may in turn stimulate further inference, and the process iterates until quiescence. To provide the theoretical basis for this approach, we review model theory and constraint processing and then we show how the satisfiability of a set of sentences in logic can be viewed as the consistency of a set of constraints in a constraint network.

8.5.1 A review of model theory

The truth of a sentence in logic is based on the interpretation of the symbols in the sentence. Consider some first-order logic language $\mathcal{L} = \langle \mathcal{P}, \mathcal{F}, \mathcal{K} \rangle$, where $\mathcal{P}$ is the vocabulary of predicate symbols, $\mathcal{F}$ is the vocabulary of function symbols, and $\mathcal{K}$ is the vocabulary of constant symbols. A theory Γ in $\mathcal{L}$ is satisfiable iff there exists some model $\mathcal{M} = \langle \mathcal{U}, I \rangle$ of the language $\mathcal{L}$ under which every sentence in Γ is true, that is, iff there exists some $\mathcal{M}$ such that $\mathcal{M} \mid= \Gamma$. The notation $\mathcal{M} \mid= \Gamma$ means that any interpretation of the symbols in $\mathcal{M}$ satisfying $\mathcal{M}$ will simultaneously satisfy Γ.

In a model $\mathcal{M} = \langle \mathcal{U}, I \rangle$ for the language $\mathcal{L} = \langle \mathcal{P}, \mathcal{F}, \mathcal{K} \rangle$, $\mathcal{U}$ is a universe of discourse while I is an interpretation function for the constant, predicate, and function symbols of $\mathcal{L}$. For every $\kappa \in \mathcal{K}$, $I(\kappa)$ is an element of $\mathcal{U}$. For every n-ary predicate symbol $p \in \mathcal{P}$, $I(p)$ is an n-ary relation over $\mathcal{U}$. For every n-ary function symbol $f \in \mathcal{F}$, $I(f)$ is an $(n + 1)$-ary relation over $\mathcal{U}$. For a functional expression $f(a_1, ..., a_n)$, where f is an n-ary function symbol and the terms $a_i, i = 1..n$, are either constant symbols or nested functional expressions, $I(f(a_1, ..., a_n))$ is an element of $\mathcal{U}$ such that $I(f)$ contains the $(n + 1)$-ary tuple $\{ I(a_1), ..., I(a_n), I(f(a_1, ..., a_n)) \}$.

Let $\mathcal{M} \models \gamma$, where γ is a sentence, means that γ is true under the model $\mathcal{M}$ and let $\mathcal{M}\{x \mapsto u\}$ mean a model containing the extended interpretation function $I \cup \{x \mapsto u\}$ The notation $x \mapsto u$ indicates that symbol x is assigned the semantic interpretation u. The rules for determining whether a sentence is true under a model are as follows:

- $\mathcal{M} \models p(a_1, ..., a_n)$ iff $\langle I(a_1), ..., I(a_n)\rangle$ is in $I(p)$.
- $\mathcal{M} \models \neg A$ iff $\mathcal{M} \not\models A$.
- $\mathcal{M} \models A \wedge B$ iff $\mathcal{M} \models A$ and $\mathcal{M} \models B$.
- $\mathcal{M} \models A \vee B$ iff $\mathcal{M} \models A$ or $\mathcal{M} \models B$.
- $\mathcal{M} \models A \Rightarrow B$ iff $\mathcal{M} \not\models A$ or $\mathcal{M} \models B$.
- $\mathcal{M} \models A \Leftrightarrow B$ iff ($\mathcal{M} \not\models A$ and $\mathcal{M} \not\models B$) or ($\mathcal{M} \models A$ and $\mathcal{M} \models B$).
- $\mathcal{M} \models (\forall X)\, A$ iff $\mathcal{M}_{\{X \mapsto u\}} \models A$ for every $u \in \mathcal{U}$.
- $\mathcal{M} \models (\exists X)\, A$ iff $\mathcal{M}_{\{X \mapsto u\}} \models A$ for some $u \in \mathcal{U}$.

8.5.2 A review of constraint processing

A constraint specifies some relationship that must be satisfied by the values, chosen from some domain, which are assumed by a group of parameters. A constraint network is a collection of constraints which are interlinked by virtue of having some parameters in common.

There has developed a considerable body of literature on the processing of constraint networks. This literature tends to fall into two broad classes. In the first kind of paper, which is where the term CSP is used, constraint networks are assumed to have finite domains [10, 13, 16, 21, 23, 25], the relationships imposed by constraints are expressed extensionally, and the CSPs are NP-complete. Although the CSPs in this literature can be solved by backtracking search, the exponential computational cost of such search algorithms has led to the development of preprocessing algorithms [12, 15, 21, 24], called consistency algorithms, which aim to reduce thrashing, that is, repeatedly exploring parameter assignments that cannot result in a solution, by *a priori* elimination of parameter to value mappings that cannot belong to any consistent valuation of the network parameters.

In the second class of literature on constraint processing, constraint networks are allowed, but not required, to have infinite domains, and the restrictions imposed by constraints can be expressed intensionally. In this literature, which includes that dealing with constraint logic programming (CLP) [8, 11, 17], the problem-solving techniques include local propagation of known states [20] and a simplex-like version of gaussian elimination [17]. Consider, for example, the CLP ($\Re$) language, which is based on Prolog and in which constraint parameters can range over the real numbers $\Re$. In this language, a constraint network is treated as a conjunctive goal which is submitted to the top-level of a Prolog-like interpreter; for example, a goal such as

?– Load = Area * Stress, Load = 30, Area = 5.

is a network containing three constraints which intensionally specify restrictions on the values that can be assumed by the three parameters, Load, Area and Stress. When the CLP ($\Re$) interpreter responds with

Load = 30   Area = 5   Stress = 6   ***Yes

it is giving an existence proof that there is a valuation for the parameters which satisfies all the constraints in the network.

In this second body of work, although constraint networks may have infinite domains, severe limits have been placed on the use of intensional specifications. Systems have restricted the types of logic connective that can be used and the situations in which they can be used. Furthermore, in most systems, only ground sentences may be used. In CLP ($\Re$), for example, the free variables in a goal are treated as being implicity subject to existential quantification, which means that effectively they are equivalent to uninterpreted constant symbols. Thus, the goal shown above can be regarded as a conjunction of three ground sentences, *load* = *area* * *stress*, *load* = 30, and *area* = 5, in which there are three uninterpreted constant symbols, *load*, *area* and *stress*. The task of the interpreter is then to determine whether there is an interpretation for these constant symbols which satisfies all three ground sentences. Constraints in CLP ($\Re$) cannot use quantification; it is not possible, for example, to submit goals such as

?–Load = Area * Stress, Load = 30,
    Area = 5, (all X : Area < X < Stress implies X > Load/2).

in which the last constraint is a universally quantified sentence. In the one language known to allow explicit quantification [7], the domain is restricted to the integers and it is unclear whether quantifiers may be arbitrarily nested.

Our work is directed towards removing these limitations and allowing arbitrary FOPC formulae to be used as constraints.

8.5.3 Definitions

The literature contains several definitions of constraint satisfaction, with varying degrees of formality. One consequence of the lack of formal definition of the concept is that many authors fail to distinguish between notions of decision, exemplification, and enumeration. In an effort to remove this ambiguity and to set the scene for a mapping between constraint processing and semantic modeling, we propose the following definitions.

Definition: Constraint Network

A constraint network is a triple $\langle \mathcal{U}, \mathbf{X}, \mathbf{C} \rangle$ where $\mathcal{U}$ is a universe of discourse, $\mathbf{X}$ is a finite tuple of q non-recurring parameters, and $\mathbf{C}$ is a finite set of r constraints. Each constraint $C_k(T_k) \in \mathbf{C}$ imposes a restriction on the allowable values for the a_k parameters in T_k, a sub-tuple of $\mathbf{X}$, by specifying that some subset of the a_k-ary Cartesian product $\mathcal{U}^{a_k}$ contains all acceptable combinations of values for these parameters.

The overall network constitutes an intensional specification of a joint possibility distribution for the values of the parameters in the network. This joint possibility distribution, called the network **intent** [2], is a q-ary relation on $\mathcal{U}^q$, as defined below.

Definition: Projection

Let R be a q-ary relation with indices $\langle x_1, x_2, ..., x_q \rangle$, and let $\mathbf{T} = \langle t_1, t_2, ..., t_m \rangle$ be an m-ary subtuple of $\langle x_1, x_2, ..., x_q \rangle$. The projection of R onto T, denoted $proj(R,T)$, is the largest set of m-tuples $\langle b_{t_1}, b_{t_2}, ..., b_{t_m} \rangle$ such that there is some q-tuple $\langle c_{x_1}, c_{x_2}, ..., c_{x_q} \rangle$ in R for which $b_{t_j} = c_{t_j}$, for all $j = 1, 2, ..., m$.

Definition: Cylindrical Extension

Let $\mathbf{X} = \langle x_1, x_2, ..., x_q \rangle$, be a q-ary tuple of indices defining a Cartesian space $\mathcal{U}^q$, let $T = \langle t_1, t_2, ..., t_m \rangle$ be a subtuple of $\langle x_1, x_2, ..., x_q \rangle$, and let $C(T)$ be a relation on the space defined by T. The cylindrical extension of $C(T)$ into the space defined by $\mathbf{X}$, denoted $E(X)$, is the largest relation R on that space such that $proj(R, T) = C(T)$.

Definition: The Intent of a Constraint Network

The intent of a constraint network $\langle \mathcal{U}, \mathbf{X}, \mathbf{C} \rangle$ is

$$\Pi_{\mathcal{U},\mathbf{X},\mathbf{C}} = E_1(\mathbf{X}) \cap ... \cap E_r(\mathbf{X}),$$

where, for each constraint $C_k(T_k) \in \mathbf{C}$, $E_k(\mathbf{X})$ is its cylindrical extension into the Cartesian space defined by $\mathbf{X}$.

The network intent is a set of q-tuples, each tuple giving, for the q parameters in $\mathbf{X}$, a valuation which is acceptable to all the constraints in $\mathbf{C}$.

A constraint network is consistent if the network intent is not the empty set. Three forms of constraint satisfaction problem (CSP) can be distinguished, as follows.

Definition: The Decision CSP

Given a network $\langle \mathcal{U}, \mathbf{X}, \mathbf{C} \rangle$, decide whether $\Pi_{\mathcal{U},\mathbf{X},\mathbf{C}}$ is non-empty.

Definition: The Exemplification CSP

Given a network $\langle \mathcal{U}, \mathbf{X}, \mathbf{C} \rangle$, return some tuple from $\Pi_{\mathcal{U},\mathbf{X},\mathbf{C}}$, if $\Pi_{\mathcal{U},\mathbf{X},\mathbf{C}}$ is non-empty, or return nil otherwise.

Definition: The Enumeration CSP
Given a network $\langle \mathcal{U}, \mathbf{X}, \mathbf{C} \rangle$ return $\Pi_{\mathcal{U},\mathbf{X},\mathbf{C}}$.

8.5.4 Networks and models

A total interpretation function for a first-order language $\mathcal{L}$ is a set of mappings, such that every predicate and function symbol of $\mathcal{L}$ is mapped onto a relation of the appropriate arity over some universe of discourse $\mathcal{U}$, and such that every constant symbol of $\mathcal{L}$ is mapped onto a member of $\mathcal{U}$. A partial interpretation function for $\mathcal{L}$ is a set of mappings which does not contain a mapping for every symbol of $\mathcal{L}$; at least one constant, predicate, or function symbol is not given a mapping.

Consider the class of problem in which, given a theory Γ written in a first-order language $\mathcal{L} = \langle \mathcal{P}, \mathcal{F}, \mathcal{K} \rangle$ and a partial interpretation function I_p for $\mathcal{L}$ in terms of a universe $\mathcal{U}$, one has to determine whether there is any total interpretation I for $\mathcal{L}$ such that $I_p \subset I$ and $\langle \mathcal{U}, I \rangle \models \Gamma$. This, the class of modeling problems, can be divided into several subclasses.

Class 1 modeling problems are those where I_p contains interpretations for all the predicate and function symbols of $\mathcal{L}$, where a finite subset of the constant symbols are uninterpreted, where every sentence in Γ references at least one of the uninterpreted constant symbols, and where each uninterpreted constant symbol is referenced by at least one sentence in Γ. Although there are several other classes (for example, those modeling problems where some subset of the function and predicate symbols lack interpretations) Class 1 problems are those that are of interest here.

Class 1 problems can be further subdivided, as follows.

Definition: The Class 1 Decision Modeling Problem
Given $\mathcal{L} = \langle \mathcal{P}, \mathcal{F}, \mathcal{K}' \cup \mathcal{K}'' \rangle$, $\mathcal{U}$, I_p and Γ, where I_p interprets, in terms of $\mathcal{U}$, all and only the symbols in $\mathcal{P}, \cup \mathcal{F}, \cup \mathcal{K}'$, $\mathcal{K}''$ is finite, each sentence in Γ references at least one symbol in $\mathcal{K}''$, and each symbol in $\mathcal{K}''$ is referenced by at least one sentence $\gamma \in \Gamma$, decide whether there exists any $\mathcal{J}$ such that $\mathcal{J}$ interprets all symbols in $\mathcal{K}''$ and $\langle \mathcal{U}, I_p \cup \mathcal{J} \rangle \models \Gamma$.

Definition: The Class 1 Exemplification Modeling Problem
Given $\mathcal{L} = \langle \mathcal{P}, \mathcal{F}, \mathcal{K}' \cup \mathcal{K}'' \rangle$, $\mathcal{U}$, I_p and Γ, as in the Class 1 Decision Modeling problem, return, if one exists, some $\mathcal{J}$ such that $\mathcal{J}$ interprets all symbols in $\mathcal{K}''$ and $\langle \mathcal{U}, I_p \cup \mathcal{J} \rangle \models \Gamma$; if no such $\mathcal{J}$ exists, return nil.

Definition: The Class 1 Enumeration Modeling Problem
Given $\mathcal{L} = \langle \mathcal{P}, \mathcal{F}, \mathcal{K}' \cup \mathcal{K}'' \rangle$, $\mathcal{U}$, I_p and Γ, as in the Class 1 Decision Modeling problem, return the set of all $\mathcal{J}$ such that $\mathcal{J}$ interprets all symbols in $\mathcal{K}''$ and $\langle \mathcal{U}, I_p \cup \mathcal{J} \rangle \models \Gamma$.

These three forms of modeling problem can be shown to correspond to the three forms of CSP defined in section 8.5.3. In what follows, let $\vec{AB}$, where A and B are tuples of the same arity, be the mapping function from the components of A onto those of B in which each component of the tuple A is mapped onto the component in the corresponding position of tuple B; for example, if $A = \langle x, y \rangle$ and $B = \langle 4, 2 \rangle$, then $\vec{AB} = \{x \mapsto 4, y \mapsto 2\}$.

Theorem 1

Given $\mathcal{L} = \langle \mathcal{P}, \mathcal{F}, \mathcal{K}' \cup \mathcal{K}'' \rangle$, $\mathcal{U}$, I_p and Γ, where I_p interprets, in terms of $\mathcal{U}$, all and only the symbols in $\mathcal{P}, \cup \mathcal{F}, \cup \mathcal{K}'$, $\mathcal{K}''$ is finite, each sentence in Γ references at least one symbol in $\mathcal{K}''$, and each symbol in $\mathcal{K}''$ is referenced by at least one sentence $\gamma \in \Gamma$. Let $\langle \mathcal{U}, \mathbf{X}, \mathbf{C} \rangle$ be a constraint network such that $\mathbf{X}$ is the lexical ordering of the elements of $\mathcal{K}''$ and such that $|\mathbf{C}| = |\Gamma|$, with each sentence $\gamma \in \Gamma$ having a corresponding constraint $C(T) \in \mathbf{C}$ such that T is the lexical ordering of those elements of $\mathcal{K}''$ which appear in γ and $\langle \mathcal{U}, I_p \cup \vec{Tt} \rangle \models \gamma$ for all tuples $t \in \mathbf{C}(T)$. Then $\langle \mathcal{U}, I_p \cup \vec{\mathbf{X}\tau} \rangle \models \Gamma$ for all tuples $\tau \in \Pi_{\mathcal{U},\mathbf{X},\mathbf{C}}$.

The proof of this theorem appears in the appendix.

Thus, the set of all $\vec{\mathbf{X}\tau}$ where $\tau \in \Pi_{\mathcal{U},\mathbf{X},\mathbf{C}}$ is the set of all $\mathcal{J}$ such that $\mathcal{J}$ interprets all symbols in $\mathcal{K}''$ and $\langle \mathcal{U}, I_p \cup \mathcal{J} \rangle \models \Gamma$. Thus the Class 1 Decision Modeling Problem corresponds to the Decision CSP, the Class 1 Exemplification Modeling Problem corresponds to the Exemplification CSP and the Class 1 Enumeration Modeling Problem corresponds to the Enumeration CSP.

Example 1

Consider a language $\mathcal{L}$, a universe of discourse $\mathcal{U}$, and a partial interpretation I_p for $\mathcal{L}$ in terms of $\mathcal{U}$, as follows:

- **Language:** $\mathcal{L} = \langle \{s, t\}, \varnothing, \{a, b, c\} \rangle$
- **Universe of discourse:** $\mathcal{U} = \{1, 2, 3, 4, 5, 6, 7\}$
- **Partial interpretation:**

$$\begin{aligned} I_p = \{ s &\mapsto \{5, 7\}, \\ t &\mapsto \{\langle 1, 4\rangle, \langle 2, 5\rangle, \langle 3, 6\rangle, \langle 4, 7\rangle, \langle 5, 1\rangle, \langle 5, 3\rangle, \\ &\quad \langle 6, 2\rangle, \langle 6, 3\rangle, \langle 6, 7\rangle, \langle 7, 1\rangle, \langle 7, 3\rangle\}, \\ b &\mapsto 7\}. \end{aligned}$$

Suppose that, given the above situation, we need to determine the satisfiability of the following theory:

$$\Gamma = \{t(a,c), (\forall X)(s(X) \Rightarrow t(X, a)), t(c, b)\}.$$

The constraint network $\langle \mathcal{U}, \mathbf{X}, \mathbf{C} \rangle$ corresponding to this is a possibility distribution, such that Γ is satisfied, for interpretations of the remaining uninterpreted constant symbols of $\mathcal{L}$, in this case for a and c. The components of the network are:

$$\mathcal{U} = \{1,2,3,4,5,6,7\}$$
$$\mathbf{X} = \langle a, c\rangle$$
$$\mathbf{C} = \{C_1\langle a, c\rangle, C_2\langle a\rangle, C_3\langle c\rangle\}.$$

Each sentence in the theory has a corresponding constraint in the network, the definition of which depends on whatever information is provided, by the partial interpretation I_p about the symbols appearing in the sentence. In this case, the tuples admitted by these constraints are as follows:

$$C_1(a,c) = \{\langle 1,4\rangle,\langle 2,5\rangle,\langle 3,6\rangle,\langle 4,7\rangle,\langle 5,1\rangle,\langle 5,3\rangle,\langle 6,2\rangle,\langle 6,3\rangle,\langle 6,7\rangle,\langle 7,1\rangle,\langle 7,3\rangle\}$$
$$C_2(a) = \{1,3\}$$
$$C_3(c) = \{4,6\}.$$

Consider, for example, $C_1(a, c)$, which corresponds to the sentence $t(a, c)$. There are two parameters in this constraint, corresponding to the two constant symbols in the sentence. The restriction imposed by the constraint is derived from the interpretation in I_p for the predicate symbol t in the sentence. Consider $C_2(a)$, which corresponds to the sentence $(\forall X)(s(X) \Rightarrow t(X, a))$; the restriction imposed by this constraint is derived from the model-theoretic rules for universal quantification and implication and from the interpretations in I_p for the predicate symbols s and t in the sentence. Consider $C_3(c)$, which corresponds to the sentence $t(c, b)$; although two constant symbols appear in the sentence, there is only one parameter in the constraint, because I_p provides an interpretation for b.

Since the constraint $C_1(a, c)$, references all parameters in **X**, the cylindrical extension of the constraint is just the constraint itself; $E_1(a, c) = C_1(a, c)$. That is,

$$E_1(a,c) = \{\langle 1,4\rangle,\langle 2,5\rangle,\langle 3,6\rangle,\langle 4,7\rangle,\langle 5,1\rangle,\langle 5,3\rangle,\langle 6,2\rangle,\langle 6,3\rangle,\langle 6,7\rangle,\langle 7,1\rangle,\langle 7,3\rangle\}$$

The constraint $C_2(a)$ admits two values for a so, since the cardinality of $\mathcal{U}$ is 7, the cylindrical extension of this constraint in the Cartesian space defined by **X** contains 14 pairs:

$$E_2(a, c) = \{\langle 1,1\rangle,\langle 1,2\rangle,\langle 1,3\rangle,\langle 1,4\rangle,\langle 1,5\rangle,\langle 1,6\rangle,\langle 1,7\rangle, \langle 3,1\rangle,\langle 3,2\rangle,\langle 3,3\rangle,\langle 3,4\rangle,\langle 3,5\rangle,\langle 3,6\rangle,\langle 3,7\rangle\}.$$

Similarly, the cylindrical extension of $C_3(c)$, also contains 14 pairs:

$$E_3(a, c) = \{\langle 1,4\rangle,\langle 2,4\rangle,\langle 3,4\rangle,\langle 4,4\rangle,\langle 5,4\rangle,\langle 6,4\rangle,\langle 7,4\rangle, \langle 1,6\rangle,\langle 2,6\rangle,\langle 3,6\rangle,\langle 4,6\rangle,\langle 5,6\rangle,\langle 6,6\rangle,\langle 7,6\rangle\}.$$

However, if we compute the intersection of these three cylindrical extensions, we find there is little overlap, the intent of the network containing only two pairs:

$$\Pi_{\mathcal{U},\mathrm{X},\mathrm{C}} = \{\langle 1, 4\rangle, \langle 3, 6\rangle\}.$$

The network intent constitutes a joint possibility distribution for a and c. By projecting this joint distribution onto the a and c dimensions, we can obtain marginal possibility distributions for the individual parameters. Thus, the marginal possibility distribution for a is $\Pi_a = \{1, 3\}$ and that for c is $\Pi_c = \{4, 6\}$.

The network intent being $\Pi_{\mathcal{U},\mathrm{X},\mathrm{C}} = \{\langle 1, 4\rangle, \langle 3, 6\rangle\}$ means that the theory Γ is satisfied under the following two models of the language $\mathcal{L}$:

$$\mathcal{M}_1 = \langle \mathcal{U}, I_p \cup \{a \mapsto 1,\ c \mapsto 4\}\rangle;$$
$$\mathcal{M}_2 = \langle \mathcal{U}, I_p \cup \{a \mapsto 3,\ c \mapsto 6\}\rangle;$$

Augmenting a theory by asserting an additional sentence often reduces the number of possible models. Suppose we augment Γ by adding the assertion $t(b,c)$ to the theory. From $I_p\ (t)$ and $I_p\ (b)$, the corresponding constraint is $C_4(c) = \{1, 3\}$. Since this constraint admits neither of the two values for c that are allowed by the existing network, the new network intent will be the empty set. Thus, there is no model $\langle \mathcal{U}, I\ \rangle, I_p \subset I$, of the language $\mathcal{L}$ under which the following theory is satisfied:

$$\Gamma' = \{t(a,c), (\forall X)(s(X) \Rightarrow t(X, a)), t(c, b), t(b, c)\}.$$

Suppose, however, we retract $t(c, b)$ and insert $t(c, a)$. The resultant theory in this case,

$$\Gamma'' = \{t(a,c), (\forall X)(s(X) \Rightarrow t(X, a)), t(c, b), t(c, a)\}.$$

is satisfiable. However, the new network intent would be a strict subset of $\Pi_{\mathcal{U},\mathrm{X},\mathrm{C}}$, namely $\Pi_{\mathcal{U},\mathrm{X},\mathrm{C}''} = \{\langle 3, 6\rangle\}$. The fact that $\Pi_{\mathcal{U},\mathrm{X},\mathrm{C}''}$ admits only one pair means that unique values 3 and 6 can be inferred for a and c, respectively.

Example 2

In Example 1, the universe of discourse was finite and the predicate symbols had finite interpretations. The following example has an infinite universe and some of the predicate and function symbols have infinite sets as their interpretations:

• **Language:** $\mathcal{L} = \langle\{positive, nonnegative, =, \leq, h, j, k, g\}, \{*, l, m\}, \mathcal{R} \cup \{a, b, c, d, e, f, g1, g2, g3, g4\}\rangle$

- **Universe of discourse:** $\mathcal{U} = \Re \cup \{\texttt{g1}, \texttt{g2}, \texttt{g3}, \texttt{g4}\}$
- **Partial interpretation:**

$$
\begin{array}{lll}
I_p = I_R \cup & & \\
\{\, \textit{positive} & \mapsto & \Re^{+} \\
\text{non}\textit{negative} & \mapsto & \Re^{\cdot +} \\
= & \mapsto & \{\langle X,Y \rangle \mid X \in \mathcal{U} \wedge Y \in \mathcal{U} \wedge \text{EQUALS}(X, Y)\} \\
\leq & \mapsto & \{\langle X,Y \rangle \mid X \in \Re \wedge Y \in \Re \wedge \text{LEQ}(X, Y)\} \\
h & \mapsto & \{\langle X \mid Y \rangle \in \Re^{+} \wedge \text{LEQ}(X, \texttt{10})\} \\
j & \mapsto & \Re^{+} \\
k & \mapsto & \Re^{-} \\
g & \mapsto & \{\texttt{g1}, \texttt{g2}, \texttt{g3}, \texttt{g4}\} \\
* & \mapsto & \{\langle X,Y,Z \rangle \mid X \in \Re \wedge Y \in \Re \wedge Z \in \Re \wedge \text{EQUALS}(Z, \text{TIMES}(X, Y))\} \\
l & \mapsto & \{\langle \texttt{g1}, \texttt{200} \rangle, \langle \texttt{g2}, \texttt{250} \rangle, \langle \texttt{g3}, \texttt{350} \rangle, \langle \texttt{g4}, \texttt{390} \rangle\} \\
m & \mapsto & \{\langle \texttt{g1}, \texttt{230} \rangle, \langle \texttt{g2}, \texttt{190} \rangle, \langle \texttt{g3}, \texttt{240} \rangle, \langle \texttt{g4}, \texttt{245} \rangle\} \\
g1 & \mapsto & \texttt{g1} \\
g2 & \mapsto & \texttt{g2} \\
g3 & \mapsto & \texttt{g3} \\
g4 & \mapsto & \texttt{g4}\}.
\end{array}
$$

- **Theory**: $\Gamma = \{positive(a), h(b), positive(c), g(e), d = a * f, j(f) \Rightarrow f \leq l(e), k(f) \Rightarrow f \leq m(e), (\forall X)(g(X) \Rightarrow m(X) \leq l(e)), (\exists X)(l(X) > l(e)), a - b \,''\, c\}$.

In the language $\mathcal{L}$, $\mathcal{R}$ is the set of constant symbols composed from the characters +, -,. and 0..9 according to a grammar for real numeric strings. $\mathcal{R}$ is distinguished from $\Re$, the set of real numbers. In this chapter, to distinguish between symbols of $\mathcal{L}$ and entities of $\mathcal{U}$, we use typewriter font for the latter. Thus $200 \in \Re$ and $g1$ are constant symbols, while $\texttt{200} \in \Re$ and $\texttt{g1}$ are in the universe of discourse. In the partial interpretation I_p, I_R is a bijection from the constant symbols in $\mathcal{R}$ onto Q_f, the set of finite-length rational numbers, $Q_f \subset Q \subset \Re$; thus I_R contains mappings such as $200 \mapsto \texttt{200}$.

The constraint network $\langle \mathcal{U}, \mathbf{X}, \mathbf{C} \rangle$ corresponding to this situation is a possibility distribution, such that Γ is satisfied, for interpretations of the remaining uninterpreted constant symbols of $\mathcal{L}$, that is, for a, b, c, d, e, and f. The components of this network are:

$$
\begin{array}{l}
\mathcal{U} = \Re \cup \{\texttt{g1}, \texttt{g2}, \texttt{g3}, \texttt{g4}\} \\
\mathbf{X} = \langle a, b, c, d, e, f \rangle \\
\mathbf{C} = \{C_1(a), C_2(b), C_3(c), C_4(e), C_5(a, d, f), C_6(e, f), C_7(e, f), C_8(e), C_9(e), C_{10}(a, b, c)\},
\end{array}
$$

Infinite cardinality means that the restrictions imposed by most of the constraints must be specified intensionally:

$C_1(a) = \Re^+$
$C_2(b) = \{X \mid X \in \Re^+ \wedge \text{LEQ}(X, 10)\}$
$C_3(c) = \Re^+$
$C_4(e) = \{g1, g2, g3, g4\}$
$C_5(a, d, f) = \{\langle X, Y, Z\rangle \mid X \in \Re \wedge Y \in \mathcal{U} \wedge Z \in \Re \wedge \text{EQUALS}(Y, \text{TIMES}(X, Z))\}$
$C_6(e, f) = (\mathcal{U} \times (\mathcal{U} - \Re^+)) \cup$
$(\{g1\} \times \{X \mid \text{LEQ}(X, 200)\}) \cup$
$(\{g2\} \times \{X \mid \text{LEQ}(X, 250)\}) \cup$
$(\{g3\} \times \{X \mid \text{LEQ}(X, 350)\}) \cup$
$(\{g4\} \times \{X \mid \text{LEQ}(X, 390)\})$
$C_7(e, f) = (\mathcal{U} \times (\mathcal{U} - \Re^-)) \cup$
$(\{g1\} \times \{X \mid \text{LEQ}(230, X)\}) \cup$
$(\{g2\} \times \{X \mid \text{LEQ}(190, X)\}) \cup$
$(\{g3\} \times \{X \mid \text{LEQ}(240, X)\}) \cup$
$(\{g4\} \times \{X \mid \text{LEQ}(245, X)\})$
$C_8(e) = \{g2, g3, g4,\}$
$C_9(e) = \{g1, g2, g3,\}$
$C_{10}(a, b, c) = \{\langle X, Y, Z\rangle \mid X \in \mathcal{U} \wedge Y \in \Re \wedge Z \in \Re \wedge \text{EQUALS}(X, \text{TIMES}(X, Z))\}$

By augmenting the theory with arbitrary sentences from $\mathcal{L}$ and observing the consequent effect on the joint and marginal possibility distributions, we can solve various problems. If we assert the three equations $b = 10$, $c = 20$, and $d = 5000$, we can determine f by observing the marginal possibility distribution which is $\Pi_f = \{25\}$. If, instead, we assert the equation $a = 10$ and the inequality $2600 \leq d$, we can observe that $\Pi_f = \{X \mid 260 \leq X \leq 350\}$, that $\Pi_d = \{X \mid 2600 \leq X \leq 3500\}$ and that $\Pi_e = \{g3\}$.

8.5.5 Consistency algorithms and semantic modeling

Although most consistency algorithms which have been published [15, 21, 24] all address only networks with finite domains and binary or unary constraints, the notions of node, arc and path consistency are not restricted to such networks [12]. They also apply to infinite-domain networks containing constraints of arbitrary arity.

Thus, the published consistency algorithms can be generalized to constraint networks corresponding to arbitrary Class 1 modeling problems. Consider the published arc consistency algorithms. These operate by storing the possibility distribution for each parameter extensionally, by treating each binary constraint as a pair of arcs and by eliminating from the possibility distribution for the parameter at the head of an arc any value that is not supported by some value in the possibility distribution for the parameter at the tail of the arc [21].

These algorithms can be generalized to constraints of arbitrary arity by treating each n-ary constraint as n hyper-arcs, each hyper-arc having one parameter at its head and a tuple containing the other $(n - 1)$ parameters at its tail. If any value in the possibility distribution for the parameter at the head of a hyper-arc is not supported by some (n – 1)-tuple of values for the other (n – 1) parameters at the tail of the hyper-arc, the value can be removed from the possibility distribution.

Algorithms for arc consistency can be generalized to infinite domains by storing infinite possibility distributions intensionally. Removing unsupported values from the intensionally-defined possibility distribution for a parameter can be done by making the intensional formula suitably more specific. Whenever it is recognized that the possibility distribution has been restricted to a finite set of suitably small cardinality, the representation can be converted to an extensional format, if necessary.

The published path consistency algorithms can be generalized in a similar fashion. These algorithms operate by storing extensionally the joint possibility distribution corresponding to a constraint and by eliminating from the extension any tuple that is not supported by all other paths between the parameters in the constraint. The algorithms can be generalized to the infinite sets of tuples that are admitted by constraints in infinite-domain networks, by storing infinite joint possibility distributions intensionally. Removing unsupported tuples from an intensionally-defined joint possibility distribution can be done by making the intensional formula suitably more specific.

Thus, treating Class 1 modeling problems as CSPs of the generalized types defined in section 8.5.3, and generalizing the published consistency algorithms appropriately, would appear to be an attractive computational companion to a representation employing the full FOPC. There is a problem, however, in that generalizing the consistency algorithms in this way means that the intensional formulae may become arbitrarily complex, which raises the specter of undecidability.[1] Even the satisfiability of formulae that are theoretically decidable may be beyond the competence of whatever inference algorithm we embed in our system.

The way to escape from this problem is to side-step it. No inference algorithm for the full FOPC can be sound, terminating and complete. Soundness is a *sine qua non* (unless unsound inferences are allowed in a principled way, by the introduction of non-monotonic common-sense assumptions) and, to be practical, an algorithm has to terminate. However, if we take the right attitude to possibility distributions, incompleteness need not be a disaster. This suggests, therefore, a framework for a research effort directed at supporting the full FOPC as a

[1] A problem is said to be undecidable if it is not possible even in theory to devise an algorithm for solving it. A class of problems is undecidable if it contains at least one specific problem instance which is undecidable. For example, while many specific satisfiability problems in FOPC have algorithmic solutions, not all do, and so satisfiability in FOPC is undecidable.

programming language for knowledge-based applications. Applications are cast as Class 1 modeling problems, with salient parameters being treated as constant symbols in an FOPC theory which represents the currently available knowledge about the application area. Users of an application program interact with the program by augmenting the theory embedded in the program with assertions about their particular problem instance and by interpreting the computed marginal possibility distributions appropriately. The results of research leading to improved inferential competence can be incorporated into the run-time system for the language over time without altering application programs, in a fashion analogous to the way in which improvements to the run-time speed of a language interpreter have no impact on the correctness of programs in the language.

8.6 GALILEO4 REVISITED

As we have seen, Galileo4 is a programming language highly applicable to supporting production. On a more formal plane, Galileo4 is firmly rooted in the notion of constraint networks and on the relationship, discussed above, between such networks and possibility distributions for semantic models of theories written in the full FOPC. The run-time system for the language is a test-bed for various approaches to computing the intent of sets of well-formed FOPC formulae and to generalizing the consistency algorithms.

8.6.1 Galileo4 programs

An application program in Galileo4 provides a declarative specification of a constraint network, analogous to the problem specifications in Example 1 and Example 2. That is, in general, an application program in Galileo4 specifies a first-order language $\mathcal{L} = \langle \mathcal{P}, \mathcal{F}, \mathcal{K} \rangle$, a theory Γ containing closed sentences from that language, a universe of discourse $\mathcal{U}$, and a partial interpretation function I_p for $\mathcal{L}$. Of these, only the theory Γ must always be specified explicitly, because, in many situations, a default universe and a default partial interpretation provided by the run-time system are adequate, while the system can determine the language $\mathcal{L}$ by computing the union of the constant symbols used in Γ with the vocabulary of constant symbols in the default language $\mathcal{L}_g$ $\langle \mathcal{P}_g, \mathcal{F}_g, \mathcal{K}_g \rangle$ provided by the run-time system. In $\mathcal{L}_g$, $\mathcal{K}_g = \mathcal{R}$ contains the real numeric strings, $\mathcal{P}_g$ contains names of standard predicates (=, =<, etc.) and $\mathcal{F}_g$, contains names of standard functions (*, +, etc). The run-time system also provides a default universe of discourse $\mathcal{U}_g = \mathfrak{R}$ and an interpretation function I_g for $\mathcal{L}_g$ giving the standard interpretations in terms of $\mathcal{U}_g$. For example, the constant symbols in $\mathcal{K}_g$ are mapped onto the intended finite-length rational numbers with interpretations such as 1 $\mapsto$ `1` and 2.1 $\mapsto$ `2.1`.

The language $\mathcal{L} = \langle \mathcal{P}, \mathcal{F}, \mathcal{K} \rangle$ defined by an application program in Galileo4 has the following components: $\mathcal{P} = \mathcal{P}_g \cup$ {predicate symbols defined in the program}; $\mathcal{F} = \mathcal{F}_g \cup$ {function symbols defined in the program}; $\mathcal{K} = \mathcal{K}_g \cup$ {constant symbols used in the program}. The universe of discourse $\mathcal{U}$ defined by a Galileo4 program is the union of $\mathcal{U}_g = \Re$ with any application-specific domains that are defined in the program. The partial interpretation function I_p defined by the program is the union of the set of mappings in I_g with any application-specific interpretations that are provided in the program.

8.6.2 The Galileo4 run-time system

The Galileo4 run-time system attempts to compute the marginal possibility distribution Π_x the projection of $\Pi_{\mathcal{U}, \mathbf{X}, \mathbf{C}}$ on the dimension x, for each parameter x in $\mathbf{X}$ in the network $\langle \mathcal{U}, \mathbf{X}, \mathbf{C} \rangle$ that is defined by a Galileo4 program. In general, it is only possible to compute $\tilde{\Pi}_x$, an approximation to Π_x such that $\Pi_x \subseteq \tilde{\Pi}_x$. However, in many networks the run-time system can compute Π_v for all x in $\mathbf{X}$.

Propagation algorithm

An important aspect of our ongoing research consists of the development of constraint propagation algorithms which compute better approximations to the marginal possibility distributions for parameters in infinite domain constraint networks. The propagation algorithm which is currently used to compute marginal possibility distributions is called **compound constraint propagation** (CCP). Its top-level is shown in Fig. 8.5.

Full details of the algorithm are beyond the scope of this chapter. However, as shown in Fig. 8.5, CCP involves interleaved application of three inference techniques: LPKS, local propagation of known states [20]; GAC, a version of arc consistency [21], generalized, along the lines suggested in section 8.5.5, to finite domains and constraints of arbitrary arity; GPC, a form of path consistency [21], generalized to infinite domains and constraints of arbitrary arity, which is only applied to small portions of the network in certain very specific circumstances.

The basic idea behind the CCP algorithm is to use a set of complementary inference techniques, with the computationally cheapest technique being used whenever possible. Thus, all constraints are initially placed in Q_{lp}, th local propagation queue, so that local propagation of known states (LPKS), a relatively inexpensive technique, is tried first. When LPKS is unable to draw any further inferences, the constraints at which LPKS reached a dead-end are placed on the generalized arc consistency queue Q_{gac}. If arc consistency results in restricting the domain of possible values

```
procedure CCP(Γ);
localvar Q_lp, Q_gac, Q_gpc;
begin
Q_lp ← Γ;
Q_gac ← ∅;
Q_gpc ← ∅;
repeat
   LPKS (Q_lp, Q_gac, Q_gpc);
   if Q_gac ≠ ∅
         then repeat
            GAC  (Q_lp, Q_gac, Q_gpc);
            if  Q_lp = ∅ ∧ Q_gpc ≠ ∅
                  then  GPC  (Q_gac, Q_gpc)
            until Q_lp ≠ ∅ ∨ Q_gac = ∅
until  Q_lp = ∅
end
```

Figure 8.5 Top level of the CCP algorithm.

for any parameter to a singleton set, then all constraints which reference this parameter are placed by GAC into Q_{lp}, GAC terminates and LPKS is re-invoked. If arc consistency fails to draw any conclusions, then GPC is invoked at those constraints where arc-consistency reached a dead-end. However, the GPC procedure only applies path consistency if certain conditions are satisfied, the details of which are beyond the scope of this chapter. Whenever path consistency succeeds in making any constraint more specific, arc consistency is re-invoked at the arcs involved in this constraint.

The various techniques are used to overcome each other's deficiencies. Thus, for example, arc consistency can push inference through inequality constraints, which is beyond the competence of local propagation of known states. Similarly, path consistency is used to overcome what is probably the best-known deficiency of local propagation of known states: its inability to penetrate so-called 'constraint cycles' [20]. Path consistency is also used to overcome an important deficiency of arc consistency: its tendency to fall into infinite loops when processing parameters which range over (infinite subsets of) $\mathfrak{R}$ [9].

Within GAC and GPC, conditional reductions [1] are used for manipulating the intensional formulae that arise during the application of arc and path consistency criteria to the sets which represent marginal and joint possibility distributions. A reduction is a rewrite rule which changes all subformulae of a certain form into a target form. A conditional reduction is a reduction with an associated guard condition; the reduction is applicable only if the guard condition is true. A reduction is **semantic**, in that its use is based not upon the syntactic structure of the form to be rewritten but upon its meaning; a guard condition is also semantic, in that it is decided by applying the model-theoretic rules for truth to the interpretations of the symbols which it references, rather than by the application of syntactic proof theory. Consider, for example, the following conditional reduction:

$$X < A \wedge X > B \xrightarrow{A \leq B} \bot$$

This can be applied to the intensional formula in the possibility distribution.

$$\{X \mid X < 5 \wedge X > 6\}$$

because model theory indicates that the guard condition $5 \leq 6$ is true; as a result of applying the reduction, it is recognized that the possibility distribution is the empty set.

It is a subject of ongoing research to determine just what is the most appropriate collection of reductions for use in GAC and GPC. Nonetheless, the sets currently in use can be proved to allow no infinite reductions. This means that, although GAC and GPC cannot be guaranteed to find the minimal representation of an intensional specification, they are guaranteed to terminate.

Monotonicity, soundness, and completeness

The CCP algorithm is a monotonic filter. It operates by removing from the marginal possibility distribution for each parameter in the network those values which it is able to prove cannot appear in any tuple of the true intent $\Pi_{u,\mathbf{X},\mathbf{C}}$ of the network. The algorithm does not treat negation as failure or make any other non-monotonic assumptions. Thus, a candidate value v or a parameter x in $\mathbf{X}$ is removed from $\tilde{\Pi}_x$ only when it can be proven that $v \notin \Pi_x$.

The CCP algorithm is sound: it can be shown, although the proof is beyond the scope of this chapter, that Π_x the true intensional marginal possibility distribution for a parameter x is guaranteed to be contained within the distribution $\tilde{\Pi}_x$ which the algorithm computes for x. If CCP were complete, the true intensional distribution for a parameter would be exactly that computed; that is, $\tilde{\Pi}_x = \Pi_x$, for all x in $\mathbf{X}$. But CCP is not guaranteed to be complete, so the intensional distribution for a parameter may sometimes be a proper subset of the computed distribution; that is, sometimes, for some x in $\mathbf{X}$, $\pi_x \subset \tilde{\Pi}_x$.

By default, CCP is the only algorithm which is applied to a constraint network. However, the user may also, at his discretion, specify that backtracking search should be employed. Thus, although the proof is beyond the scope of this chapter, it can be shown that, in problems where each x in $\mathbf{X}$ is subject to at least one positive literal whose predicate symbol has a finite interpretation, the Galileo4 run-time system can guarantee to compute not just $\tilde{\Pi}_x$. but Π_x for all x, albeit with expontential computational cost.

User interaction

Galileo4 programs are interactive. The user is allowed to augment the theory defined in a program with additional sentences and, whenever he does so, the Galileo4 run-time system uses CCP to compute revised possibility distributions.

The system maintains a set of dependency records and, at any time, the user can request a justification for the current computed marginal possibility distribution $\tilde{\Pi}_x$ of any parameter x in the network.

Whenever, as a result of a sequence of user assertions, the system computes that $\tilde{\Pi}_x = \varnothing$ for any x, the system reports a violation, since $\tilde{\Pi}_x = \varnothing$ implies $\Pi_x = \varnothing$, which implies that $\Pi_{\mathcal{U},\mathbf{X},\mathbf{C}} = \varnothing$.

If this happens, the user can recover from the situation by retracting one of his earlier assertions. This ability to unpeel the consequences of a previous assertion is enabled by the same dependency records which enable the provision of justifications. By making assertions and retractions in any order he pleases, the user can explore various 'what if' scenarios.

Backward- and forward-chaining

The system supports both forward- and backward-chaining reasoning. The user can volunteer assertions to augment a theory, in which case the system will use CCP to forward-chain on the consequences of each assertion. Alternatively, the user can ask the system to determine the value of some parameter in the network, in which case the system will backward-chain through the sentences that correspond to the constraints in the network, asking the user for information about other parameters in the network.

During backward-chaining, whenever the system asks for information about a parameter, the user can input any well-formed assertion. At its simplest, this could be a ground equation giving the exact value for the parameter; e.g. the assertion $a = 2$ enables the system to infer the interpretation $a \mapsto 2$, since I_g contains both the interpretation $2 \mapsto 2$, and the standard interpretation for equality. Alternatively, the user's assertion could be any well-formed sentence (atomic, compound or quantified) in the first-order language $\mathcal{L}$. In particular, since the equality predicate need not be used; the user can provide whatever information he has about the parameter, even if he does not know the parameter's value.

The fact that inference in Galileo4 is based on the computation of possibility distributions for parameters means that the system's inferential competence degrades gracefully whenever insufficient information is obtained during backward-chaining. Thus, if the user invokes backward-chaining by asking the system to determine the value of some parameter x, but he is then unable to specify values for other parameters related to x, the system may not be able to infer a singleton value for x; however, the system may still be able to compute a more restricted possibility distribution for x from whatever lower-grade information the user is able to provide about the other parameters.

Mixed-initiative interaction

The ability to accept any well-formed sentence during backward-chaining has an important added benefit: it supports mixed-initiative interaction. Whenever the system, during backward-chaining, asks for information about one parameter, the user can, if he sees fit, input an assertion that does not even reference the parameter which the system asked about. Since the system forward-chains on the consequences of each assertion before asking further questions, an experienced user who follows the system's line of reasoning can short-circuit the process by providing information that the system has not asked for but which the user knows is relevant.

The user can determine the system's line of reasoning in the usual way, by asking the system to justify itself whenever it asks a question during backward-chaining. To provide a justification, the system merely outputs the sentences which link the parameter about which information is being sought to the parameter whose value the user asked the system to determine.

8.6.3 Simple example programs

In this section we provide a practical application program written in Galileo4. Initially, however, we build on the discussion in section 8.5.4 by providing programs that correspond to the abstract problems discussed there.

Program 1

The Galileo4 program corresponding to Example 1 is shown in Fig. 8.6.

The last three statements in this program define the theory given in Example 1. The first three statements in the program, in conjunction with the set of constant symbols used in the theory, define the first-order language $\mathcal{L} = \langle \mathcal{P}, \mathcal{F}, \mathcal{K} \rangle$, whose components are as follows: $\mathcal{P} = \mathcal{P}_g \cup \{s, t\}$; $\mathcal{F} =$

```
relation s =::= {5.7}.
relation t =::= {(1,4),(2,5),(3,6),(4,7),(5,1),
                  (5,3),(6,2),(6,3),(6,7),(7,1),(7,3)}.
constant b =::= 7.
t(a,c).
all X : s(X) implies t(X,a).
t(c,b).
```

Figure 8.6 Galileo4 program 1.

$\mathcal{F}_g$; $\mathcal{K} = \mathcal{K}_g \cup \{a, b, c\}$. Since the program defines no application-specific domains, the universe of discourse is just the default; that is $\mathcal{U} = \mathcal{U}_g = \Re$. As well as contributing to the definition of $\mathcal{L}$, the first three statements in the program also specify that the partial interpretation function I_p is the union of the default partial interpretation I_g with the set of mappings for s, t and b given in the problem statement of Example 1.[2]

If the Galileo4 run-time system were given this program, it would report that $\tilde{\Pi}_a$ was {1, 3} and that $\tilde{\Pi}_c$ was {4, 6}. As in Example 1, if the user were to assert $t(b, c)$ when interacting with this program, the run-time system would report a constraint violation, since now $\tilde{\Pi}_c$ would be $\varnothing$. If the user then retracted $t(b, c)$ and replaced it by $t(c, a)$, the system would compute that $\tilde{\Pi}_a$ was {3} and that $\tilde{\Pi}_c$ was {6} and would report that $a \mapsto 3$, and $c \mapsto 6$.

If instead of volunteering assertions, the user were to invoke backward-chaining by asking the system to determine the value of a, the system would respond by asking the user for the value of c. If, in response, the user were to assert that $c = 4$, the system would infer that $c \mapsto 4$ and $a \mapsto 1$; if, however, the user were to assert that $c > 5$, the system would infer that $c \mapsto 6$ and $a \mapsto 3$.

Program 2

A Galileo4 program corresponding to Example 2 is shown in Fig. 8.7.

The first statement in the program defines an application-specific

[2] Actually, because of the default language $\mathcal{L}_g$ defined by the Galileo4 run-time system, there are two differences between the situation defined by this program and that defined in Example 1. First, the language $\mathcal{L}$ defined by this program is a superset of the language in Example 1; second, the partial interpretation function defined by this program is a superset of the function defined in Example 1. Despite these differences, however, the constraint networks of Example 1 and Program 1 are identical, for three reasons: first, the theories are identical; second, these theories contain the same set of uninterpreted constant symbols; third, the partial interpretation functions provide the same mappings for all symbols in the theory that they do interpret.

domain. The universe discourse $\mathcal{U}$ is the union of $\mathcal{U}_g$ with all application-specific domains, so

$$\mathcal{U} = \Re \cup \{\texttt{g1,g2,g3,g4}\}$$

As well as introducing an entity to the universe of discourse, each member of an application-specific domain also introduces a constant symbol to the first-order language and introduces to the partial interpretation a mapping from the constant symbol to the entity. Thus, as well as introducing the four entities `g1,g2,g3` and `g4` to $\mathcal{U}$, the first statement in this program also introduces the four symbols *g*1, *g*2, *g*3 and *g*4 to the constant symbol vocabulary of $\mathcal{L}$ and the four mappings $g1 \mapsto \texttt{g1}$, $g2 \mapsto \texttt{g2}$, $g3 \mapsto \texttt{g3}$, and $g4 \mapsto \texttt{g4}$ to I_p.

A domain declaration also introduces the domain name into the first-order language, as a predicate symbol, and specifies that the set of domain contents constitute the interpretation for this symbol. Thus, the first statement in this program also introduces the following mapping to I_p.

$$g \mapsto \{\texttt{g1,g2,g3,g4}\}.$$

The next five statements define application-specific function and predicate symbols; both extensional and intensional specifications are used.

```
domain   g =::= {g1,g2,g3,g4}.
relation j -::= {X : positive (X)}.
relation k =::= {X : negative (X)}.
relation h =::= {X : 0 < X = < 10}.
function l(g) - > positive
                  =::= {(g1,200),(g2,250),(g3,350),(g4,390)}.
function m =::= {(g1,230),(g2,190),(g3,240),(g4,245)}.
positive (a).
h(b).
positive(c).
g(e).
d = a * f.
j(f) implies f = <l(e).
k(f) implies f = < m(e).
all X : g(X) implies m(X) = <l(e).
exists X : l(X) >l(e).
a = b * c.
```

Figure 8.7 Galileo4 program 2.

The last ten statements specify the following theory:

$$
\begin{aligned}
\Gamma \mapsto \{ \quad & positive\ (a), h\ (b), positive\ (c), g\ (e), \\
& d = a * f, j\ (f) \Rightarrow f \leq l\ (e), \\
& k\ (f) \Rightarrow f \leq\ m\ (e), \\
& (\forall X)(g(X)\ \Rightarrow m(X) \leq\ l\ (e)), \\
& (\exists X)(g(X)\ \wedge l(X) >\ l\ (e)), \\
& a = b * c \qquad\qquad\qquad\qquad \}.
\end{aligned}
$$

Note that, as shown in the definition for the function l, definitions for predicate and function symbols optionally can include type information about the arguments (and, for functions, about the value returned). This type information facilitates processing the theory specified by a program. Thus, for example, the type information that the function l takes arguments of type g enables the system, when implementing the statement

exists X : 1(X) > 1(e)

to introduce a relativizing predicate:

$$(\exists X)(g(X)\ \wedge l(X) >\ l\ (e)).$$

This relativizing predicate enables the system to treat the quantified variable as ranging over the finite set {g1, g2, g3, g4}, instead of over the infinite set $\mathcal{U}$.

As a consequence of the domain, function and predicate symbol definitions, and of the constant symbols used in the theory, the language $\mathcal{L}$ and the partial interpretation I_p defined by this program are as follows:

$$
\begin{aligned}
\mathcal{L} = \ & \langle \quad \mathcal{P}_g \cup \{g, j, k, h\}, \mathcal{F}_g \cup \{l, m\}\ \mathcal{K}_g \cup \{a, b, c, d, e, f, g1, g2, g3, g4\} \quad \rangle \\
I_p = \ & I_g \cup \ \{ \quad g \mapsto \{\mathtt{g1,g2,g3,g4}\}, j \mapsto \Re^+, \\
& \qquad k \mapsto \Re^-, h \mapsto \{X \mid X \in \Re^+ \wedge \text{LEQ}\ (X, 10)\}, \\
& \qquad l \mapsto \{\langle \mathtt{g1,200}\rangle, \langle \mathtt{g2,250}\rangle, \langle \mathtt{g3,350}\rangle, \langle \mathtt{g4,390}\rangle\}, \\
& \qquad m \mapsto \{\langle \mathtt{g1,230}\rangle, \langle \mathtt{g2,190}\rangle, \langle \mathtt{g3,240}\rangle, \langle \mathtt{g1,245}\rangle\}, \\
& \qquad g1 \mapsto \mathtt{g1},\ g2 \mapsto \mathtt{g2},\ g3 \mapsto \mathtt{g3},\ g4 \mapsto \mathtt{g4} \qquad \}.
\end{aligned}
$$

As with Program 1, a user interacting with this program can solve a variety of problems by making various assertions and observing the consequent marginal possibility distributions, or by invoking backward-chaining to determine the value or restricted possibility distribution for some parameter.

8.7 HUMAN–MACHINE INTERACTION IN GALILEO4

Human–machine interaction in Galileo4 can be described in terms of the view of machine-interlocutor model in Fig. 8.8. In this view, the universe of discourse (set of all possible values for all parameters) defined in a Galileo4 application program constitutes the World. User inputs to the machine contain descriptions about the state of the World, expressed as sentences in the first-order language $\mathcal{L}$ which is defined in the Galileo4 application.

In our constraint-based approach to logic, the machine state M at any time is characterized by a set of possible conceptualizations of the World. A conceptualization of the World comprises the universe of discourse $\mathcal{U}$ and those interrelationships between the entities in $\mathcal{U}$ that the machine finds interesting. There may be many other interrelationships in which the machine is uninterested. Indeed, since, as a first-order language, $\mathcal{L}$ has finite predicate and function vocabularies, this has to be the case when $\mathcal{U}$ is infinite: the number of possible q-ary relations between the entities of $\mathcal{U}$ is $2^{|\mathcal{U}|^q}$.

If the machine has a total model $\mathcal{M}$, $= \langle \mathcal{U}, I \rangle$ for the language $\mathcal{L}$, (that is, if I provides a mapping for each symbol in the union of the symbol vocabularies in $\mathcal{L}$) then the machine knows everything worth knowing about (that is, it has a total conceptualization of) the World.

Giving inputs to the machine about the state of the World has a purpose only if the machine does not know all the interrelationships that exist between the entities of $\mathcal{U}$, that is, if the machine does not have a total model of the language $\mathcal{L}$ or if its model is not an accurate depiction of the state of the World.

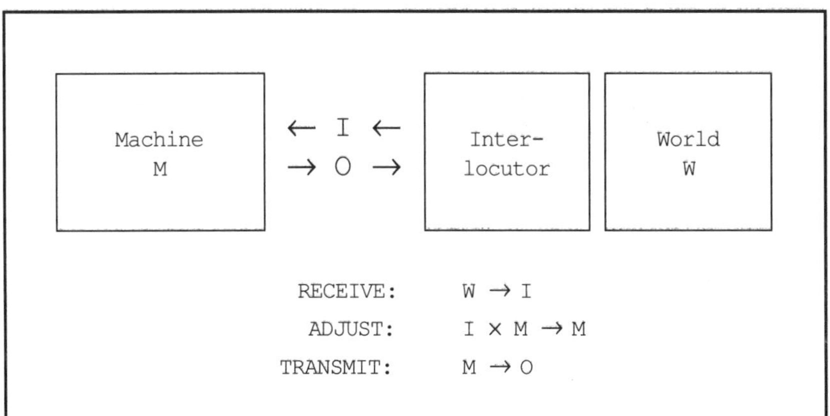

Figure 8.8 Machine–interlocutor interaction.

When the machine does not have a total model for $\mathcal{L}$, the state of the machine (the set of possible conceptualizations of the World) is the set of possible models for $\mathcal{L}$. That is, if at some time t the machine's partial model of $\mathcal{L}$ is $\mathcal{M} = \langle \mathcal{U}, I_t \rangle$ then the machine state M_t is the set of all total models $\mathcal{M}_t = \langle \mathcal{U}, I \rangle$ for $\mathcal{L}$, $I_t \subset I$, which are consistent with the information about the state of the World that the machine has been given. Thus, if at time t the information that the machine has been given about the state of the World is a set Γ_t of sentences, then the state M_t of the machine is the set of all total models $\langle \mathcal{U}, I \rangle$, $I_t \subset I$, which are consistent with the truth of all sentences in Γ_t.

Since the state of the World, or the interlocutor's perception of it, may change, the information about the World which the machine is given may not be monotonic. One input to the machine may say that a certain sentence γ of $\mathcal{L}$ should be taken as true and a later input may say that γ should no longer be taken as true. This means that the inputs to the machine are not sentences of $\mathcal{L}$; rather, they are **about** sentences of $\mathcal{L}$. The two inputs just mentioned could not be just γ and $\neg\gamma$; that would lead to a contradiction in the theory that the machine has about the World. Rather, the two inputs must be meta-level sentences, which might be expressed imperatively as ASSERT γ and RETRACT γ or declaratively as $\Gamma_{t+1} = \Gamma_t \cup \{\gamma\}$ and $\Gamma_{t+2} = \Gamma_{t+1} - \{\gamma\} = \Gamma_t$.

This form of machine-interlocutor interaction is what happens when a user interacts with a Galileo4 application program. The program defines the language $\mathcal{L}$, the universe of discourse $\mathcal{U}$ and an initial partial interpretation I_p. The program also primes the machine with an initial set Γ of $\mathcal{L}$ sentences about the World.

8.8 SUMMARY

In this chapter, we have introduced a generalized notion of constraint processing. We have shown that this generalized notion corresponds to a class of modeling problem in logic. We have then used this correspondence as the theoretical basis for a computational approach to supporting the full first-order predicate calculus as a knowledge representation language. In this approach, constraint propagation is used to treat inference as semantic entailment rather than syntactic theorem-proving.

We have introduced Galileo4, a new constraint programming language based on these ideas and shown how short constraint-based application programs can have high problem-solving versatility.

We have discussed a new algorithm for constraint propagation which generalizes and integrates several of the best existing techniques. The algorithm interleaves local propagation of known states with forms of arc and path consistency which have been generalized to infinite domains and constraints of arbitrary arity. In the case of infinite domains, semantic reductions are used for entailment. However, we regard this algorithm as only the first in a sequence of progressively more competent and efficient algorithms. We regard Galileo4 as a test-bed for further developments in this area. Although no terminating constraint propagation algorithm can ever be both sound and complete for the full first-order predicate calculus, an incomplete algorithm is adequate for some applications, as we have illustrated by means of examples from the realms of design for testability and equipment selection.

We have shown how the man–machine interface presented by Galileo4 is a special case of Nilsson's view of machine–world interaction. We believe that the lexical economy offered by the full first-order predicate calculus and the problem-solving versatility offered by a constraint-based approach to inference mean that the generalized form of constraint-based programming that we have introduced in this chapter is destined to become the declarative paradigm of choice.

However, this constraint-based approach to supporting the full first-order predicate calculus is still relatively immature and much remains to be done. Notable open problems include: the development of constraint propagation algorithms which are more efficient and which compute better approximations to the true intent of infinite domain constraint networks; extending the object-level language to include modal operators; supporting application-specific sets of reductions and the use of results such as the Knuth–Bendix theorem [18] to compute complete reduction sets from the application-specific sets proposed by application programmers.

Several other related issues also present themselves. One involves the extension of the constraint-based approach to problems having non-scalar universes of discourse, where constant symbols can be interpreted to sets or sequences; work here might include looking at issues of non-monotonic inference. Other work might be directed at using second-order and meta-level notions to ease the programming of large-scale applications. The specification of partial interpretations can be eased by allowing the interpretations of predicate symbols to be specified as the reflexive, symmetric or transitive closure of interpretations for other symbols. The specification of a theory in a program, or of augmentations to the theory by a user, can be eased by using meta-level notions to allow these theories and augmentations to be specified intensionally rather than extensionally; in addition, using such meta-level notions in a constraint-based approach to logic makes it possible to treat (the declarative aspects of) frame-based programming as a special case of theory specification.

REFERENCES

[1] Bledsoe, W. (1977) Non-resolution Theorem Proving. *Artificial Intelligence*, **9**, 1–35.
[2] Bowen, J. and Bahler, D. (1991) *Conditional Existence of Variables in Generalized Constraint Networks*. Proc. 9th. Nat. Conf. on Artificial Intelligence (AAAI-91), pp. 215–20.
[3] Bowen, J. and Bahler, D. (1991) Supporting cooperation between multiple perspectives in a constraint-based approach to concurrent engineering. *Journal of Design & Manufacturing*, **1**, 89–105.
[4] Bowen, J. and Bahler, D. (1992) Frames, quantification, perspectives and negotiation in constaint networks for life-cycle engineering. *Int. Journal of AI in Engineering*, **7**, 199–226.
[5] Bowen, J. and Bahler, D. (1993) Constraint-based software for concurrent engineering. *IEEE Computer*, Special Issue on Computer Support for Concurrent Engineering, **26** (1), 66–8.
[6] Borning, A. (1981) The programming language aspects of ThingLab, a constraint-oriented simulation laboratory. *ACM Trans. on Prog. Lang. and Sys.*, **3** (4), 353–87.
[7] Chronaki, C. and Baldwin, D. (1990) A Front End for CONSUL. Technical Report 319, Department of Computer Science, University of Rochester.
[8] Colmerauer, A. (1990) An introduction to Prolog III. *Commun. ACM*, **28** (4), 412–18.
[9] Davis, E. (1987) Constraint propagation with interval labels. *Artificial Intelligence*, **32**, 281–331.
[10] Dechter, R. (1990) Enhancement schemes for constraint processing: Backjumping, learning, and cutset decomposition. *Artificial Intelligence*, **41**, 273–312.
[11] Dincbas, M., van Hentenryck, P., Simonis, H., Aggoun, A., Graf, T. and Berthier, F. (1988) *The Constraint Logic Programming Language CHIP*. Proc. Fifth Gen. Comp. Sys. '88.
[12] Faltings, B. (1994) Arc-consistency for continuous variables. *Artificial Intelligence*, **65** (2), 363–76.
[13] Freuder, E. (1982) A sufficient condition for backtrack-free search. *Journal of the ACM*, **29** (1), 24–32.
[14] Gosling, J. (1983) Algebraic Constraints. Technical Report CS-83–132, Carnegie-Mellon University, Pittsburgh.
[15] Han, C.-C. and Lee, C.-H. (1988) Comments on Mohr and Henderson's path consistency algorithm. *Artificial Intelligence*, **36**, 125–30.
[16] Haralick, R. and Elliot, G. (198) Increasing tree search efficiency for constraint satisfaction problems. *Artificial Intelligence*, **14**, 263–313.
[17] Jaffar, J., Michaylov, S., Stuckey, P., and Yap, R. (1992) The CLP($\mathfrak{R}$) Language and System. *ACM Trans. on Prog. Lang. and Systems*, **14** (3), 339–95.
[18] Knuth, D. and Bendix, P. (1970) Simple word problems in universal algebras, in *Computational Problems in Abstract Algebra*, (ed. J. Leech), Pergamon Press, New York.
[19] Lambert, K. and van Fraassen, B. (1972) *Derivation and Counterexample: An Introduction to Philosophical Logic*, Dickenson Publishing Company, Enrico, CA.
[20] Leler, W. (1988) *Constraint Programming Languages*, Addison-Wesley, Reading, MA.
[21] Mackworth, A. (1977) Consistency in networks of relations. *Artificial Intelligence*, **8**, 99–118.

[22] Maher, M. and Stuckey, P. (1989) *Expanding Query Power in Constraint Programming Languages*. Proc. North American Conf. on Logic Programming, pp. 20–36.
[23] Mittal, S. and Falkenhainer, B. (1990) *Dynamic Constraint Satisfaction Problems*. Proc. 8th. Nat. Conf. on Artifical Intelligence (AAAI-90), pp. 25–32.
[24] Mohr, R. and Henderson, T. (1986) Arc and path consistency revisited. *Artificial Intelligence*, **28**, 225–33.
[25] Nadel, B. (1989) Constraint satisfaction algorithms. *Computational Intelligence*, **5** (4), 188–224.
[26] Nilsson, N. (1991) Logic and artificial intelligence. *Artificial Intelligence*, **47**, 31–56.
[27] Paramasivam, M. (1991) An Axiom Based Approach to Constraint Satisfaction. MS Thesis, Computer Science Department, North Carolina State University, Raleigh.
[28] Sussman, G. and Steele, G. (1980) Constraints: A language for expressing almost-hierarchical descriptions. *Artificial Intelligence*, **14**, 1–39.
[29] Sutherland, I. (1963) *SKETCHPAD: A Man–Machine Graphical Communication System*. Proc. Spring Joint Comp. Conf., IFIPS.

APPENDIX: PROOF OF THEOREM 1

The proof of this theorem depends on the following lemma:

Lemma

Given $\mathcal{L} = \langle \mathcal{P}, \mathcal{F}, \mathcal{K}' \cup \mathcal{K}'' \rangle$, $\mathcal{U}, I_p$ and γ, where I_p interprets, in terms of $\mathcal{U}$, all and only the symbols in $\mathcal{P} \cup \mathcal{F} \cup \mathcal{K}'$, $\mathcal{K}''$ is finite, and the sentence γ references at least one symbol in $\mathcal{K}''$. Let $\mathbf{X}$ be the lexical ordering of the elements of $\mathcal{K}''$ and let $C(T)$ be a constraint such that T is the lexical ordering of those elements of $\mathcal{K}''$ which appear in γ and such that $\langle \mathcal{U}, I_p, \cup T\vec{t} \rangle \models \gamma$ for all tuples $t \in C(T)$. Then $\langle \mathcal{U}, I_p, \cup \mathbf{X}\vec{r} \rangle \models \gamma$ for all tuples $\tau \in E(\mathbf{X})$ where $E(\mathbf{X})$ is the cylindrical extension of $C(T)$ into the space defined by $\mathbf{X}$.

Proof

Take any tuple $\tau \in E(\mathbf{X})$. The projection of the singleton set $\{\tau\}$ on T will be a singleton set containing some tuple τ'; that is, $proj(\{\tau\}, T) = \{\tau'\}$. Then, by definition of cylindrical extension $\tau' \in C(T)$. The mapping function $\mathbf{X}\vec{r}$ provides a mapping for each constant symbol in $\mathcal{K}''$. However, the truth of γ depends only on the interpretation of those members of $\mathcal{K}''$ which appear in γ. Thus, the truth of γ depends only on the mappings provided by $T\vec{r}'$, which is a subset of $\mathbf{X}\vec{r}$. Since $\tau' \in C(T)$, $\langle \mathcal{U}, I_p, \cup T\vec{r}' \rangle, \models \gamma$. Since $\langle \mathcal{U}, I_p, \cup T\vec{r}' \rangle, \models \gamma$, $\langle \mathcal{U}, I_p, \cup \mathbf{X}\vec{r} \rangle, \models \gamma$, because the additional mappings in $(\mathbf{X}\vec{r} - T\vec{r}')$ have no impact on the truth of γ. Since τ is any tuple in $E(\mathbf{X})$, $\langle \mathcal{U}, I_p, \cup \mathbf{X}\vec{r} \rangle, \models \gamma$ for any tuple $\tau \in E(\mathbf{X})$.

Q.E.D.

Proof of Theorem 1

Since $\langle \mathcal{U}, I_P, \cup T_1\vec{t}\rangle, \vDash \gamma_1$ for all tuples $t \in C_1(T_1)$ then, by the lemma, $\langle \mathcal{U}, I_P, \cup \mathbf{X}\vec{r}\rangle \vDash \gamma_1$ for all tuples $\tau \in E_1(\mathbf{X})$, the cylindrical extension of $C_1(T_1)$. Similarly, $\langle \mathcal{U}, I_P, \cup \mathbf{X}\vec{r}\rangle \vDash \gamma_2$ for all tuples $\tau \in E_2(\mathbf{X})$, the cylindrical extension of $C_2(T_2)$. Therefore, $\langle \mathcal{U}, I_P, \cup \mathbf{X}\vec{r}\rangle \vDash \{\gamma_1, \gamma_2\}$ for all tuples $\tau \in E_1(\mathbf{X}) \cap E_2(\mathbf{X})$. Similarly, $\langle \mathcal{U}, I_P, \cup \mathbf{X}\vec{r}\rangle \vDash \{\gamma_1, \gamma_2, ..., \gamma_r\}$ for all tuples $\tau \in E_1(\mathbf{X}) \cap E_2(\mathbf{X}) \cap ... \cap E_r(\mathbf{X})$. That is, $\langle \mathcal{U}, I_P, \cup \mathbf{X}\vec{r}\rangle \vDash \Gamma \in \Pi_{u,\mathbf{X},\mathbf{C}}$.

Q.E.D.

CHAPTER 9

Scheduling methodology: optimization and compu-search approaches I

M.-C. Portmann

ABSTRACT

This chapter focuses on the scheduling problem level of the hierarchical approach of production management. It proposes, on one hand, a scheduling problem classification and, on the other hand, various scheduling methods of solution. After a presentation of the most classical scheduling problems, the great variety of solving schemes is considered: constructive methods, neighborhood methods, splitting-up methods, model changing methods, AI and mixed methods. Neighborhood methods in particular are developed at the end of the chapter.

9.1 INTRODUCTION

Scheduling problems arise in many fields of human activities: products in a workshop, activities of a project, classes at a university, deliveries by a number of vehicles, flights of a fleet of airplanes, applications in a computer system, patients in an emergency room, and so forth. **Scheduling** implies the definition of the starting and finish times and the assignment of resources to each task of the given set, while restricted by several constraints which may involve the tasks and/or the resources. When attention is focused on the **order** of the tasks rather than their (absolute) times of operation, then the problem is called a sequencing problem.

This chapter will focus on scheduling problems that arise in the production/manufacturing context. The first step in solving a problem is to model it. Then the model must be analyzed and a solution schema must be chosen. The scheduling problems concerned are generally NP-hard, that is, it is probably impossible to secure optimal solutions using fast algorithms (i.e. algorithms that run in polynomial time in the 'size' of the problem), though some problems are in fact 'easy', in the sense that they are solvable to optimality by fast algorithms. Generally, for NP-hard problems, an approximation schema is constructed for the whole problem. It uses polynomial algorithms for its sub-problems, and establishes specific dominance properties in order to reduce the solution set. When building a specific approximation schema, several methods and techniques may be used together such as hierarchical and iterative approaches, exact polynomial methods for solving particular sub-problems, relaxation methods, constructive methods, local search improvement techniques and genetic algorithms.

In section 9.2 (which is completed by the appendix), we illustrate the diversity of manufacturing scheduling problems (concrete and theoretical) and present two classifications. In section 9.3 we focus on the modelling task and on solving potential schemes (optimal or otherwise). Some local search improvement methods are described in section 9.4. Chapter 10 (on compu-search methodologies II) is devoted to two specific approaches: genetic algorithms and artificial neural network approaches.

9.2 CLASSIFICATION OF SCHEDULING PROBLEMS

9.2.1 Problem contexts and hypotheses

We consider the so-called 'short term scheduling' problems in a production system. The objects to be scheduled are individual products. A product can be a single unit or a batch of units to be processed, and generally transported, together. It can therefore be considered in the model as 'a product'. In order to manufacture a given product, we need what we will call 'machines: a machine can be an actual machine, or a more complicated resource which is modeled as a single resource required by the product to perform a given operation. To obtain a finished product, a set of operations must be performed on one or several units (when you have to assemble several units to make the product). In manufacturing scheduling problems, pre-emption is generally forbidden. That is, it is prohibited to interrupt the operation on a product and move the residual processing to a different time interval.

The problem is defined as follows:

- the finished product set to be produced and, eventually, either the dates by which they have to be imperatively finished (**deadlines**) or the target dates at which they are supposed to be completed (the

product is 'early' if it is completed before the target and 'late' if it is completed after the target date) with possible associated penalty cost for both earliness and lateness (**due dates**);
- the date of availability of the components used in the production process (release date);
- the set of 'routes' which can be used to obtain each finished product with (i) the corresponding processing time (including set-up times if necessary), (ii) the corresponding machine used for each operation in each route (with possibility of multiple choice for the machines), and (iii) the set of precedence constraints in each route;
- the time intervals when a given machine is available;
- the measure of quality of the completed schedule. Generally this is a function of the completion time of each finished product but it is possible also to integrate a function of the completion time of each operation.

Some of the data of these general problems can be simplified. For example:

(c1) all components are available at time 0,
(c2) the machines are always available throughout the scheduling period,
(c3) there exists only one route for each product defined by a set of successive operations,
(c4) each operation of any route is assigned to only one machine,
(c5) the successive operations of every route visit the machines in the same order: machine 1, machine 2, . . ., machine m,
(c6) the set of precedence constraints is empty except the precedence constraints between successive operations included in the routes,
(c7) there exists only one machine (or at least only one machine which needs to be scheduled first because the other machines are not bottleneck resources).

Depending on the set of hypotheses of a given problem, we can classify the more classical scheduling problems (see section 9.2.1) which have been extensively studied in the literature (see, for instance, GOThA (1993); Lawler *et al.*, (1992)).

We may divide scheduling problems into two subsets depending on the workshop structure:

- basic manufacturing problems (section 9.2.2),
- non-basic manufacturing problems and other problems (section 9.2.3).

We may also divide the problems into different subsets depending on their theoretical difficulty (section 9.2.4 and appendix).

9.2.2 Basic manufacturing problems

In all basic problems, conditions (c1) to (c4) are present. The set of classical workshop problems is traditionally divided into three subsets:

- one machine scheduling problems,
- parallel machine scheduling problems, and
- 'shop' scheduling problems (flow-, job- and open-shop).

In the first subset, there is only one machine and the routes consist of only one operation performed on this machine. In the second subset, the routes consist of only one operation, but there are several machines which are able to perform this operation with processing times which may be:

- independent of the machine (**identical** parallel machines);
- inversely proportional to the speed of the machine (**uniform** parallel machines);
- dependent on the product and of the machine (**unrelated** parallel machines).

In the third set, the routes consist of a set of operations to manufacture each product, there is only one route per product (it is here that traditionally a 'product' is called a 'job'); and the order of the machines used by the operations may be:

- the same for the whole set of jobs (**flow-shop** scheduling problem);
- given and specific for each job (**job-shop** scheduling problem);
- without any precedence constraints (**open-shop** scheduling problem).

9.2.3 Non-basic manufacturing problems

Non-basic manufacturing problems use other combinations of the list of hypotheses listed in section 1. For example:

- there are parallel machines inside a 'shop' scheduling problem;
- a route is only defined by a partial sequence of operations (this is a median situation between the total operation order in job-shops or in flow-shops, and an empty set of precedence constraints, such as in open-shops);
- particular constraints exist upon the storage of the inventories;
- precedence constraints exist between the jobs (assembly lines);
- additional resources must be considered (tools, transportation devices);
- deadlines and release dates exist;
- machines are not available all the time;
- set-up times exist between operations on a machine.

In this case, in order to describe the nature of the considered scheduling problem, the nearest basic workshop is used and details are given about the particular extensions or hypotheses relative to the non-basic workshop problems. For instance, we consider a flow-shop problem with sequence dependent set-up times between the operations on a given machine, or where the level of the storage between two machines is limited to one product.

Until now, we have supposed that the problem data are precise and deterministic. We have in fact accepted the following assumption: if some input data are stochastic, we substitute for them in the model an estimated 'certainty equivalent', for example the average value, without a great loss in the performance results. When stochastic phenomena must be taken into account explicitly to solve scheduling problems, then stochastic approaches must be adopted. These use Markov chain, stochastic Petri net, queuing theory and Gittins indices. We do not develop the stochastic methods in this chapter; the reader can find a good survey of stochastic methods in Weiss (1992).

9.2.4 Notation

In order to facilitate the reference to scheduling problems, researchers have proposed the standardized classification $\alpha|\beta|\gamma$ (Conway *et al.* (1967); Rinnooy Kan (1976)) where α described the machine environment, β the job characteristics and γ the criteria (see the appendix). The scientific literature presents scheduling problems using this classification: see for example the survey of complexity results by Lawler *et al.* (1992).

When considering a given scheduling problem, it is very useful to search its classification $\alpha|\beta|\gamma$ and its complexity (extracts of such lists are given in the appendix). If you are fortunate, your particular problem has already been studied and has been proved to be 'easy', and one or several polynomial algorithms have been proposed for solving it. In less favorable instances, the particular problem has been studied but was proved to be NP-hard; then exact or approximation solving schemes generally exist. Exact schemes are restricted to relatively 'small' problem sizes. Approximation methods are more or less efficient depending on the particular example considered. In most of the remaining cases, the problem considered is a generalization of one or several already studied problems (through the addition of specific constraints or conditions); then it is NP-hard if it contains sub-problems which are themselves NP-hard. This is very often the case for industrial problems. In this case, it is necessary to develop specific solving schemes. This chapter and Chapter 10 are dedicated to helping you build new solution schemes.

9.3 METHODS OF SOLUTION

Our aim is to list the various techniques found useful in designing algorithms for scheduling problems. We first emphasize the specific properties of some problems. This allows us either to prove the optimality of some algorithms or to improve them by decreasing the number of feasible solutions to be considered. We then present a non-exhaustive list of methods of solution, most of them leading to approximate algorithms, since most scheduling problems are NP-hard. A special emphasis will be given to local search methods in this chapter, as well as in the following one.

9.3.1 Some useful properties

Some useful definitions for building efficient algorithms are given below (see, for instance, Baker 1974).

Regular measure of performance

A measure of performance (or criterion) of a scheduling problem that is a function of job completion times $F(C_1, C_2, \ldots, C_n)$ is said to be regular if

$$\forall_i, C_i \leqslant C'_i \rightarrow F(C_1, C_2, \ldots, C_n) \leqslant F(C'_1, C'_2, \ldots, C'_n).$$

Criteria generally used in scheduling problems are regular, except those which are computed using the 'earliness' measure. In this case there is a cost to take into account when a job is finished strictly before its due date. This particular criterion must be considered in the just-in-time (JIT) manufacturing systems. There are relatively few interesting results concerning this criterion in the literature, where tactical delays are included inside the schedule in order to eventually be nearer the due dates (Kanet and Sridharan, 1992).

Dominant subset

A subset of solutions of a problem is said to be a dominant subset for a given criterion if it contains at least one optimal solution (Baker, 1974).

For every solution outside the dominant subset, there exists at least one solution in the dominant set which is at least as good. There may remain some optimal solutions outside a dominant subset. This does not matter if the only goal is to find an optimal solution.

For instance, if we consider the one-machine basic scheduling problem with a regular criterion, the set of sequences without pre-emption or idle time is a dominant subset. As a consequence, for these scheduling problems, the solution schemes can consider only sequences without pre-emption and without idle times on the machine, which will considerably simplify them. If we do the same for a criterion which is not regular, such as a weighted sum of earliness and tardiness, then not only we can miss the optimal solution set, but also the considered solution subsets may contain only very poor values for this criterion. The knowledge of the regularity of a criterion and the knowledge of any dominant subset help in building efficient solving schemes.

It is well known that:

- for flow-shop problems, the set of schedules which have the same order of jobs on machines 1 and 2 is a dominant subset for any regular criterion;
- for flow-shop problems, the set of schedules which have the same order of jobs on machines 1 and 2, or M − 1 and M (the last two machines) is a dominant subset for the makespan criterion (minimize the total duration of the schedule).

Generally, the scheduling algorithms work only upon some traditional sets of schedule which are defined below.

Semi-active schedule or left-justified schedule

In a **semi-active** schedule, it is not possible to begin an operation earlier without changing the order of the operations on at least one machine. Operations are **left-justified** either on the previous operation of the job route (after an eventual transportation time represented in the following figures by an arrow) or on the previous operation performed on the machine. The notion of left-justified operation is illustrated in Fig. 9.1. The operations are represented as dotted rectangles. Other rectangles correspond to operations of other jobs already scheduled on the machine.

Active schedule

An **active** schedule is a semi-active schedule in which it is impossible to initiate an operation earlier without some other operation beginning later (Fig. 9.2).

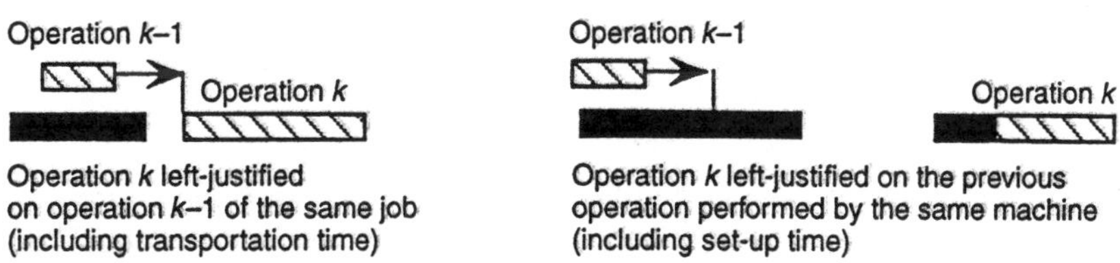

Figure 9.1 Semi-active or left-justified schedule.

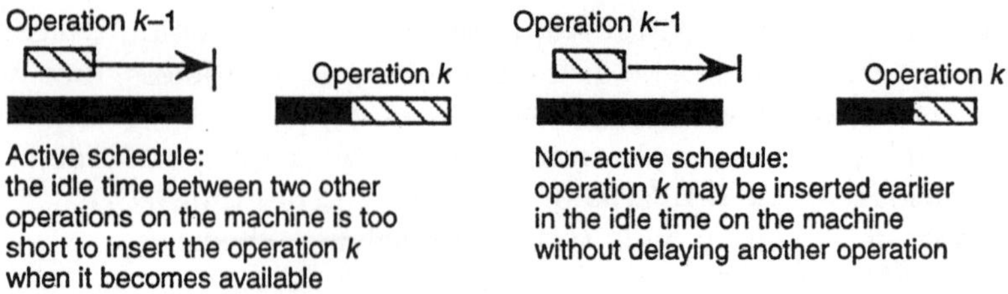

Figure 9.2 Active schedule.

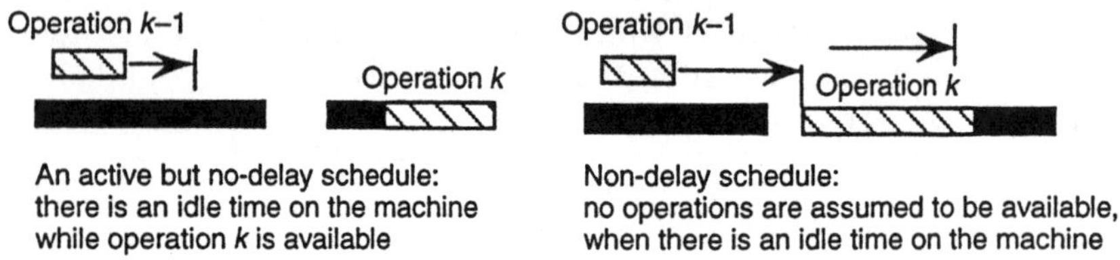

Figure 9.3 Non-delay schedule.

Non-delay schedule

A non-delay schedule is an active schedule in which idle times on a machine exist only if no operation is waiting for this machine (Fig. 9.3).

The set of semi-active schedules and the set of active schedules are dominant subsets for the job-shop problem with any regular criterion. But the set of non-delay schedules is not a dominant subset. Nevertheless many approximation schemes use non-delay schedules, as illustrated in the next section.

When considering a particular problem, it is always useful to prove some dominant properties in order to define dominant subsets. For example, for the $1//\bar{T}$ problem (basic one-machine scheduling problem with the mean tardiness criterion), a dominant subset may be built by using the dominant properties of Emmons (1969). The dynamic programming approach may be improved by exploring only the corresponding subsets. This is implemented in the algorithm of Srinivasan (1971).

9.3.2 The design of algorithms

We propose to regroup the solution schemes in five families, already presented in GOThA (1993) and Carlier and Portmann (1992), but not illustrated by examples:

- progressive building methods or constructive methods (often 'greedy' methods),
- neighborhood methods (or 'local search' methods).
- splitting-up methods (or 'decomposition' methods),
- model-changing methods (generally 'relaxation' methods),
- miscellaneous methods, including artificial intelligence (AI) approaches.

It is obvious that in order to solve NP-hard scheduling problems efficiently, several approximation schemes must be used either independently or concurrently in mixed approaches.

Constructive (progressive building) methods

In this family of methods, we include any iterative method which begins with an empty (or partial) set of decisions concerning the values of the schedule unknowns and which, at each step, adds one or several decisions to the decision set. The process ends when a complete schedule is obtained.

A person with no particular knowledge about operational research and/or scheduling theory will generally use this class of methods to try to solve scheduling problems.

At each step, a partial schedule is completed by a new decision: the position in the sequence, or the beginning time of a single operation (or of a set of operations, such as the whole set of operations of a job). The schedules can be of semi-active, active or non-delay type. The one-operation-at-each-step methods generally build non-delay schedules. The better known part of this family consists in the simulation methods using dispatching rules (Panwalkar and Iskander, 1977; Haupt, 1989). They build active and generally non-delay schedules by adding, at each step, an accessible operation (whose predecessors are already scheduled) on a machine.

Constructive methods can be deterministic or stochastic. In the latter case some random numbers are generated and used in order to make the decision at each step. The solution obtained by a stochastic method can differ from one execution to another. When the constructive methods are approximation schemes, then it is possible to improve the solution by applying a stochastic method several times or by using a set of different deterministic or stochastic constructive methods and keeping the best of the solutions obtained.

Constructive methods are generally 'greedy' algorithms, i.e. the decision made is the most advantageous in the short term, and the partial schedule or the partial set of decisions obtained at any step is never modified by subsequent steps. So, if the complexity function of the algorithm used to make the decisions of each step is polynomial, then the complexity of the complete process is also polynomial. However, when deadlines exist or when the resource constraints are difficult to verify and finding a feasible schedule is at least as difficult as finding an optimal one, then it is necessary to use backtracking. In this case, there is a branching tree corresponding to the decision set, and the number of nodes of this tree increases exponentially with the size of the problem (for example:

$$2^{(n(n-1)/2)}$$

where n is the number of operations to be scheduled). Then, even if the complexity function of each step is polynomial, the complete method can be very time consuming.

A constructive method can be an optimal one. To illustrate this, we will consider some well-known methods and a numerical example (Table 9.1).

For the basic one-machine problem, Smith's rule (Smith, 1956), also called the **SPT rule** (for shortest processing time rule), which consists of ordering the operations by non-decreasing order of processing times, gives an optimal schedule for the mean flow time criterion ($1//\bar{C}$) (Table 9.2).

For the basic one-machine problem, Jackson's rule (Jackson, 1955), also called the **EDD rule** (for earliest due date rule), which consists of ordering the operations by non-decreasing order of due dates, gives an optimal schedule for the maximum tardiness criterion ($1//T_{max}$) (Table 9.3).

For the basic one-machine problem, Moore's algorithm (Moore, 1968), whose implementation was improved by Hodgson, gives an optimal

Table 9.1 Input data 1

i	1	2	3	4	5	6	7	8
p_i	1	5	7	2	4	3	3	3
d_i	2	5	25	26	7	10	20	14

Table 9.2 SPT order: $\bar{C} = 12$

i	1	4	6	7	8	5	2	3
p_i	1	2	3	3	3	4	5	7
C_i	1	3	6	9	12	16	21	28

Table 9.3 EDD order: $T_{max} = 3$

i	1	2	5	6	8	7	3	4
p_i	1	5	4	3	3	3	7	2
C_i	1	6	10	13	16	19	26	28
d_i	2	5	7	10	14	20	25	26
T_i	0	1	3	3	2	0	1	2

schedule for the number of tardy jobs ($1//N_T$). At each step, this algorithm increases the set of definitively late jobs at the end of the sequence, the job chosen to be definitively late is the longest between the first job scheduled and the first job which is late in the beginning sequence sorted by EDD. In our numerical example, the first job which is sent at the end of the sequence is the job 2 (first late in the EDD sequence and longer than job 1).

For the basic m-identical parallel machine problem the SPT rule which assigns the shortest remaining operation to the first idle machine remains optimal for minimizing the mean flow time ($P//\bar{C}$).

For the basic flow-shop problem with two machines ($F2//C_{max}$), Johnson's rule (Johnson, 1954) is optimal for minimizing the makespan (i.e. the total duration, or the greatest completion time of the jobs: C_{max}). It consists of considering at each step one of the remaining jobs with the shortest processing time either on the first (condition 1) or on the second (condition 2) machine, and placing it either at the end of a beginning schedule list if condition 1 is verified or at the beginning of an ending schedule list if condition 2 is verified (with initial empty lists).

Concerning approximation-constructive methods, we will limit our illustrations to the most famous designed for the 'permutation' flow-shop problem (in which no 'passing' is permitted – the order of the jobs is the same on the whole set of machines, even if this restriction does not give a dominant subset).

For the flow shop problem with the makespan criterion when there are more than two machines ($M > 2$), many constructive methods have been developed under different hypotheses (see, for example, Dannenbring (1977), Proust *et al.* (1991)). The more famous and still one of the most efficient is the CDS method (for Campbell, Dudek and Smith (Campbell *et al.*, 1970)). The CDS method builds only permutation schedules. It uses Johnson's rule on $M - 1$ fictitious flow-shops composed of two machines in order to obtain different job permutations. For the kth possibility, one fictitious machine is obtained by summing up the processing times of machines 1 through k and the other one by summing up the processing times of machines $M - k + 1$ through M. The makespan of the flow-shop problem corresponding to each permutation is determined, and the best solution is kept.

In most of the software packages currently available, the scheduler makes scheduling decisions with the help of the software program. A constructive method is proposed which works on a partial schedule that is completed at each step by the whole set of operations of a given job. The scheduler can ask the program to classify the set of jobs into classes of different priorities and inside each class to order the jobs using a given rule. Using these results, the scheduler can choose a job to be scheduled. He or she then has to choose between several placement policies.

Neighborhood search methods

These methods work on complete schedules and try to improve them at each step of the method (see, for instance, Dannenbring (1977)). They are also called 'local search methods' because they normally stop when they find a local optimum, and because at each step they study the local neighborhood of a given complete solution. To build a local search method three procedures have to be designed: a procedure which provides an initial (feasible) schedule, a procedure which provides a schedule in the neighborhood of a given schedule, and a function which provides the performance measure of a given schedule.

Neighborhood methods are classified depending on the use of these three procedures inside some general schemes. We can distinguish between **stochastic** and **deterministic** schemes, depending on whether there is use of random decisions within the solution schema; and between **no-null** or **null** threshold schemes, depending on whether a neighbor can be selected that is worse, but only if the loss is smaller than a given computed threshold (Vaessens *et al.*, 1994).

The better-known neighborhood methods are the following.

- If the neighborhood of each schedule is a finite set, and if we choose the best schedule in the neighborhood and stop the search when the best schedule is not strictly better than the one in hand, then we obtain a 'best ascendant' method (deterministic and null threshold).
- If the chosen neighbor is one of the neighbors that is obtained randomly among those which improve the value of the criterion, and stop the search when there are no better values in the neighborhood than the one in hand, then we obtain a 'hill climbing' method (stochastic and null threshold).
- If we choose randomly one of the neighbors and accept it, always when it is better, or with a probability exp(–degradation/T) when it is worse, where T is a 'temperature' which is controlled from within the

method, then we obtain a 'simulated annealing' method (by analogy with the simulation of annealing of solids, which explains why parameter T is called temperature (Kirkpatrick *et al.*, 1982, 1983; Cerny, 1985) (stochastic and no-null threshold). This approach is explained in more detail in section 9.4.

- If we use a 'best ascendant' or a 'hill climbing' method, with the possibility of accepting a worse schedule in order to leave a local optimum, but with Tabu schedules (for instance the last k visited schedules are forbidden) or Tabu modifications of the working schedule (for instance the last k used modifications to go from a schedule to its neighbor are forbidden), then we obtain one of the numerous variants of a Tabu search method (Glover, 1987, 1989, 1990; Glover *et al.*, 1993; Hertz and De Werra, 1990) (deterministic or stochastic, but no-null threshold). This approach is given in more detail in section 9.4.

Various variants of local search methods have been proposed in order to help move away from a local optimum, such as using perturbations (Charon and Hudry, 1992) or modifications of the neighborhood when the probability of being in a local optimum reaches a given threshold (Fleury, 1993).

When using a **population** of schedules (rather than only one schedule) which we try to improve at each step, we obtain a method of evolutionary strategies such as genetic algorithms (Holland, 1975; Goldberg, 1989). This method works on a population using **reproduction**, **cross-over** and **mutation** techniques, similarly to the life of animal populations (see Chapter 10). This family of methods uses, in fact, an hyper-neighborhood technique on a population of solutions instead of on a single solution (Vaessens *et al.*, 1994).

Splitting-up methods

This is one of the most natural approaches that come to mind: split the original problem to yield sub-problems that are smaller in size and/or simpler to solve. Generally speaking, the global solution built at the end of this solution schema may not be optimal even if the sub-problems were optimally solved. But the final (assembled) solution may be better than solutions obtained by global approximation schemes to the original problem.

There are many approaches for building algorithms through splitting the original scheduling problem in order to construct a feasible, and hopefully near-optimal, schedule for the original problem using the partial schedules of the sub-problems.

One may consider **hierarchical** methods (Axsater and Johnson, 1984; Bitran *et al.*, 1982, Erschler *et al.*, 1985) which split the decision problem into several **levels**, the higher levels working on aggregate data in simplified models; with the decision at a higher level constraining the lower levels, which leaves fewer solutions to be considered at the lower level.

The **structural splitting** method was introduced by Roy (1970). It consists of splitting the constraint set into two families (for instance, resources usage constraints and temporal constraints) and solving iteratively two kinds of problems: one in which the unknown values of one family are fixed and the values of the other family are considered as unknown, and vice-versa.

If we split the **set of solutions**, we obtain decomposition methods, such as the branch-and-bound (BaB) methods which are well known in operations research (Lenstra, 1977). We obtain a heuristic method when a BaB method is truncated before having ascertained that the best schedule found is the optimum.

The splitting method can use a **temporal** or a **spatial decomposition** of the problem. In temporal decomposition (Portmann, 1988), the best schedule is sought after only a restricted subset of the jobs (e.g. those which have arrived before a given date $\Delta 1$). Then the beginning of the schedule is fixed, and the end is mixed with a new subset of the jobs (those arriving between dates $\Delta 1$ and $\Delta 2$, etc.). In spatial decomposition (Portmann, 1988; Chu *et al.*, 1992), the shop is split into sub-shops using cross decomposition methods or group technology methods, and an iterative approach is built to obtain a global feasible schedule, using local feasible sub-schedules. These methods are strongly related to FMS scheduling problems where the sub-shops can exist effectively inside the 'parent' shop. There are several ways to cross spatial and temporal decomposition techniques (Chryssolouris *et al.*, 1991; Portmann, 1988; Meguelati, 1988).

Model changing methods

This class of methods modifies the model of the problem in order to simplify it, either to (i) insert the simplified model inside a branch-and-bound approach in order to find an optimal schedule, or (ii) to obtain a restricted set of schedules which will be easier to explore.

All the relaxation methods – integer relaxation, Lagrangean relaxation (Fisher, 1973; Van De Velde, 1991), surrogate relaxation (Glover, 1975; Fisher *et al.*, 1983) – belong to the first approach. Examples of methods that belong to the second approach include the linearization of the objective function and the addition of restrictive constraints. An example of the latter approach is working only on the set of permutation flow-shop schedules (which contains $n!$ schedules) rather than on the set of flow-shop schedules (which contains $(n!)^{m-1}$ schedules, dominant set with the same order of the jobs on machines 1 and 2), where n is the number of jobs and m the number of machines. In this last case, we may not obtain the optimal solution, but we hope that the gap between the best solution of permutation schedules and the best solution of the whole set of schedules is not too large.

To illustrate this family of methods, we take one example from the thesis of Van De Velde (1991). We consider the one-machine problem under the criterion of minimizing the sum of weighted flow times when a graph of precedence constraints is added to the basic problem. This problem is NP-hard.

Let C_j denote the completion time of job j, the parameters p_j and w_j denote, respectively, the processing time and the weight of job j. Let E denote the set of edges of the precedence constraint graph. The problem may be formulated as follows

$$\min \left(\sum_j w_j C_j \right)$$

$$C_k \geqslant C_j + p_k \textbf{ or } C_j \geqslant C_k + p_j \qquad \forall j, \forall k, j = k \qquad (9.1)$$

$$C_j - p_j \geqslant 0 \qquad \forall j \qquad (9.2)$$

$$C_j \geqslant C_k + p_j \qquad \forall j, \forall k, (k,j) \in E \qquad (9.3)$$

If we use a vector λ of Lagrangean coefficients $\lambda_{k,j}$ in order to relax the third constraint set, then we obtain the L_λ problem:

$$\sum_{j=1}^{n} \left[\left(w_j + \sum_{(j,k) \in E} \lambda_{j,k} - \sum_{(k,j) \in E} \lambda_{k,j} \right) \cdot C_j + \sum_{(j,k) \in E} \lambda_{j,k} \cdot p_k \right]$$

$$C_k \geqslant C_j + p_k \textbf{ or } C_j \geqslant C_k + p_j \qquad \forall j, \forall k, j = k \qquad (9.1)$$

$$C_j - p_j \geqslant 0 \qquad \forall j \qquad (9.2)$$

We note that

$$\sum_{(j,k) \in E} \lambda_{j,k} \cdot p_k$$

is a constant; and if we define W_j by

$$W_j = \left(w_j + \sum_{(j,k) \in E} \lambda_{j,k} - \sum_{(k,j) \in E} \lambda_{k,j} \right)$$

we note that we obtain the same form as the original problem except that the set of constraints (9.3) has disappeared. This problem is denoted by $1//\bar{C}_w$ in the $\alpha|\beta|\gamma$ classification and is a polynomial one. A polynomial solving algorithm consists in scheduling the jobs in non-decreasing order of the processing times divided by the weights.

It is then easy to obtain a solution of the L_λ problem; the value of the corresponding criterion is denoted by $L(\lambda)$. This solution is a lower bound for the solution of the original problem. To obtain the greatest lower bound it is necessary to solve the Lagrangean Dual problem which consists in finding the vector λ which maximizes $L(\lambda)$. Efficient polynomial approximation or exact methods can be developed to solve this Lagrangean Dual problem (Van De Velde, 1991). This lower bound can be used in a BaB method. An approximation schema is developed by Van De Velde (1991) using particular properties (dual block decomposition) of the optimal schedules of L_{λ}.

AI approaches and other methods

We consider here all the methods and tools used and developed in the field of artificial intelligence (AI) which do not belong to the previous classes. This includes **constraints programming**, **expert systems** and **knowledge-based systems** with or without learning.

To illustrate this family of methods, we briefly present a knowledge-based system (KBS) which is called 'Artificial Memory' (Chu and Portmann, 1993). Similar approaches are developed in different approaches building KBS: they are based on the same main idea, that is, to help a scheduler making a choice among a set of approximation methods in order to solve a given scheduling problem instance. The various KBSs differ only in the technical and theoretical choices made to build them.

An Artificial Memory is dedicated to one family of problems. For this family of problems, we suppose that a set of approximation algorithms has been designed and a large set of problem instances has been generated. The whole set of approximation algorithms is applied on the whole set of problem instances by a simulation phase and the corresponding values of performance measures are collected. For a given instance, due to the variety of the approximation algorithms, each measure varies from its worst value to its best value and we can cluster these values into five subsets: 'very bad', 'bad', 'medium', 'good', 'very good'. Using this data, it is possible to build clusters which are sets of instances for which the application of a given algorithm provides a given value ('very bad', 'bad', . . .) to a given measure. A distance is designed to measure the similarity between elements of each cluster (Chu and Portmann, 1993). The Artificial Memory is built off-line and contains the clusters and the elements which are necessary to compute the different distances. The Artificial Memory is used on-line in order to choose the 'best' approximation algorithm for each new encountered instance. For a chosen measure, the cluster(s) nearest to the new instance is (are) selected and the one that provides the best performance for the measure gives the algorithm which must be chosen. If there is no cluster in the proximity of the new instance, then a standard default algorithm is chosen (the best in average for the whole set of instances) and this instance is added to a list of 'poorly represented' instances. If the list of poorly represented instances becomes too large, then a new simulation phase is called off-line which integrates the unsatisfied instances and which gives a new version of the Artificial Memory (memory refreshment).

Mixed approaches

Some approaches mix several of the previous approaches. For instance, when considering the job shop problems with the criterion of minimizing the makespan, a method called the 'shifting bottleneck method' is typically a mix of several approaches. It is an iterative method that improves at each macro-step the criterion value and may be considered as a local search method. Inside each macro-step, some relaxed problems are solved by considering a series of one-machine problems, so it uses relaxation and splitting up methods. This method is explored further in section 9.4.

9.4 LOCAL SEARCH APPROACHES

The discussion will now focus on three approaches: simulated annealing, tabu search and shifting bottleneck.

9.4.1 The simulated annealing (SA) approach

This was first proposed independently by Kirkpatrick *et al.* (1983a and b) and by Cerny (1988) by analogy with the annealing of a solid (Var Laarhoven and Aarts, 1987):

> Annealing denotes a physical process in which a solid in a heat bath is heated up by increasing the temperature of the heat bath to a maximum value at which all particles of the solid randomly arrange themselves in the liquid phase, followed by cooling through slowly lowering the temperature of the heat bath. In this way, all particles arrange themselves in the low energy ground state of a corresponding lattice, provided the maximum temperature is sufficiently high and the cooling is carried out sufficiently slowly.

To illustrate this family of methods, we discuss two issues. The first is not problem specific, but method specific: it gives the general schema of simulated annealing. The second subsection gives some examples of the procedures which have to be written in order to complete the general schemes of simulated annealing for some specific problems.

General schema of SA

```
fb:=+∞;
for k varying from 1 by 1 to P do     % P trials
    xo:=INITIAL SOLUTION;             % Initial schedule
    fm:=f(xo);                        % Initial best value of criterion
    J:=0;                             % Initial value of stopping counter
    T:=To;                            % Initial value of temperature
    while J≤Q do
        I:=0;                         % Initial value of temperature
                                        counter
        while I≤N & J≤Q do            % Testing a neighbor
        x1:=NEIGHBOR(xo);
        I:=I+1; J:=J+1;
        if f(x1)<f(xo) then
                   xo:=x1;
               if f(xo)<fm then
                   xm:=xo; fm:=f(xo); J:=0
                   endif;
    else
              Δf:=f(x1)−f(xo); U:=Unif(0,1);
           if U ≤ exp (−Δf/T) then xo:=x1 endif
    endif;
    endwhile
 T:=T*α;                              % Temperature modification
    endwhile
 if fb>fm then write fm,xm; fb:=fm
   endif;
 endfor (k)
```

where:

$f(x)$: the criterion value of solution x.

f_b: the criterion value of the best found solution since the beginning of the algorithm.

f_m: the criterion value of the best found solution since the beginning of the current trial (for a given k).

P: the number of trials with randomly generated different initial solutions (obtained by the procedure INITIAL SOLUTION); a given parameter.

x_o: the current solution.

x_1: the potential future current solution (obtained by the procedure NEIGHBOR(x_o)).

Q: the maximal number of micro-steps (trying to go from x_o to x_1) without improvement of the criterion value; a given parameter.

J: the counter of micro-steps without criterion improvement.

N: the maximal number of micro-steps with the same temperature value; a given parameter.

T: the current temperature.
T_o: the initial temperature; a given parameter.
Δf: the gain or the loss in the criterion value between x_o and x_1.
Unif: generates randomly a real value between two bounds (here 0 and 1).
exp: exponential function.
α: a real value such that $0 < \alpha < 1$, which decreases exponentially the value of the temperature.

The previous pseudo code is a general schema of simulated annealing, and is also the simplest one. Many authors have proposed different alternative schemes. As an example, we present the approach proposed by (Van Laarhoven and Aarts, 1987) which is one of the most utilized.

In this particular schema, the description of the cooling process of the temperature is as follows: the initial temperature is not arbitrarily chosen, but computed by the formula:

$$T_o = \frac{\Delta \bar{C}^{(+)}}{ln(X_o^{-1})}$$

where X_o is a given percentage of acceptance of the obtained solutions and $\Delta \bar{C}^{(+)}$ is the average value (obtained by experimentation: random generations of solutions and their neighbors) of criterion degradation. The stopping test is given by the temperature which becomes smaller than a threshold T_f obtained by the formula:

$$T_f = \varepsilon . \frac{\bar{C}(T_o)}{(\partial \bar{C}(T)/\partial T)_{T=T_f}}$$

where ε is a given parameter, and numerator and denominator use the average value and the derivative value of the cost with respect of the temperature (obtained again by experimentation). The coefficient α is given by the formula:

$$\alpha = \left(1 + \frac{ln(1 + \delta).T_k}{3\sigma(T_k)}\right)^{-1}$$

where δ is a given parameter and $\sigma(T)$ is the standard deviation of the cost for the temperature T.

Utilization of SA in some specific problems

In order to implement the simulated annealing method for a specific problem, one has to write two procedures and one function:

- the procedure 'INITIAL SOLUTION',
- the procedure 'NEIGHBOR',
- the function f(x).

To obtain an initial solution, one can use any approximation schema which is able to provide a complete solution, such as constructive methods. These methods are among the simplest approximation schemes in which one can introduce random decisions so that the initial solution can be different for each trial.

To write the NEIGHBOR procedure, one must examine the structure of the complete solutions and design simple modifications which transform a complete schedule into another complete one. The more simple NEIGHBOR procedures are obtained for 'one-machine' problems or for permutation flow-shop. In this case, transpositions of jobs on the machines give a very simple neighborhood. For more complicated problems, the neighborhood is described by a set of possibilities; for instance, when there are several parallel machines, an operation can be moved from one machine to another, or the position of two operations can be swapped. For the job-shop scheduling problem with makespan criterion, a given solution is defined by a feasible selection of edges from the 'disjunctive' graph (set of edges between two operations using the same machine which must be oriented in order to decide which operation will be executed before the other). On the 'conjunctive' graph CS, consisting of the precedence constraints, initial graph completed by the selected edges, the critical paths are computed (longest paths). In order to build a neighbor, only selected edges of a critical path are considered and are reversed. In this case Van Laarhoven *et al.* (1992) have proved that no cycle is created in the conjunctive graph CS and that, if the reversal of a non-critical edge leads to an acyclic graph CS′, the longest path in this graph CS′ cannot be shorter than the longest path in CS. This property limits the neighbors of any given solution to interesting ones.

The function $f(x)$ is needed whatever the solution schema may be. There is little to add here except that, if it is difficult to find feasible solutions of the given problem, then we may use non-feasible solutions in local search methods. In this case, the error in the unsatisfied constraints must be included in the criterion with large weights so that the criterion decreases appreciably each time a previously violated constraint becomes satisfied.

9.4.2 The tabu search (TS) method

As stated in Hertz *et al.* (1993), 'the roots of tabu search go back to the 70s; it was first presented in its present form by Glover (1987); the basic ideas have also been sketched by Hansen (1986)'. The main initial idea of this method is that it is possible to leave local optima by accepting a degradation of the criterion value but without using previous visited solutions which are stored in a list of forbidden, or 'tabu', solutions.

Since the specific problem procedures are similar for all neighborhood methods, we focus here only on the general TS schemes, i.e. the method specific to TS.

Tabu search uses a 'best ascendant' method, with the possibility of accepting a worse schedule in order to leave a local optimum, but with tabu schedules (for instance, the last k visited schedules are forbidden) or tabu modifications of the working schedule (for instance, the last k used modifications to go from a schedule to its neighbor are forbidden). Then we obtain one of the numerous variants of the TS method (Glover, 1987, 1989, 1990; Glover *et al.*, 1993; Hertz and De Werra, 1990).

A simplified description of tabu search method is as follows.

1. Choose an initial solution x_o; $x_b = x_o$; $f_b = f(x_0)$.
2. Generate a subset V^* of solutions included in the neighborhood of x_o (where tabu solutions are excluded).
3. Find the best solution x_1 in V^* (minimal value for the criterion).
4. $x_o := x_1$.
5. If $f(x_o) < f_b$ then $x_b = x_o$; $f_b = f(x_o)$.
6. Go to (2) if no stopping condition is met.

The same stopping condition as in SA schema may be used, that is a maximal number of steps without improvement of the best criterion value f_b.

Building the subset V^* is the main step. V^* is the set of neighbors of x_o minus the non-authorized tabu solutions. For instance, a tabu file of constant length L which contains the last L reversed movements π_1, $\pi_2, \ldots, \pi_L$ performed to go from $x_n + 1 - L$ to x_n (x_n is the current point called x_o in the procedure) is memorized. Normally $v = \pi_k(x_n)$ is tabu and not accepted in V^* except if $f(v) \leqslant a(f(x_n))$ where a is an 'aspiration function' (generally a is defined by $a(x) = x - \Delta x$ with $\Delta x = 1$ or the smallest variation of the cost function $f(x)$, but the aspiration function may be dynamically modified for different values of x during the execution of the algorithm).

Some other techniques are included in tabu search in order to intensify the search in some part of the solution space and/or to diversify the search inside the solution space. The interested reader may consult Barnes and Chambers (1994), Dell'Amico and Trubian (1993), Glover (1987, 1989, 1990), Glover *et al.* (1993), Hertz and De Werra (1990), Hertz *et al.* (1993), Laguna *et al.* (1991), Nowicki and Smutnicki (1993), Taillard (1994), where methods for building tabu search on parallel processors are explained, and Widmer (1989).

9.4.3 The shifting bottleneck approach

This is an iterative improvement method which uses again and again different relaxations of the problem into one-machine scheduling problems. It was first proposed by Adams *et al.* (1988) and then improved by Dauzere-Perez and Lasserre (1993, 1994a and b) by taking into account more general precedence constraints in the relaxed problem. We will present here the original version. The problem considered here is the basic job shop problem with the makespan criterion.

In this approach, machines are sequenced one at a time, consecutively, and each machine will be considered several times.

At the *k*th macro-step of the method, a set M_o of k–1 machines ($M_o = \varnothing$ at the first macro-step) are supposed to be scheduled (i.e. the operation order is given for these machines) and M–M_o machines are still free (the constraints between operations due to the corresponding machines are supposed to be relaxed, i.e. machines are considered as having infinite capacity).

Every machine m in the set M–M_o is successively considered. Considering that the sequences are fixed in the set M_o, that the machine capacity in M–M_o–$\{m\}$ is infinite, and that the capacity of machine m is 1 and the order on this machine is free, we obtain a one-machine scheduling problem with, for each operation, a 'tail' (identical to release dates) and 'queue' (delay between completion time and makespan value), and the makespan criterion to be minimized. The obtained one-machine scheduling problem is solved optimally with known methods (Carlier, 1982; McMahon and Florian, 1975).

In the first step, every machine of M–M_o is considered; the corresponding relaxed one-machine problem is solved and the corresponding makespan is obtained. Machines are then ranked in decreasing order of the obtained makespan values. The machines corresponding to the greatest value of the makespan are bottleneck machines. The first bottleneck machine m_b is included in M_o with its corresponding sequence.

In the second step, every machine of M_o is re-optimized. This means that for every k of M_o, the one-machine problem with operations of M_o–$\{k\}$ sequenced, free operation order on k and machines of M–M_o with infinite capacity, is optimally solved and the corresponding new order for machine k is kept.

This approximate schema stops when M_o is equal to M. This method is presented as one of the most efficient approximation schemes for the considered problem.

9.5 CONCLUSIONS

Our aim was to present a guide for the practitioner to the great variety of exact and approximation schemes with a classification of the problems and a classification of the methods.

A final comment on solving real-life industrial scheduling problems is in order. The prime difficulty in practice is to secure the 'correct' set of constraints imposed on the scheduler. Typically, opinions differ radically on what is essential and what is optional, which constraints are an integral part of the process and which are included to improve the performance of the resulting schedule. The sifting of priorities in this respect is demanding in time and expertise, but is essential.

In general, it is found that the use of the hierarchical approach for modeling and designing the decision support system is most advantageous. After the first division of the decision hierarchy, one usually encounters several scheduling problems, depending on the division and the time unit adopted. For each of these problems one searches for the specific characteristics that define it, and (with the participation of the decision maker) one usually combines several of the solution approaches described in this chapter to construct efficient algorithms. Finally, every new decision tool must be thoroughly tested (normally off-line, but eventually on-line) before it is integrated into the production process as an everyday tool of decision making.

APPENDIX

Notations

In order to present the main results about complexity, we will use the well-known classification $\alpha|\beta|\gamma$ for scheduling problems first introduced by Conway *et al.* (1967) and Rinnooy Kan (1976) and definitively stated in Lawler *et al.* (1992).

α describes the machine environment

$\alpha 1$ = $\emptyset$: single machine
= *P*: identical parallel machines
= *U*: uniform parallel machines
= *R*: unrelated parallel machines
= *O*: open shop
= *F*: flow shop
= *J*: job shop

$\alpha 2$ = 1, 2, . . .: the number of machines is fixed to the given number
= *m*: the number of machines is fixed inside the definition of the problem
= $\emptyset$: the number of machines is variable depending on the instance of the problem

β describes job characteristics

$\beta 1$ = *pmtn*: operations can be pre-empted
= $\emptyset$: pre-emption is forbidden

ß2 = ∅: there are no precedence constraints (except the routes for job shop and flow shop)
= *tree*: there are additional precedence constraints for which the graph forms trees
= *prec*: there are additional precedence constraints for which the graph is general (an acyclic directed graph)

ß3 = ∅: all release dates are identical and can be taken equal to 0
= r_j: release dates that may differ per job are specified

ß4 = p_j=1 or p_{ij}=1: each operation has unit processing time
= p_j=p or p_{ij}=p (with ß3 = r_j): the processing time of each operation is identical to p, p not equal to 1, and at least one of the r_j is not divisible by p
= ∅: all processing times are arbitrary non negative integers

ß5 = ∅: set-up times are constant and included inside the processing time
= S_{nsd}: setup times are assigned to each operation, are independent of the sequence on the machine, but are not included inside the processing time (the setup can be made before arriving of the job in the queue of the machine)
= S_{sd}: setup times are not included inside the processing time and are sequence dependent
(by replacing S with R, we obtain the same distinction for the reset times after the operations)

ß6 = d_j=d: all due dates are identical (common due date)
= ∅: due dates are arbitrary non negative integers
= Δ_j: due dates are replaced by imperative deadlines

γ *describes optimality criteria*

The meaning of the used letters are the following:

C = completion time,
L = lateness = completion time minus due date,
T = tardiness = lateness if it is positive and 0 if not,
E = earliness = lateness if it is negative and 0 if not,
G = a general non-decreasing function of completion time
∃ = decision problem: does a feasible solution exist?

- First family of criteria
  C_{max}, L_{max}, T_{max}, E_{max}, G_{max}:
  minimizing the biggest value of a given function of the completion times for the whole set of jobs
- Second family of criteria
  $\bar{C}$, $\bar{L}$, $\bar{T}$, $\bar{E}$:

minimizing the average value of a given function of the completion times for the whole set of jobs

- Third family of criteria

  $\overline{C_w}$, $\overline{L_w}$, $\overline{T_w}$, $\overline{E_w}$:
  minimizing the weighted sum of a given function of the completion times for the whole set of jobs
- Miscellaneous family
  N_T: the number of tardy jobs

Most famous results of problem complexity

Problem family	*Polynomial problems*	*NP-hard problems*
Single machine and $\bar{T}$ or T_w	$1/p_j = p/\bar{T}_w$ $1/r_j, p_j = 1/\bar{T}_w$	$1//\bar{T}$ $1/r_j/\bar{T}$ $1//\bar{T}_w$ $1/r_j/\bar{T}_w$ $1/prec/\bar{T}$ $1/prec, p_j = p/\bar{T}_w$ $1/prec/\bar{T}_w$
Single machine and N_T	$1/prec/\exists\, ?N_T = 0$ $1//N_T$ $\begin{cases} 1/r_j/N_T \\ r_i \leqslant r_j \Rightarrow d_i \leqslant d_j \end{cases}$ $\begin{cases} 1//N_{Tw} \\ w_i \leqslant w_j \Rightarrow d_i \leqslant d_j \end{cases}$	$1//N_{Tw}$ $1/prec/N_T$ $1/chains, p_j = 1/N_T$ $1/r_j/N_T$
Single machine and L_{max} or G_{max}	$1//L_{max}$ $1/prec/L_{max}$ $1//G_{max}$ $1/prec/G_{max}$ $1/r_j, p_j = p/L_{max}$ $1/r_j, preempt/L_{max}$ $1/r_j, preempt, prec/L_{max}$ $1/r_j, preempt/G_{max}$ $1/r_j, preempt, prec/G_{max}$	$1/r_j/L_{max}$
Single machine and $\bar{C}$ or C_w	$1/forest/\bar{C}_w$ $1/series\text{–}parallel/\bar{C}_w$ $1/\Delta_j/\bar{C}$ $1/prec, preempt/\bar{C}$ $1/r_j, preempt/\bar{C}$	$1/r_j/\bar{C}$ $1/r_j, preempt/\bar{C}_w$ $1/prec/\bar{C}$ $1/prec, p_j = 1/\bar{C}_w$ $1/r_j/\bar{C}_w$ $1/\Delta_j/\bar{C}_w$ $1/prec/\bar{C}$
Single machine and S_{sd}		$\begin{cases} 1/S_{sd}, \Delta_j/\exists \\ \lvert\Delta_j\rvert \leqslant D \end{cases}$ $1/S_{sd}, \Delta_j/\exists$ $1/S_{sd}/C_{max}$

Most famous results of problem complexity (*cont.*)

Problem family	*Polynomial problems*	*NP-hard problems*
Parallel machines and $\bar{C}$ or C_w	$P//\bar{C}$ $Q//\bar{C}$ $R//\bar{C}$ $P/p_j = p/\bar{C}_w$	$\left\{\begin{array}{l}P_m//\bar{C}_w\\ m \geqslant 2\end{array}\right.$ $P_m/preempt/\bar{C}_w$ $P/preempt/\bar{C}_w$ $P_2/prec/\bar{C}_w$ $P/prec/\bar{C}$ $P/in\text{-}tree$ or $out\text{-}tree/\bar{C}_w$ $P/prec/\bar{C}_w$ $P//\bar{C}_w$
Parallel machines and C_{max}	$P/p_j = p/C_{max}$ $Q/p_j = p/C_{max}$ $P_2/prec,p_j = 1/C_{max}$ $P/forest,p_j = 1/C_{max}$ $P_2/prec,preempt/C_{max}$ $U_2/prec,preempt/C_{max}$ $P/forest,preempt/C_{max}$ $U/preempt/C_{max}$ $R_m/preempt/C_{max}$	$P_2//C_{max}$ $P_m//C_{max}$ $P/prec,p_j = 1/C_{max}$ $P/prec,preempt/C_{max}$ $P_2/prec,p_j \in \{1,2\}/C_{max}$ $P//C_{max}$
Parallel machines and Δ_j	$P/in\text{-}tree,p_j = 1,\Delta_j/\exists$ $P_2/prec,p_j = 1,\Delta_j/\exists$ $P_2/r_j,prec,p_j = 1,\Delta_j/\exists$ $P/p_j = 1,\Delta_j/\exists$ $P/r_j,p_j = 1,\Delta_j/\exists$ $P/preempt,\Delta_j/\exists$ $U_2/r_j,preempt,\Delta_j/\exists$	$P/tree,p_j = 1,\Delta_j/\exists$ $P/prec,p_j = 1,\Delta_j/\exists$
Flow shop and C_{max} or $\bar{C}$	$F_2//C_{max}$ $F_2/no\ wait/C_{max}$ $F/preempt/C_{max}$	$\left\{\begin{array}{l}F_m/no\ wait/C_{max}\\ m \geqslant 4\end{array}\right.$ $F_2/limited\ buffer > 0/C_{max}$ $F_2/r_j,preempt/C_{max}$ $F/r_j,preempt/C_{max}$ $F3//C_{max}$ $F//C_{max}$ $F2//\bar{C}$ $F//\bar{C}$
Job shop and C_{max}	$J_2/n_j \leqslant 2/C_{max}$ $J_2/p_j = 1/C_{max}$ $J/n \leqslant 2/C_{max}$	$J_2/p_j \in \{1,2\}/C_{max}$ $J_2/n_j \leqslant 3/C_{max}$ $J_2//C_{max}$ $J_3/p_j = 1/C_{max}$ $J_3/n_j \leqslant 2/C_{max}$ $J/p_j = p/C_{max}$ $J//C_{max}$
Open shop	$O_2//C_{max}$ $O_2/r_j/C_{max}$ $O/preempt/C_{max}$ $O/p_j \in \{p_1,p_2\},preempt/C_{max}$ $O/preempt,r_j,\Delta_j/\exists$	$O/p_j \in \{0,1\},preempt/\bar{C}$ $O//\bar{C}$ $O//\bar{C}_w$ $O_3//C_{max}$ $O_m//C_{max}$ $O//C_{max}$ $O_2//N_T$ $O//N_T$ $O_2/\Delta_j/\exists$ $O/\Delta_j/\exists$

REFERENCES

Adams, J., Balas, E. and Zawack, D. (1988) The shifting bottleneck procedure for job shop scheduling. *Management Science*, **34**, 391–401.

Axsater, S. and Johnson, H. (1984) Aggregation and disaggregation in hierarchical production planning. *EJOR*, **17**, 338–50.

Baker, K.R. (1974) *Introduction to Sequencing and Scheduling*, John Wiley.

Barnes, J.W. and Chambers, J.B. (1994) Solving the job shop scheduling problem using tabu search. *IIE Trans.*, to appear.

Biegel, J.E. and Davern, J.J. (1990) Genetic algorithms and job shop scheduling. *Computers Ind. Eng.*, **19**(1–4), 81–91.

Campbell, H.G., Dudek, R.A. and Smith, M.L. (1970) A heuristic algorithm for the n-job, m-machine sequencing problem. *Management Science*, **16**, B630–B637.

Carlier, J. (1982) The one machine problem. *EJOR*, **11**, 42–7.

Carlier, J. and Portmann, M.C. (1992) *Machine scheduling problems*. Summer School on Scheduling Theory and its Applications, Bonas (France), Sept 28–Oct 2, pp. 240–8.

Cerny, V. (1985) Thermodynamical approach to the traveling salesman problem. *J. Optim. Theory Appl.*, **45**, 41–51.

Charon, I. and Hudry, O. (1992) *Le bruitage: une méthode prometteuse d'optimisation combinatoire*. Journée du groupe combinatoire de l'AFCET, Nancy, 4 juin.

Chu, C., Portmann, M.C. and Proth, J.M.P. (1992) A splitting-up approach to simplify job-shop scheduling problems, *Int. J. Prod. Res.*, **30**(4), 859–70.

Chu, C. and Portmann, M.C. (1993) Application of artificial memory in scheduling problems. *J. Intell. Manuf.*, **4**, 151–61.

Conway, R.W., Maxwell, W.L. and Miller, L.W. (1967) *Theory of Scheduling*. Addison-Wesley.

Dannenbring, D.G. (1977) An evaluation of flow-shop sequencing heuristics. *Management Science*, **23**(11), 1174–82.

Dauzere-Perez, S. and Lasserre, J.B. (1993) A modified shifting bottleneck procedure for job-shop scheduling, *Int. J. Prod. Res.*, **31**, 923–32.

Dauzere-Perez, S. and Lasserre, J.B. (1994) An integrated approach in production planning and scheduling, *Lecture Notes in Economics and Mathematical Systems*, **411**, Springer-Verlag.

Dauzere-Perez, S. and Lasserre, J.B. (1994) A procedure for one-machine sequencing problem with dependent jobs, *EJOR*, to appear.

Dell'Amico, M. and Trubian, M. (1993) Applying tabu search to the job-shop scheduling problem. *Ann. Oper. Res.*, **41**, 231–52.

Emmons, H. (1969) One-machine sequencing to minimize certain functions of job tardiness, *Oper. Res.*, **17**(4), 701–15.

Erschler, J., Fontan, G. and Merce, C. (1985) Consistency of the disaggregation process in hierarchical planning. *Oper. Res.*, **34**, 464–69.

Fisher, M.L. (1973) Optimal solution for scheduling problems using Lagrange multipliers, part 1. *Oper. Res.*, **21**, 1114–27.

Fisher, M.L., Lageweg, B.J., Lenstra, J.K. and Rinnoy Kan, A.H.G. (1983) Surrogate duality relaxation for job shop scheduling. *Discrete Appl. Math.*, **5**, 65–75.

Fleury, G. (1993) Méthodes stochastiques et déterministes pour les problèmes NP-difficiles. Application à des problèmes d'ateliers et de placements, Thèse de l'Université de Clermont Ferrand II, 25 février.

Glover, F. (1975) Surrogate constraint duality in mathematical programming. *Oper. Res.*, **23**, 434–51.

Glover, F. (1987) Tabu search methods in artificial intelligence and operations research. *ORSA, Artif. Intell. Newsletter*, **1**.

Glover, F. (1989) Tabu Search – Part I. *ORSA J. Comput.*, **1**, 190–206.
Glover, F. (1990) Tabu Search – Part II. *ORSA J. Comput.*, **2**, 4–32.
Glover, F., Taillard, E. and De Werra, D. (1993) A user's guide to tabu search. *Ann. Oper. Res.*, **41**, 3–28.
Goldberg, D.E. (1989) *Genetic Algorithms in Search, Optimization and Machine Learning*, Addison-Wesley.
GOThA (1993) Les problèmes d'ordonnancements. *RAIRO Oper. Res.*, **27**(1), 77–150.
Hansen, P. (1986) *The Steepest Ascent Mildest Descent Heuristic for Combinatorial Programming*, presented at Congress on Numerical Methods in Combinatorial Optimization, Capri, Italy.
Haupt, R. (1989) A survey of priority rule-based scheduling. *OR Spektrum*, **11**, 3–16.
Hertz, A. and De Werra, D. (1990) The tabu search metaheuristic: how to use it. *Annals of Math. and Artif. Intell.*, **1**, 111–21.
Hertz, A., Taillard, E. and De Werra, D. (1992) Tabu search, in *Local search in combinatorial optimization*, (ed. J.K. Lenstra). Working paper ORWP 92/18, DMA-EPFL, Switzerland.
Holland, J.H. (1975) *Adaptation in Natural and Artificial Systems*, MIT Press, Cambridge, Mass.
Jackson, J.R. (1955) Scheduling a production line to minimize maximum tardiness. Research Report 43, Management Science Research Project, University of California, Los Angeles.
Johnson, S.M. (1954) Optimal two- and three-stage production schedules with set-up times include. *Naval. Res. Logist. Quart.*, **1**, 61–8.
Kanet, J.J. and Sridharan, V. (1992) Tactically delayed scheduling: problem taxonomy and literature review. Conference International POMS-92, Orlando, October 19–21.
Kirkpatrick, S., Gelatt, C.D., Vecchi, M.P. (1982) Optimization by simulated annealing. Res. Rep. R.C. 9335, IBM Thomas J. Watson Center, Yorktown, NY.
Kirpatrick, S., Gelatt, C.D. and Vecchi, M.P. (1983) Optimization by simulated annealing. *IEEE Trans. on Computer-Aided Design*, **CAD2** (4).
Kirkpatrick, S., Gelatt, C.D. and Vecchi, M.P. (1983) Optimization by simulated annealing. *Science*, **220** (4598), 671–80.
Laguna, M., Barnes, J.W. and Glover, F. (1991) Tabu search methods for a single machine scheduling problem. *J. Intell. Manuf.*, **2**, 63–74.
Lawler, E.L., Lenstra, J.K., Rinnooy Kan, A.H.G. and Shmoys, D.B. (1992) Sequencing and scheduling: algorithms and complexity. Report BS-R8909, Centre for Mathematics and Computer Science, Amsterdam, and Summer School on Scheduling Theory and its Applications, Bonas (France), Sept 28–Oct 2.
Lenstra, J.K. (1977) Sequencing by enumerative methods. Math. Cent. Tracts 69, Centre for Mathematical and Computer Science, Amsterdam.
McMahon, G.B. and Florian, M. (1975) On scheduling with ready times and due dates to minimize maximum lateness. *Oper. Res.,*. **23**, 475–82.
Moore, J.M. (1968) An n-job, one machine sequencing algorithm for minimising the number of late jobs. *Management Science*, **15**, 102–9.
Nowicki, E. and Smutnicki, C. (1993) A fast taboo search algorithm for the job shop problem. Preprint, 8/1993, Institute of Engineering Cybernetics, Technical University of Wroclaw.
Panwalkar, S.S. and Iskander, W. (1977) A survey of scheduling rules. *Oper. Res.*, **25**(1), 45–61.

Portmann, M.C. (1988) Méthodes de décomposition spatiales et temporelles en ordonnancement de la production. *RAIRO-APII*, **22**(5), 439–451.
Proust, C., Gupta, J.N.D. and Deschamps, V. (1991) Flowshop scheduling with setup, processing and removal times separated. *Int. J. Prod. Res.*, **29**(3), 479–93.
Rinnooy Kan, A.H.G. (1976) *Machine Sequencing Problems: Classification, Complexity and Computation*, Nijhoff, The Hague.
Roy, B. (1970) *Algèbre moderne et théorie des graphes*, tome 2, Dunod, Paris.
Smith, W.E. (1956) Various optimizers for single-stage production. *Naval. Res. Logist. Quart.*, **3**, 59–66.
Srinivasan, V. (1971) A hybrid algorithm for the one-machine sequencing problem to minimize total tardiness. *Naval Res. Logist. Quart.*, **18**(3).
Taillard, E. (1994) Parallel taboo search technique for the jobshop scheduling problem. *ORSA J. Comput.*, to appear.
Vaessens, R.J.M., Aarts, E.H.L. and Lenstra, J.K. (1994) Job shop scheduling by local search. Working Paper, Eidhoven University, February.
Van De Velde, S.L. (1991) Machine scheduling and Lagrangian relaxation, Doctoral Thesis, CWI, Amsterdam.
Van Laarhoven and Aarts, E.H.L. (1987) *Simulated Annealing: Theory and Applications*, D. Reidel Publishing Company, Dordrecht, Holland.
Van Laarhoven, P.J.M., Aarts, E.H.L. and Lenstra, J.K. (1992) Job shop scheduling by simulated annealing, *Oper. Res.*, **40**, 113–25.
Weiss, G. (1992) A tutorial in stochastic scheduling, in Summer School 'Scheduling theory and its application', Bonas, France, Sept 28–Oct 2.
Widmer, M. (1989) Job-shop scheduling with tooling constraints: a tabu search approach. OR working paper 89/22, Departement de mathematiques, EPFLS.

FURTHER READING

Aarts, E.H., Van Laarhoven, P.J.M., Lenstra, J.K. and Ulder, N.L.J. (1994) A computational study of local search algorithms for job shop scheduling. *ORSA J. Comput.*, to appear.
Baker, K.R. and Su, Z.S. (1974) Sequencing with due-dates and early start times to minimize maximum tardiness. *Naval Res. Logist. Quart.*, **21**, 171–6.
Bitran, G.R., Haas, E.A. and Hax, A.C. (1982) Hierarchical production planning: a two stage system. *Oper. Res.*, **30**, 232–51.
Bratley, P., Florian, M. and Robillard, P. (1973) On sequencing with earliest starts and due dates with application to computing bounds for the (n/m/G/Fmax) problems. *Naval Res. Logist. Quart.*, **20**, 57–67.
Carlier, J. (1987) Scheduling with release dates and tails on identical machines to minimize makespan. *EJOR*, **29**, 298–306.
Carlier, J. and Pinson, E. (1989) An algorithm for solving the job shop problem. *Management Science*, **35**(2), 164–76.
Carlier, J. and Pinson, E. (1990) A practical use of Jackson's preemptive schedule for solving the job-shop problem. *Ann. Oper. Res.*, **26**, 269–87.
Chryssolouris, G., Dicke, K. and Lee, M. (1991) An approach to short interval scheduling for discrete parts manufacturing. *Int. J. Comput. Integrated Manufacturing*, **4**(3), 157–68.
Chu, C. and Portmann, M.C. (1992) Some new efficient methods to solve the $n/1/r_i/\Sigma T_i$ scheduling problems. *EJOR*, **58**(3), 404–13.

Davis, L. (1987) *Genetic Algorithms and Simulated Annealing*, Morgan Kaufmann Publishers.

Falkenauer, E. and Bouffouix, S. (1991) *A Genetic Algorithm for Job Shop*. Proceedings of The IEEE International Conference on Robotics and Automation, Sacramento, **1**, pp. 824–9.

Ford, L.R. and Fulkerson, D.R. (1956) Maximal flow through a networks, *Canad. J. Math.*, **8**, 399–404.

Ford, L.R. and Fulkerson, D.R. (1962) *Flows in networks*, Princeton University Press, NJ.

Fox, M.S. and Smith, S.F. (1984) ISIS, A knowledge based system for factory scheduling, *Expert Sys. J.*, **1**(1), 25–45.

Fulkerson, D.R. (1961) A network flow computation for project cost curves. *Manage. Sci.*, **7**(2).

Jackson, J.R. (1956) An extension of Johnson's results on job lot scheduling. *Naval Res. Logist. Quart.*, **3**, 201–3.

Matsuo, H., Suh, C.J. and Sullivan, R.S. (1988) A controlled search simulated annealing method for the general jobshop scheduling problem. Working paper 03–04–88, Graduate School of Business, University of Texas, Austin.

Meguelati, S. (1988) Méthodes de classification pour la constitution d'îlots de fabrication et l'ordonnancement, Thèse de Doctorat de l'INSA (Toulouse).

Nawaz, M., Enscore, E. and Ham, I. (1983) A heuristic algorithm for the m machine, n job flow shop sequence problem. *Omega*, **11**(1), 91–5.

Nuitjen, W.P.M., Aarts, E.H.L., Van Erp Taalman Kip, D.A.A. and Van Hee, K.M. (1993) Job shop scheduling by constraint satisfaction. Computing Science Note 93/39, Department of Mathematics and Computing Science, Eindhoven University of Technology.

Portmann, M.C. and Camps, J.J. (1993) Steel coils scheduling in a flow shop with parallel tools and tight constraints, in *Advances in Factories of the Future, CIM and Robotics* (eds M. Cotsaftis and F. Vernadat), Elsevier, pp. 489–95.

Talbi, E.G. and Muntean, T. (1992) *Hill-climbing, Simulated Annealing and Genetic Algorithms: A Comparative Study*. Int. Conf. on Parallel and Vector Processing CONPAR92-VAPP V, Lyon, Sept.

Yamamoto, M. (1977) An approximate solution of machine scheduling by decomposition method, *Int. J. Prod. Res.*, **15**(6), 599–608.

CHAPTER 10

Compu-search methodologies II: Scheduling using genetic algorithms and artificial neural networks

F. Alexandre, C. Cardeira, F. Charpillet, Z. Mammeri and M.-C. Portmann

ABSTRACT

This chapter complements the previous chapter 'Scheduling Methodology: Optimization and Compu-search Approaches I' about the scheduling level of a production manufacturing hierarchical approach. It presents various ways of using genetic algorithms and artificial neural networks to solve scheduling problems. Genetic algorithms are used for scheduling problems without assignment unknown values (solutions are completely decribed by the list of job sequences on each resource). The potential use of artificial neural networks for solving scheduling problems is illustrated with a simple multiprocessor scheduling problem.

10.1 INTRODUCTION

Scheduling problems are found in a wide range of real-world applications. Therefore, these problems have attracted the attention of various research communities belonging to different domains, such as real-time systems, manufacturing processes and artificial intelligence. We address in this chapter the outcomes of two research areas that we expect to be applied soon to scheduling problems with promising results: genetic algorithms and neural networks.

Scheduling problems were presented in detail in the previous chapter; only their general definition is given here. The scheduling process provides the timing for executing a set of constrained tasks. In most cases, a large number of schedules satisfied the constraints. Additional criteria for measuring the performance of a schedule make it possible to select the most appropriate one. The following exact and approximation methods, among others, were introduced in Chapter 9: iterative methods, such as greedy algorithms; neighborhood methods of which simulated annealing and tabu search are two examples; splitting-up methods; constraint-based approaches; and artificial intelligence (AI) methods. We dedicate the remainder of this chapter to the use of two AI methods, that is, artificial neural networks (ANNs) and genetic algorithms (GAs).

Artificial neural networks have been applied to a large spectrum of artificial intelligence applications such as signal processing, speech recognition, scene interpretation and process control. Since the publication of Hopfield and Tank's paper (1985) on the application of Hopfield ANN for solving the Traveling Salesman Problem (TSP), a lot of attention has been given to the application of ANNs to optimization problems. A chief rationale of this approach relies on the capability to obtain fast and parallel implementations over analog or digital hardware. The possibility of getting good solutions to large intractable problems within a few microseconds is very attractive, especially for real-time applications. This allows the ANN to be run several times within a short time period, and to obtain good solutions even if some of them are not applicable. The success in using Hopfield ANNs for optimization problems relies on the dynamic of this kind of network. This leads to the minimization of an energy function that can be mapped to the optimization problem to be solved.

Both techniques were created by using analogy with biological sciences in the 1950s for ANNs and in 1975 for GAs. However, their current use as approximation schemes for solving NP-hard optimization problems is relatively recent (1985 for both of them). Most of the earlier research on these methods was conducted in the artificial intelligence field. Recently the operational research community has begun to integrate this approach with more traditional ones. ANNs and GAs were developed in two different contexts: pattern recognition and classifier system for the former and optimization methods for the latter. In this chapter, we will only consider the latter case, with the particular objective of solving scheduling problems. Each section of the chapter begins with a general presentation of the approach and ends with considerations or examples related to scheduling problems.

10.2 GENETIC ALGORITHMS

Genetic algorithms (GAs) were developed by John Holland and his colleagues at the University of Michigan (Holland, 1975). They are search algorithms which explore a solution space and try to mimic the processes observed in the natural evolution of a living population. The main features of GAs are (randomized) **creation of an initial population** of chromosomes,[1] **natural selection of best genitors** from the population of chromosomes, **genetic operators** applied to the genitors whose role is to create new members in the population by crossing the genes of two genitors (cross-over operators) or by modifying the genes of one genitor (mutation operators), **natural selection of the (best) members of the population** who will survive, and **natural convergence** of the whole population that is globally improved at each step of the algorithm. Depending on the choice of different parameters describing every detail of a particular algorithm, there is in fact a great variety of GA implementations applied to a great variety of optimization or multicriteria decision problems. In section 9.2.1 we present genetic algorithms independently of specific applications, and in section 9.2.2 we focus on the variety of GAs linked with scheduling problem solving.

10.2.1 General genetic algorithm presentation

In the following sections we present the different global structures or implementations of GA, we consider the various types of chromosome contents, and the different types of genetic operators. We then describe selection procedures, and some well-known theoretical results are given in outline; the possibility of using parallel computing for GA is then briefly considered.

Various GA implementations

The simple genetic algorithm schema (Goldberg, 1989)

First we present the simple genetic algorithm schema (SGA) proposed by Goldberg (1989). We assume here that every solution of a considered problem can be described by a string of n binary values. The string is called a **chromosome** and each binary value of the string is called a **gene** or an allele. The position of a gene in the string is called **string position** or locus.

[1] A chromosome contains either the description of one solution of the considered problem or data sufficient to build one solution of the considered problem. The initial population is obtained, for example, by applying random constructive methods as presented in the previous chapter.

At the beginning of the algorithm, a population of chromosomes is created by some specific algorithm or by randomized generation, and the criterion value of every chromosome is computed. One of the chromosomes corresponding to the best criterion value (also called 'best chromosome') is kept in a special memory. The criterion value is normalized in order to give a fitness value in order to always maximize the fitness value and to obtain chromosomes with similar criterion value quality and also quite near similar fitness values (for normalization techniques see the section below, 'Fitness value computation and normalization').

Every step of the SGA consists of a generation phase that creates a new population and kills the old one. Any population has a given constant size *N* (*N* is an even integer). For each generation phase, the following steps are executed:

- selection in the old population of *N*/2 couples of parents or mates using a roulette wheel with slots sized according to fitness values (selection techniques are described later);
- creation of a new population of *N* offsprings (two for each couple of parents) by applying a genetic operator called cross-over (for binary cross-over description, see section on cross-over operators, below);
- with a very small probability, every gene value of the new population may be modified by mutation: in this case with probability 1/2 the considered gene takes the value 1 or the value 0 (so it can be modified or not by the mutation process);
- the new modified population is evaluated (computation of criterion and fitness values) and the best chromosome of the population is kept in the special memory if the best criterion value is improved;
- the old population is totally replaced by the new modified one;
- the algorithm is stopped when the maximal number of trials is reached.

The result of SGA is the best created chromosome kept in the special memory.

General genetic algorithm schema

According to Goldberg (1989), De Jong in 1975 was the first to test overlapping between populations of two consecutive steps. He defined a parameter G of overlapping and analyzed the influence of G on the performance of SGA. In contrast with SGA, most of the GAs found in the literature use overlapping techniques. We give here the general overlapping GA schema (Davis, 1991):

1. Initialize a population of chromosomes.
2. Evaluate each chromosome in the population.
3. Create new chromosomes by mating chromosomes or by applying mutation operator.

4. Delete members of the population to make room for the new chromosomes.
5. Evaluate the new chromosomes and insert them into the population.
6. If the stop condition is verified (for instance time is up), stop and return the best created chromosome; if not, go to 3.

In the SGA, whether identical chromosomes are present in the generated populations is never tested. Therefore, good chromosomes which exist in several exemplars may be often chosen as genitors and may increase more and more the number of identical chromosomes. In the general GA schema, already existing chromosomes (either in the old population or in the new created chromosomes) are generally not inserted in the population. In this case, less chromosomes need to be deleted in order to make room for the new chromosomes, and two separate chromosomes of a given population are never identical.

Chromosome type classification

Originally, GAs used essentially binary strings. But when strings need to represent solutions of complex problems, binary representations are less convenient (or inadequate for some cross-over designs). Therefore, symbolic representations are used. Goldberg (1989) noted that the position of the genes inside the string may be important. He proposed to complete binary strings by position strings and noted that, for sequencing problems, only the position strings present some interest. For the traveling salesman problem (TSP) or for scheduling problems where a solution can be described by a sequence or a permutation, the content of the string is the corresponding sequence. So, most of the GA literature concerning TSP or scheduling problems design and use genetic operators (cross-over and mutation) applied to strings which contain job or operation sequences. When some operations may be processed by several machines or some jobs belong to different feasible routes inside the workshop (Uckun *et al.*, 1993), then chromosomes contain sequences, assignment data and routes data. There are other string possibilities specific or not for scheduling problems which will be presented later.

Genetic operator classification

Cross-over operators

We assume here that couples of mates have been already selected. We focus our attention on how to create one or two offsprings from a given couple of parents. The simplest cross-over operator used in SGA is of the **one-point** type cross-over. In the simple case of binary strings, the first step of 'one-point' cross-over operation consists in randomly choosing a cross-over point which cuts each mate string into two sub-strings, the beginning string and the terminal string, and then builds offspring 1 using the beginning string of mate 1 and the terminal string of mate 2 (and the reverse for offspring 2). The same process can be used for symbolic chromosomes, but then each sub-string may be modified in order to verify specific constraints of the problem (we present a one-point cross-over applied to permutation chromosomes below).

The immediate generalization of one-point cross-over operators is **two-points** cross-over operators and then **k-points** cross-over operators. The first step of 'k-points' cross-over operation consists in randomly choosing k cross-over points: $j_1, j_2, \ldots, j_k$ with $0 < j_1 < j_2 < \ldots < j_k < n$ which cuts each mate string into $k + 1$ sub-strings. In the simple case of binary strings, the offspring 1 is obtained by taking the even numbered sub-strings of mate 1 and the odd numbered sub-strings of mate 2 (and the reverse for offspring 2).

According to Goldberg (1989), De Jong defined **uniform** cross-over as k-points cross-over with k equal to $n/2$. For Syswerda in Davis (1991), uniform cross-over operators consist of n steps: at step k, the binary value of the kth locus of mate 1 (resp. 2) is put in the locus k of offspring 1 (resp. 2) or of offspring 2 (resp. 1) with probability 1/2. We will use this latter definition of uniform cross-over operator.

Almost every cross-over operator found in the literature is either of the type k-points or uniform cross-over.

Mutation operators

By always marrying people of the same initial population, we obtain in real life the phenomenon of degeneration. Doing the same in GAs, we take the risk of always remaining in a limited subset of the whole set of solutions. In order to avoid this drawback, a mutation phase is included after the cross-over phase: it consists of modifying arbitrarily one or several members of the population. We have already presented the mutation technique used in the simple genetic algorithm SGA. In the case of permutation chromosomes, mutation operators are very similar to creating a neighbor in local search methods. The most frequently used of them consists of:

- randomly choosing a position k between 2 and n and exchanging genes in position k and $k-1$ (called **transposition of two consecutive genes**),
- randomly choosing two different positions $k1$ and $k2$ between 1 and n and exchanging their genes (called **transposition of two genes**),
- randomly choosing two positions $k1$ and $k2$ between 1 and n and reversing the order of the gene values between position $k1$ and $k2$ (called **inversion** operator[2]).

[2] This genetic operator, depending on the authors, is considered either as a mutation operator or as a 'two-points' cross-over operator which gives only one offspring. In 1975, Holland considered it as a third family of genetic operators.

Selection methods in genetic algorithm schema

There are two selection phases inside a general GA. One selection consists of choosing the genitors or couples of mates. Another consists of deciding which chromosomes of the old population must be killed. Generally, every selection phase uses what is called the chromosome fitness value which is computed using the criterion value of the chromosomes.

Fitness value computation and normalization

If the considered criterion has to be maximized, then the criterion value may be used as a fitness value. It is the simplest way for obtaining it. If the considered criterion has to be minimized, then we need an upper bound UBC of the criterion value, either a static one given, for instance, as an input coefficient or computed by a procedure for the particular data instance of the problem that the GA is solving, or a dynamic one, generally the greatest value of the criterion in the considered population.

If $C(x)$ is the criterion value of chromosome x and $F(x)$ the fitness value, then the simplest way to compute $F(x)$ is the following one: $F(x) = UBC - C(x)$ (if UBC is not an upper bound and the obtained value for $F(x)$ is negative, then $F(x)$ is put equal to 0).

When the exact value of the criterion matters less than the rank of each chromosome inside the population, then a linear normalization is used (Davis, 1991):

- two arbitrary input integer coefficients are given: A and a with $A \geqslant N.a$ (N is the population size);
- the population is sorted by decreased values of $F(x)$ (ties randomly broken): $F(x_1) \geqslant F(x_2) \geqslant \ldots \geqslant F(x_N)$;
- the element x_i of rank i gets the fitness: $F(x_i) = A - (i - 1).a$.

Some linear scaling procedures have also been defined and tested (Goldberg, 1989, pp. 75–9).

Selections of mates

Mates may be randomly chosen inside the population. Only couples randomly created with mate 1 identical to mate 2 are deleted. In natural selection, it is assumed that better genitors will create better young; so the mates are selected with a probability dependent on their fitness. The technique uses a roulette wheel with slots sized according to fitness. Described by an algorithm, the corresponding selection procedure is the following:

- compute TF, the sum of chromosome fitness for the whole considered population,

- consider the chromosome in any order $x_1, x_2, \ldots x_N$,
- generate uniformly a real number π between 0 and 1,
- select the chromosome q such that:

$$\frac{\sum_{k=1}^{q-1} F(x_k)}{TF} \leqslant \pi < \frac{\sum_{k=1}^{q} F(x_k)}{TF}$$

(the left inequality becomes $\leqslant$ when $q = N$).

Selections linked to the population evolution
Several overlapping techniques exist to delete P chromosomes of a population of size N in order to make room for P new created ones. The deterministic selection consists in sorting the chromosomes according to decreasing fitness values (or even criterion values) and keeping the $N - P$ best chromosomes of the population. The stochastic selection uses the procedure described above (like a roulette wheel) in order to select successively $N - P$ different (best) chromosomes from the old population which will be added to the set of newly created chromosomes.

General theorems and properties

We consider here binary (resp. p-ary) gene values and the corresponding one-point cross-over. Holland (1975) has defined the notion of schema. A schema is a 3-ary (resp. $p + 1$-ary) string of n characters, each character being equal to 0, 1 or * (resp. 1, 2, . . . p or *).

A 'schema' S (string $s_1 s_2 \ldots s_n$) matches a chromosome C (string $c_1 c_2 \ldots c_n$) if and only if, for each k, $s_k = c_k$ or $s_k = *$. If a schema S matches a chromosome C, then C is said to be a representative of schema S. A schema S is said to be good if the average value of fitness (or criterion) of its representatives is good. The 'order' of a schema S is the number of fixed values 1 or 0 (resp. 1, 2, . . ., p) inside the schema. The 'defining length' of a schema S is the difference between the position of the last and first fixed values inside the schema. For instance, the schema *01***10** has an order of 4 and a defining length of 6.

The fundamental schema theorem is the following:

Assuming that a 'one-point' cross-over is used (beginning of a mate and ending of another mate), that a roulette wheel selection procedure is used so that the best genitors are chosen and the best chromosomes are kept (or a generation process is used like in SGA), and that identical chromosomes are kept inside the population, then the number of representatives of good schemata with small order and small defining length is not decreasing along the steps of the genetic algorithms.

This theorem explains partially the fact that the whole population is globally improving along the execution of GA. Defining length is linked to gene positions inside the chromosomes. If these positions are badly chosen, one-point cross-over may be bad relatively to the fundamental theorem, so it could be a reason for using uniform cross-over. For global convergence of GA proved by Markov chain analysis, see Eiben *et al.* (1991) and Cerf (1996).

Genetic algorithms and parallel computing

It is very easy to implement GAs on parallel processors. For m processors, generally, the population is split up into $m - 1$ sub-populations assigned to processors 2 to m, processor 1 being kept as central processor. Every K steps, the processors exchange (best) chromosomes of their populations under the control of the central processor which always keeps the best found solution.

10.2.2 Genetic algorithm and scheduling problems

GAs for solving a scheduling problem begin with a population of schedules. As in the simulated annealing approach, these schedules may be obtained by any approximation schema or by random generations. Several codes may be used for scheduling problems. Cross-over and mutation operators depend on the particular code. We consider successively different codes for the genes and examine for each of them which genetic operators may be used.

Symbolic codes and scheduling solutions

Correspondence between symbolic codes and scheduling solutions
In most of the GAs found in the literature, researchers or practitioners are using symbolic codes. In this case, the content of each chromosome is a list of sequences (or permutations) of operations (or jobs). In the simplest case, the one-machine problem, a chromosome contains the sequence of operations to be performed in this order on the machine. In a similar simple case, the permutation flow-shop problem, a chromosome contains a sequence of jobs, the same one for each machine. In this case, every semi-active (or left-justified) schedule is also an active one.

For the general flow-shop problem, when the order of the jobs may be different on each machine, a chromosome contains a list of job sequences, one sequence for each machine, and the cross-over operator is applied successively for each sequence inside the chromosome corresponding to a given machine (we cross the sequence of machine 1 of mate 1 with the sequence of machine 1 of mate 2, then sequence of machine 2 of mate 2 with sequence of machine 2 of mate 2 . . . until machine m). Every created chromosome corresponds then to a feasible schedule, but every semi-active schedule is not necessarily an active one and so its performance may be bad. Therefore, to solve general flow-shop problems, techniques developed for job-shop problems may be preferred.

For the job-shop problem, we may always use operation sequences for each machine. But in this case, some chromosomes may give non-feasible schedules. For instance, if the kth operation of job i_1 is performed on the machine m_1 and the $(k + 1)$th operation of job i_1 is performed on the machine m_2, while the jth operation of job i_2 is performed on the machine m_2 and the $(j + 1)$th operation of job i_2 is performed on the machine m_1, then if the $(j + 1)$th operation of job i_2 is scheduled in the sequence of operations before the kth operation of job i_1 on machine m_1 and the $(k + 1)$th operation of job i_1 is scheduled before the jth operation of job i_2 on machine m_2, we have a cycle in the graph of precedence constraints as shown in Fig. 10.1, and there is no feasible schedule that verifies these sequences of operations: an operation must be performed strictly before itself because of the cycle!

As operation sequences randomly generated or given by mutation or cross-over operators do not always give a feasible schedule for the job-shop problem, not all authors use the contents of the chromosome as the order of operations on the different machines but as 'preference lists' that are used as priority rules in a constructive method. The position of operations in the sequence corresponding to a machine is used as the priority value for each operation. The smaller the position of an operation in the sequence is, the greater the priority of this operation. This priority is used in order to build either a non-delay schedule (frequently found in the GA literature) or an active schedule (rarely found in the GA literature).

Cross-over operator for permutations or sequences

Partially mapped cross-over or PMX cross-over As a first example, we describe the cross-over operator on permutations used by Biegel and Davern (1990) for solving a job-shop problem. This cross-over was originally designed for the traveling salesman problem by Goldberg and Lingle (1985). It is a two-points cross-over. An example where the

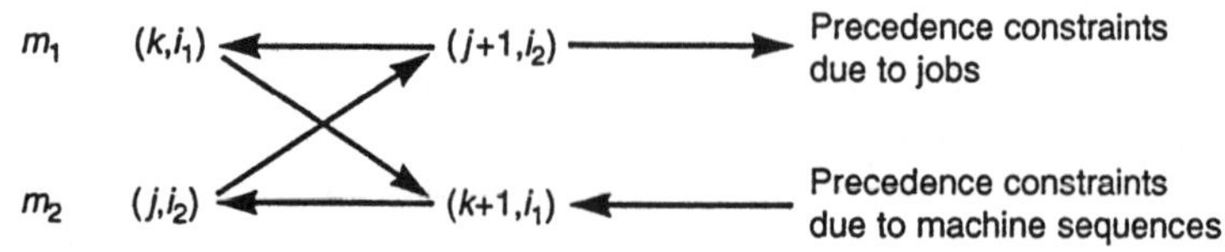

Figure 10.1 Graph of precedence constraints.

operations are denoted by letters is given to illustrate it. A cross-area is randomly chosen by two generated points, from the 4th to the 6th position, and the value of mate 1 and of mate 2 are exchanged inside the cross-area as shown in the following table.

mate 1	i	h	d	e	f	g	a	c	b	j
mate 2	h	g	a	b	c	j	i	e	d	f
offspring 1 in preparation	i	h	d	b	c	j	a	[c]	[b]	[j]
offspring 2 in preparation	h	[g]	a	e	f	g	i	[e]	d	[f]

After this modification, we no longer have a permutation: b, c and j exist two times in the first list. This is the case for e, f and g in the second list. To correct the offspring 1, in the non-cross-over region, b is replaced by e, c by f and j by g which are the lost values and which are in the mate 2 in front of these letters in the initial table. To correct the offspring 2, e is replaced by b, f by c and g by j. It gives:

offspring 1	i	h	d	b	c	j	a	f	e	g
offspring 2	h	j	a	e	f	g	i	b	d	c

Many different cross-over operators have been developed for permutation chromosomes. Most of them are now described using only one example and giving some remarks about their relative interest for scheduling problems. In order to simplify the presentation, we will always use the following sequences for mate 1 and mate 2:

mate 1	a	b	c	d	e	f	g	h	i	j
mate 2	c	b	a	g	h	i	j	f	d	e

Edge recombination cross-over This cross-over operator was proposed by Whitley *et al.* (1989). It has been designed especially for the traveling salesman problem, but may also be useful (deleting the fact that positions n and 1 are considered as neighbors), for instance, for one machine scheduling problem with sequence dependent set-up times.

First, an edge table is built: it contains for each operation a list of operations which are immediate neighbors either in mate 1 or in mate 2 (a sign – is put when an operation is neighbor twice, and position n and 1 are considered as neighbor positions)

a	–b	g	j		f	d	e	g	j
b	–a	–c			g	a	f	–h	
c	–b	d	e		h	–g	–i		
d	c	–e	f		i	–h	–j		
e	c	–d	f		j	a	f	–i	

Only one offspring is built. An arbitrary first operation is chosen among the operations with the smallest list of neighbors, for instance, and is called the current operation. Then the following iterative algorithm is used:

1. Select an operation which is a neighbor of the current one and which has the fewest remaining neighbors (breaking ties randomly); select an arbitrary remaining operation if the current operation has no remaining neighbor.
2. The operation is added to the sequence and becomes the new current one.
3. Until n operations are selected go to 1.

We obtain the following offspring:

[b] a [j] i h g f [d] [e] c

where □ = random breaks of ties

Cycle cross-over This is another cross-over designed for the traveling salesman problem by Oliver *et al.* (1987). In a first phase, a cycle using both mates is copied in offspring 1 (resp. 2). In a second phase, the remaining empty positions of the considered offspring are filled with the gene of mate 2 (resp. 1).

The description of the first phase (for offspring 1) is as follows:

1. Select randomly a starting position (here position 4) which is called the current position.
2. Copy the value of the current position of mate 1 in offspring 1.
3. Search the position of this value in mate 2, this position becomes the current position.
4. If the current position is yet empty in offspring 1, go to 2 else stop first phase.

mate 1	a	b	c	d	e	f	g	h	i	j	
mate 2	c	b	a	g	h	i	j	f	d	e	
offspring 1	$\underline{c}$	$\underline{b}$	$\underline{a}$	d	e	f	g	h	i	j	without underlining = phase 1
offspring 2	$\underline{a}$	$\underline{b}$	$\underline{c}$	g	h	i	j	f	d	e	with underlining = phase 2

Order cross-over (OX, LOX . . .) A first two-points order cross-over was proposed by Davis (1985). A modified version called OX cross-over was proposed by Oliver *et al.* (1987). It is a two-points cross-over. The middle cross-area of mate 2 is used for offspring 1 and the middle cross-area of mate 1 is used for offspring 2. Then, beginning with the third area and followed by the first area, the missing values in offspring 1 (resp. 2) are written in the order they each have in mate 1 (resp. 2).

mate 1	a	b	c	d	e	f	g	h	i	j
mate 2	c	b	a	g	h	i	j	f	d	e
offspring 1	e	f	j	g	h	i	a	b	c	d
offspring 2	h	i	j	d	e	f	c	b	a	g

The linear OX cross-over (or LOX cross-over) was proposed by Falkenauer and Bouffouix (1991) as an improvement of OX for job shop scheduling problems. The first steps are identical, but when empty positions are filled in the order of mate 1 (resp. 2), the first extreme area is filled before the last extreme area, so operations are put in the order of mate 1 (resp. 2).

An uniform order cross-over was proposed by Syswerda in Davis (1991). It is a 'uniform'-like cross-over, but more complex. For offspring 1 (resp. 2), the randomly chosen positions considered in mate 2 (resp. 1) give a partial order of the corresponding operations. The remaining operations are kept to their positions in mate 1 (resp. 2), and the holes are completed following the partial order.

Position based cross-over The position based cross-over was also proposed by Syswerda in Davis (1991). It is a modified uniform cross-over operator. Here, the chosen position values of mate 1 (resp. 2) are copied in the same position in offspring 1 (resp. 2). Then the empty positions of offspring 1 (resp. 2) are filled with the missing values following the order of mate 2 (resp. 1).

mate 1	a	b	c	d	e	f	g	h	i	j
mate 2	c	b	a	g	h	i	j	f	d	e
	*	*		*		*	*			
offspring 1	a	b		d		f	g			
			c		h			i	j	e
offspring 2	c	b		g		i	j			
			a		d			e	f	h

k-points simple cross-over A one-point cross-over was recently designed for a project by an engineering student (Molongo) of the École des Mines de Nancy, and Portmann (Djerid *et al.*, 1995) proposed the corresponding 'k-points' cross-over. *k* may be an input coefficient or randomly chosen between 1 and *n*/2. *k* cross-points are generated. Four offspring are generated (identical chromosomes are deleted). Offspring 1 (resp. 2) consists in copying the odd numbered sub strings of mate 1 (resp. 2) and reordering in the order of mate 2 (resp. 1) each even numbered substring of mate 1 (resp. 2). Offspring 3 (resp. 4) is obtained in the same manner by reversing even and odd substring roles.

mate 1	a	b	c	d	e	f	g	h	i	j	
mate 2	c	b	a	g	h	i	j	f	d	e	
offspring 1	~~a~~	~~b~~	~~c~~	~~d~~	~~e~~	~~f~~	~~g~~	~~h~~	~~i~~	~~j~~	= mate 1
offspring 2	c	b	a	g	h	i	j	d	e	f	
offspring 3	b	a	c	d	e	g	f	h	i	j	
offspring 4	b	c	a	g	h	i	j	f	d	e	

This k-points simple cross-over has the drawback that created offspring are too similar to the mates (especially if *k* is great). But it possesses the following property 1, that is not true for the other cross-over operators previously presented in the literature except for the one with binary values briefly presented below and proposed in Djerid *et al.* (1995).

- **Property 1**
  If *i* precedes *j* in both mate 1 and mate 2, then *i* precedes *j* in the generated offspring.

This property is useful for scheduling problems. Actually, precedence constraints may exist (and must be verified) between the operations of the sequence in the definition of the scheduling problems. In some scheduling problems, selected dominant properties may be proved for couples of jobs (*i*,*j*): there exists at least one optimal solution in which job *i* precedes job *j*; the application of dominance properties gives a set of new precedence constraints which may be added to the definition of the problem and thus may reduce considerably the size of the solution set to be explored in any optimal solution search (for example, for the one-machine scheduling problem with mean tardiness criterion a set of precedence constraints may be *a priori* obtained by applying Emmons dominance properties (Emmons, 1969).

The mutation operators for permutation or sequence chromosomes have been presented earlier in this chapter. There are few of them. They have been directly presented because it is easy to disturb a permutation slightly when keeping a correct permutation. Under the hypothesis that precedence constraints must be maintained between pairs of operations, mutation operators must be slightly modified (forbidding when *i* must precede *j* to reverse *i* and *j*). Some techniques presented for simulated annealing and tabu search using critical operations may be also used here.

n-ary code, production and priority rules

Goldberg (1989) shows how genetic algorithms may be crossed with expert systems using production rules ('if condition of the rule is verified then perform the corresponding action') to build some learning classifier systems. We will present here genetic algorithms which use rules better known in scheduling methodology as priority rules. The literature concerning this technique is surprisingly very poor – Dorndorf and Pesch (1995), Chiu and Yih (1995) and Portmann and Ghedjati (1994), whose idea was initially proposed by Storer *et al.* (1992) for general local search methods.

The idea is as following: many constructive methods using priority rules exist in order to build active or non-delay schedules. Some are very general and may be applied to a great variety of scheduling problems. Others are developed for specific problems (Portmann and Ghedjati, 1994). We consider a given problem and a generator of solutions for this problem which, at each step, adds an operation to a partial schedule by applying a rule chosen among a given set of rules. We number the different rules.

We assume now that there are n operations to be scheduled and p available rules. A chromosome (of length n) contains in position k an integer i whose value is between 1 and p and which represents the number of the rule to be applied in order to schedule the operation that must be scheduled in the kth step of the constructive method. The initial population may be filled by chromosomes whose genes are all identical (in this case, the chromosome is analogous to applying only one simple priority rule) and by chromosomes which are totally randomly filled (in this case, each gene value is randomly generated between 1 and p). Every k-point and uniform cross-over operator designed for binary values may be used here for integer values and schemata are probably significant (it can be assumed for instance that a rule such as SPT will be more efficient at the beginning of the schedule construction and rules that use due dates such as SLACK will be more efficient at the end of the schedule construction).

Binary code and scheduling properties

It is generally considered that for permutation problems direct representation with binary codes needs the development of complicated genetic operators. This possibility is rarely considered in the literature, except in Nakano and Yamada (1991).

Considering a given permutation of length n, it is clear that to obtain an equivalent binary representation, simply consider the corresponding anti-symmetric and transitive precedence matrix **A** defined by $a_{ij} = 1$ if i precedes j in the permutation and 0 otherwise ($a_{ii} = 1$). This matrix has the property that there is one and only one line which contains k values equal to 1, k varying from 1 to n and there is one and only one column which contains k values equal to 0, k varying from 0 to n–1. The size of matrix **A** is n^2, but only $n(n-1)/2$ may be kept in the chromosome, using the anti-symmetry of the matrix; the elements below the diagonal are given by complementarity from the elements above it. Nevertheless, in order to define cross-over operators on this binary representation, it is easier to work on a complete matrix **A**.

The Nakano and Yamada cross-over is complex. They use a classical binary cross-over and obtain the upper triangular significant part of matrix **A**. The complementary part of matrix **A** is obtained by anti-symmetry property with or without creating this corresponding part in the memory. This matrix may not represent a permutation. In this case, they develop an iterative 'repairing' procedure in order to transform the square matrix **A** into a correct permutation matrix. At a step of the procedure, they compute the number of 1s in each line of the current matrix and sort these numbers by decreasing values: $k_1, k_2, \ldots, k_n$. The matrix is correct if $k_1 = n$, $k_2 = n-1$, $\ldots$, $k_n = 1$. They consider the smallest j such that $k_j \neq n + 1 - j$ and they force some binary values to be equal to 0 or 1 to obtain the correct value.

The k-points simple cross-over operators presented earlier may be easily presented as very simple binary cross-over using the precedence matrix **A**. For instance, for the one-point simple cross-over operator (k = 1), if we denote by **A1** and **A2** the precedence matrices of mate 1 and mate 2, the procedure for computing the precedence matrix **O1** of offspring 1 (it is very easy to modify the procedure in order to obtain **O2**, **O3** or **O4**) is the following:

1. A cross-point q is randomly chosen between 1 and n–1.
2. The lines of matrix **A1** are numbered $l_1, l_2, \ldots, l_n$ by decreasing values of total number of values l on the line.
3. The lines $l_1, l_2, \ldots, l_q$ of mate 1 are put in offspring 1.
4. The remaining empty positions of the columns $l_1, l_2, \ldots, l_q$ of offspring 1 are filled with the corresponding values of mate 1.
5. The remaining empty positions for $i \notin (l_1, l_2, \ldots, l_q)$ and $j \notin (l_1, l_2, \ldots, l_q)$ are filled with the corresponding values of mate 2.

Very recently Portmann (Djerid *et al.*, 1995) designed a new binary cross-over operator of 'uniform' type whose design is simpler than the one of Nakano and Yamada. The main ideas of this operator are the following (no example is given due to lack of space):

1. Let O be the Boolean matrix obtained by applying the 'and' Boolean operator to matrices mate 1 and mate 2 (1 = 'true' and 0 = 'false').
2. Keep 'definitively' every coefficient (i,j) such that $O(i,j)$ or $O(j,i) = 1$.
3. Choose randomly with probability 1/2 (or π and $1-\pi$) from mate 1 or mate 2 the value of the first (for example, smallest number line and then smallest number column) non-definitive coefficient in O which becomes definitive.
4. Compute the transitive closure of O and put it in O.
5. If there remains a non-definitive coefficient go to 2.

This binary 'uniform' cross-over, like the 'k-points' simple cross-over, possesses the Property 1. Furthermore, it is a very powerful operator which may be extended to 3-ary codes and used to build GAs or to cross GAs and branch and bound methods for any disjunctive scheduling problems and not only permutation scheduling problems (Djerid and Portmann, 1995).

Performance of genetic algorithms for scheduling problems

As most of the randomized approximation schemes that explore randomly the solution spaces, GAs improve the results obtained by applying successively a lot of specific or not specific constructive methods. In our own numerical experiments (Portmann and Ghedjati (1994), for instance), we improve the average value of the criterion between 8 and 14% depending on the chromosomes representation (priority rules or permutations) and the parameters of the general schema of the used GA (with population overlapping). Compared with exact branch-and-bound methods for NP-hard problems, GAs are obviously more efficient and faster when the size of the problem increases and are not too bad for small sizes. Many GA users have compared different schemes of genetic algorithms, but not with other approximate approaches. When GA is compared to tabu search or simulated annealing by specialists of these methods, and they but not the GA approach are optimized, then it evidently loses out. But when parallel computing is used, then GA becomes a better approach than the others. GA is said to be a robust approach (Goldberg, 1989), that is, a specific method better than GAs may exist for a particular instance, but on average GAs are never bad for a very large variety of problems. The published permutation cross-over was designed more for the traveling salesman problem than for scheduling problems; new permutation cross-overs that are especially designed for particular scheduling problems (for example with precedence constraints) are promising for future implementations of GAs or crossing of GAs with operations research methods.

A small practical example

We consider the sequencing problem of minimizing the sum of weighted tardiness for the basic one-machine problems. The n operations are ready to be processed at time 0. Each operation i must be processed on a single machine always available and possesses a processing time p_i, a due date d_i and a penalty cost w_j for every unit time of tardiness. We will use a genetic algorithm as an approximation solving schema.

T(σ) is a function which provides us with the sum of the tardiness penalties when the n operations are scheduled 'left justified' following the sequence σ.

FT is an array where the fitness value of each element of the population is stored before using a selection procedure.

RAND(N) gives randomly and equiprobably any integer between 1 and N.

RANDSEQ is a function which creates with probability $1/n!$ any of the $n!$ permutations of 1, 2, . . ., n.

EDD is a function which creates the earliest due date sequence.

WHEEL(V) is a function which gives randomly the rank i with the probability

$$\frac{V(i)}{\sum_{i'=1}^{n} V(i')}.$$

OFFSPRINGS(σ_1,σ_2) gives two new permutations by applying one of the permutation cross-over operators presented in this chapter.

MUT(σ) gives a new permutation by applying one of the mutation operators presented in this paper.

SORTED(P) gives a list (π_1, π_2, . . ., π_N) of the population P sorted by increasing values of the tardiness criterion.

BEST is the better found weighted tardiness and BESTSEQ is a sequence which gives this best value to the criterion.

MBEST(σ) is the following procedure which up-dates BEST and BESTSEQ each time a better solution is encountered:

CT:=T(σ);
if CT < BEST then BESTSEQ:=σ; BEST:=CT endif

The corresponding GA program may be the following:

Initialization:
BESTSEQ:=EDD; BEST:=T(BESTSEQ);
for i varying from 1 by 1 until N do σ_i:=RANDSEQ and MBEST(σ_i);
K generation phases

```
for k varying from 1 by 1 until K do
begin
  computation of the fitness for the old population
  π:=SORTED(σ_i, σ_2, . . ., σ_N)
  FT(σ_1):≠N;
  for i varying from 2 by 1 until N do FT(π_i):=FT(π_{i-1})-1;
  selection of the mates and generation of offsprings
  for i varying from 1 by 1 until N/2 do
  begin
    repeat
      μ_{i,1}:=WHEEL(FT); μ_{i,2}:=WHEEL(FT);
    until μ_{i,1} ≠ μ_{i,2};
    (σ'_{2i-1},σ'_{2i}):=OFFSPRINGS(π_{μ_{i,1}},π_{μ_{i,2}})
    MBEST(σ'_{2i-1}); MBEST(σ'_{2i});
  end;
  mutation phase
  j:=RAND(N); σ'_j:=MUT(σ'_j); MBEST(σ'_j);
  (σ_1, σ_2, . . ., σ_N):=(σ'_1, σ'_2, . . ., σ'_N);
end;
print BEST,BESTSEQ.
```

10.3 ARTIFICIAL NEURAL NETWORKS

This section focuses on the use of artificial neural networks (ANNs) for solving scheduling problems. First, ANNs are presented. Then their basic components, the formal neurons, are described together with the main structures of these networks. A specific model, the Hopfield model, is presented first as a content-addressable memory, second as a possibly adapted model for optimization and scheduling problem. Finally, an original application is presented, where a Hopfield model is used for a real-time scheduling of tasks.

10.3.1 Artificial neural networks presentation

ANNs must be studied in the framework of parallel distributed processing. They can be defined as a set of interconnected elementary units sharing fundamental adaptative properties. This restricted definition and the very name of this field (also called connectionism or neurocomputation) indicates strong links with biological nervous systems. Nevertheless, ANN models are rare where this analogy relies on strong evidences. Anyway, it is not necessary to refer to neurosciences to work in the ANN field. Indeed, ANNs can also be viewed as belonging to graph or cellular automata theory. In this case, neurosciences are only considered as a 'vocabulary provider' and we will make extensive use here of biologically inspired terms (e.g. neurons, synaptic weight) only as analogical references and not with precise justifications, as is usual in the ANN literature.

The formal neuron

ANNs have been studied for a long time. In 1943, MacCulloch and Pitts first introduced a simple model of a neuron, as a binary threshold unit. More precisely, the neuron i computes a weighted sum of its inputs from other units j (A_i: activation function) and outputs one if the sum is above a certain threshold and zero otherwise (O_i: output function).

$$\begin{cases} A_i(t+1) = \sum_j W_{ij} O_j(t) \\ O_i(t) = f_i(A_i(t)) \end{cases}$$

Time t is taken as discrete. W_{ij} is the synaptic weight or the strength of the connection between neuron i and neuron j. $f_i(x)$ is the threshold function for unit i and is defined as:

$$f_i(x) = \begin{cases} 1 \ \textit{if}\ x \geqslant t_i \\ 0 \ \textit{otherwise} \end{cases}$$

where t_i is the threshold defined for neuron i (Fig. 10.2).

This McCulloch–Pitts neuron is also called a binary neuron. Other formal neurons can be defined in a similar way: linear neurons taking values in $\mathcal{R}$ (the threshold function is identity), continuous neurons taking values in $]-a;a[$, In this case, the threshold function can be sigmoidal:

$$f_i(x) = a\frac{(e^{kx}-1)}{(e^{kx}-1)}$$

These kinds of elementary units were used in the 1960s to build feed-forward networks for certain computational tasks (Rosenblatt, 1962;

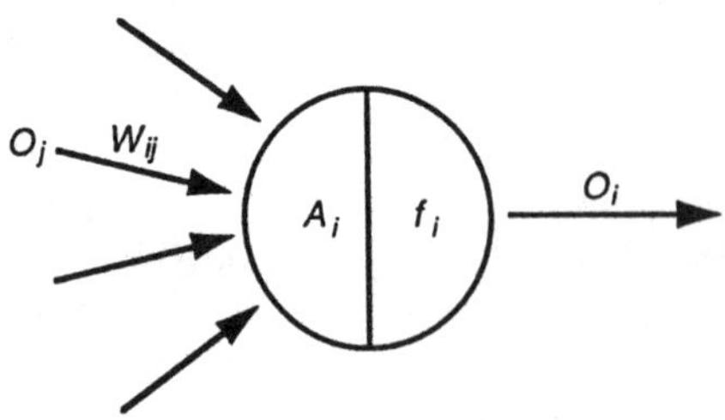

Figure 10.2 McCulloch–Pitts neuron.

Widrow and Hoff, 1960). One fundamental issue in these models was the definition of learning algorithms, i.e. ways to iteratively modify the weights so that the desired computational task was performed. Several algorithms were proposed, but only in the 1980s did really efficient learning algorithms appear (mainly the backpropagation learning algorithm, based on gradient descent techniques: Rumelhart and MacClelland (1986); LeCun (1987)).

Since then, there has been an extraordinary resurgence of interest in ANNs. Nowadays there are several successful applications of ANNs to a variety of domains. Widrow and Lehr (1990) give a very good historical review of the use of ANNs. Hertz *et al.* (1991) give an excellent introduction to the ANN theory and its various models. Basically, we can divide this set of models into two parts. First, there are feed-forward models. They include an input layer, hidden layers and an output layer. These models learn to link the input space to the output space in a supervised manner and are mainly used for classification or matching tasks. Second, there are feedback models, in which a signal can propagate in various directions (not only from the input to the output neurons, as for feed-forward models). Here, we often have to wait for stabilization of the network which learns in an unsupervised manner to retrieve specific low energy states. These models are mainly used as content-addressable memories.

The Hopfield model

While most researchers despairing of finding a good ANN theory have abandoned ANNs, some pioneers as Grossberg, Kohonen and Hopfield have laid down the foundations of ANNs. Hopfield has proposed a model whose specific extensions are useful for optimization procedures. We first sum up the main characteristics of the Hopfield model. Then, we explain its use for optimization problems.

Basically, the Hopfield model (Hopfield, 1982) is a perfect prototype of a feedback model used as a content-addressable memory. A Hopfield network is a set of totally interconnected binary neurons. The easiest way to describe these units is to use an equivalent model with {1;–1} output states (instead of {0;1} in the original model) and with a threshold $t_i = 0$ for each neuron. This means that:

$$O_i(t+1) = \sum_j W_{ij} O_j(t) = \begin{cases} 1 \; if \; O_i(t+1) \geqslant 0 \\ -1 \; otherwise \end{cases}$$

Later, Hopfield proposed a new version of his model in which neurons are continuous units with a sigmoidal threshold function f_i. Their activation function A_i (yielding $O_i = f_i(A_i)$) verifies the differential equation:

$$\frac{dA_i}{dt} = -\frac{A_i}{n_i} + \sum_j W_{ij}O_j + I_i$$

Where

$$-\frac{A_i}{n_i}$$

is a decay term and I_i is the external input of neuron i.

The functioning of the network consists in iteratively choosing a neuron at random and updating its output until stabilization of the network. The learning of the network consists in updating the weights so that the network should be able to retrieve a set of prototypal patterns that are given to the network. This learning can be done in an iterative way, using the Hebbian learning rule (Hebb, 1949). For each prototypal pattern p and for each weight:

$$\Delta W_{ij} = \varepsilon O_i^p O_j^p$$

In fact, the prototypal patterns (Fig. 10.3) can be also considered as stabilization states for the network, which must correspond to minima of energy. From this global point of view, Hopfield showed that the weights could be directly computed as follows:

$$W_{ij} = \sum_p O_i^p O_j^p$$

and $W_{ii} = 0$

This global approach is very important because it directly leads to the use of such models for optimization. It starts from Cohen–Grossberg stability rules (Cohen and Grossberg, 1983). These rules state four conditions under which an ANN converges to a stable state and includes a Lyapunov function. These conditions are:

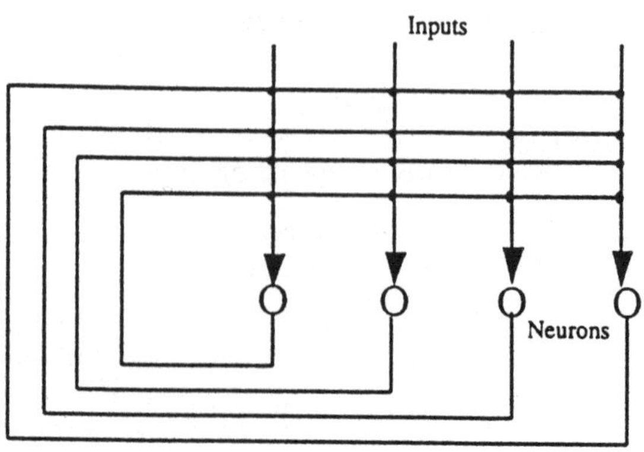

Figure 10.3 Prototypal patterns.

1. There is no self feedback (i.e. $W_{ii} = 0$).
2. The connection matrix is symmetric (i.e. $W_{ij} = W_{ji}$).
3. There is no positive feedback (i.e. $W_{ij} \leqslant 0$).
4. The transfer function of each neuron is a non-decreasing function.

Concerning his model, Hopfield derived its Lyapunov function that he called an energy function with analogy to physical systems:

$$E = -\frac{1}{2}\sum_i \sum_j W_{ij}O_iO_j - \sum_i I_iO_i \tag{10.1}$$

This quantity decreases toward a minimum which corresponds to one of the prototypal patterns in the better case, but which can also correspond to spurious states, combination of the prototypal pattern if these patterns are not orthogonal. In fact, a lot of research (mainly linked with the physical system analogy) has been conducted to evaluate the performances and the capacities of the Hopfield model (Hertz *et al.*, 1991) and is not enthusiastic about them, as will be discussed later. Many extensions of this model have also been proposed (Hertz *et al.*, 1991), including its adaptation to optimization problems, which will be discussed.

10.3.2 ANNs for optimization

The Hopfield model is not the only model that has been used for optimization problems. The same can be said for instance about Kohonen maps (Angéniol *et al.*, 1988). Nevertheless, the case of the Hopfield model is certainly the most famous and the most typical.

The energy function that we just described is at the basis of the work presented in Hopfield and Tank (1985) which explains how this model can perform optimization tasks. We have seen that the energy function is a decreasing function that leads toward a stable state. That is, if a Hopfield model has weights W_{ij} and external inputs I_i, then it will stabilize on (or it will find) those outputs O_i that minimize the energy function E. Now, consider an optimization problem where you try to find some coefficients C_i that minimize a function F, depending on some parameters P_{ij} and Q_i. The adaptation of the Hopfield model to optimization problems is basically the following. If the computational formula of F can be mapped into the following shape:

$$F = -\frac{1}{2}\sum_i \sum_j P_{ij}C_iC_j - \sum_i Q_iC_i$$

then a Hopfield model with weights P_{ij} and external inputs Q_i will stabilize on (or will give you) those outputs C_i that minimize F. This analogy is the fundamental point of this technique, but a lot can be said about its validity, as will be discussed later.

Compared with other neighborhood methods (Gotha, 1993), ANNs do not require to test whether the last solution is better than the previous. One of the properties of these ANNs is the guarantee that the last solution is always better than or at least equal to the previous one. The main difficulty is to translate the cost functions into an energy function.

This approach has been applied to several applications including the famous traveling salesman problem (Hopfield and Tank, 1985), the graph partitioning problem (Ramanujam and Sadayappan, 1988), the number partitioning problem (Ramanujam and Sadayappan, 1988), and a weapon-to-target assignment problem (Tagliarini *et al.*, 1991).

We now present an original application of the Hopfield model to an optimization problem, which is moreover a scheduling problem. It is also an opportunity to illustrate how such a problem can be expressed in terms of energy function and how the analogy can apply.

10.3.3 Using Hopfield neural nets for task scheduling

In this section, we show how Hopfield neural networks can be used to solve multiprocessor task scheduling problems. We study the task constraints which can be translated easily into ANNs (i.e. ready times and deadlines constraints). Task constraints (such as resource constraints and pre-emptive scheduling) which are under research are briefly discussed. Also, the complexity of the ANN-based scheduling algorithm is analyzed.

About the task scheduling problem

Let us take the following problem: a set of n tasks is to be scheduled over a set of m processors during a finite number of time units (L). Each task has a given processing time (p_i), and must not begin its execution before a given date (ready time, r_i) and must terminate before a given date (deadline, Δ_i). All processors are identical.

Let us associate an ANN with the set of tasks. This ANN is composed of $n*L$ neurons; the neuron states correspond to the states of the tasks during each time unit. An inactive neuron means that the corresponding task is not running during the corresponding time unit, and an active neuron means that the corresponding task has the processor during the corresponding time unit. For instance in Fig. 10.4, task T1 is active during time units 3 and 6. For sake of simplicity, the ANN is represented as a grid indicating only the neuron outputs (i.e. active and inactive neuron states).

To obtain an ANN for solving the scheduling problem, one must define rules to compute the connection weights and the external inputs of the neurons, according to the tasks constraints and the number of

	L=10									
T1	o	o	x	o	o	x	o	o	o	o
T2	o	o	o	o	x	o	o	o	o	o
T3	o	o	o	x	o	o	o	o	o	o
T4	x	x	o	o	o	o	o	o	o	o
T5	o	o	o	o	o	o	o	o	o	x
T6	o	o	o	o	o	o	x	o	o	o
T7	o	o	o	o	o	o	o	x	x	o

x active neuron
o inactive neuron

if neuron (i,j) is active then
Task i is active during time unit j

Figure 10.4 Equivalence between neuron states and task states.

processors, so that the ANN leads to an approximate solution of the scheduling problem, regardless of the initial state of the neurons.

A scheduling algorithm is characterized by the task constraints it can handle (timing constraints, precedence constraints, resource constraints, etc.). In this section we will deal with the problem of pre-emptive scheduling of time constrained and independent tasks, over a multiprocessor architecture with identical processors.

A lot of research has been carried out to solve this type of problem especially since Liu and Layland (1973) have presented the first algorithms for pre-emptive scheduling of real-time periodic tasks on a monoprocessor. Liu and Layland's algorithms (i.e. Rate Monotonic and Deadline Driven) are optimal[3] for pre-emptive scheduling of periodic and independent tasks.

The Earliest Deadline algorithm can also be used to handle precedence constraints. It is proved that it remains optimal when the deadlines are modified by a rule (Chetto *et al.*, 1990) that takes into account the tasks precedence constraints.

However the Earliest Deadline optimality proof does not hold if tasks are to be executed over a multiprocessor environment (Dertouzos and Mok, 1989). Dertouzos and Mok proposed the Least Laxity algorithm, which is not optimal but is better than the Earliest Deadline Algorithm because it can schedule all the tasks accepted by the Earliest Deadline algorithm and, moreover, it can successfully schedule sets of tasks for which the Earliest Deadline algorithm fails.

This scheduling problem may also be translated into a problem of maximizing the flow through a directed graph (Horn, 1974), which can be solved in a pseudo-polynomial time by the algorithm of Ford and Fulkerson (1962).

For the non-pre-emptive case, there is no polynomial optimal algorithm (even for the monoprocessor case): the problem belongs to the class of NP-Complete problems (Garey and Johnson, 1975).

[3] These algorithms are optimal when finding feasible solutions. In other words, if the Earliest Deadline or the Rate Monotonic algorithm fail to find a feasible schedule, no other algorithm will succeed to do so.

The k-out-of-N Rule

The k-out-of-N rule (Tagliarini *et al.*, 1991) was introduced to build ANNworks which satisfy constraints of the type 'exactly k neurons among N' should be active when the net reaches a stable state.

A cost function satisfying this constraint is:

$$E = \left(k - \sum_{i=1}^{n} O_i\right)^2 + \sum_{i=1}^{n} (O_i)(1 - O_i)$$

This function is a 'cost' function because its zeros correspond to configurations where the number of active neurons is equal to k (minimization of the first term) and when the neuron outputs are either zero or one (minimization of the second term). After some algebraic manipulation, this function may be written as:

$$E = -\frac{1}{2} \sum_{i=1}^{n} \sum_{\substack{j=1 \\ j \neq i}}^{n} (-2) O_i O_j - \sum_{i=1}^{n} (2k - 1) O_i$$

which is an energy function like (10.1) with:

$$W_{ij} = \begin{cases} 0 & \text{if } i = j \\ -2 & \text{otherwise} \end{cases}$$

and

$$I_i = (2k - 1)$$

The k-out-of-N rule means that an ANN composed of N neurons, with all the connection weights (but the self connections) set to -2, and the external inputs set to $2k - 1$, will lead to states with k active neurons, regardless the initial neuron states.

Moreover, the k-out-of-N rule may be linearly independent if applied to overlapping sets of neurons. The new weights and external inputs just need to be added to the existing ones. The net will evolve to a state satisfying all the constraints simultaneously. If there is no solution satisfying all the constraints, the net will converge to a solution that minimizes the energy function even if it does not reach zero. However, it is not guaranteed that the net will reach the global optimal of the energy function due to the presence of local optima on the energy function.

A neural net-based algorithm for task scheduling

We present now an approach for translating real-time task constraints into an energy function obtained by the application of several k-out-of-N rules.

Let us go back to the ANN introduced in Fig. 10.4: it is composed of a number of neurons equal to the number of tasks (n) times the number of time units (L) of the schedule length.

We said that the ANN has L neurons for each task. The task timing constraints mean that a task T_i must begin its execution after r_i and must terminate it before Δ_i. This means that the net should converge to a stable state with exactly p_i active neurons among the neurons corresponding to time units greater than r_i and lower than Δ_i.

The external inputs and connection weights among the neurons can now be computed by the application of the k-out-of-N rule to the concerned neurons, replacing N by the number of concerned neurons ($\Delta_i - r_i$ neurons) and k by the processing time of the task (p_i).

Moreover, the task should not be executed before its ready time or after its deadline this means that exactly 0 neurons should be active among the neurons corresponding to time intervals $[0..r_i]$ and $[\Delta_i..L]$. Again, we apply the k-out-of-N rule to satisfy these constraints (see rules R1, R2 and R3 below).

To meet the tasks timing constraints, several k-out-of-N rules are applied. The connection weights and external inputs produced by the current rule are added to the previously computed ones, because these rules are linearly independent.

Finally, we must impose the mutual exclusion of the processor use. This means that the number of active tasks during the same time unit cannot be greater than the number of processors.

If the tasks fully utilize the processors (i.e. $\mu = 1$, equation (10.2)) the number of active neurons during each time unit must be equal to the number of processors. This constraint may be solved by applying the m-out-of-n rule among neurons of the same column (see rule R4). Unfortunately the assumption of full utilization of the processors would be very restrictive. To ease this restriction out, we must extend the ANN by some slack neurons (Tagliarini *et al.*, 1991).

If one has to deal with a constraint which states that 'at most' k neurons should be active when the net reaches the stable state, it is still possible to apply the k-out-of-N rule by extending the initial ANN by k neurons. As the total number of active extended neurons cannot be greater than k, the number of active neurons among the original ones will be in the interval $[0,k]$. These extended neurons are called slack neurons as they are hidden and their state should not be taken into account when retrieving the solution of the scheduling problem.

To solve the mutual exclusion of the processor use, the net should be extended by m lines (rule R5), and the m-out-of-n rule should be applied to all the neurons of each column (rule R6). In practice, this corresponds to adding fictive tasks to the system (one fictive task for each processor) that will be executed when all real tasks do not need the processor. The convergence rate of the net is improved if we give to the fictive tasks fictive execution times so that the processors will be fully utilized (rule R7).

The previous argument leads to the following rules for translating task scheduling problems into ANNs:

- For the line associated with task T_i:
  - (R1): Apply p_i-out-of-(Δ_i–r_i) rule to the neurons corresponding to time units after r_i and before Δ_i.
  - (R2): Apply 0-out-of-r_i rule to the neurons corresponding to time units before r_i.
  - (R3): Apply 0-out-of-(L–Δ_i) rule to the neurons corresponding to time units after Δ_i.
- If the tasks fully utilize the processors (i.e. if $\mu = 1$):
  - (R4): Apply m-out-of-n rule to each column of the net (R4).
- If the tasks do not fully utilize the processors (i.e. if $\mu < 1$):
  - (R5): Extend the net by m lines.
  - (R6): Apply m-out-of-(n + m) rule to each column of the extended net.
  - (R7): For each new line, compute the corresponding fictive task processing time (p_i) using equation (10.3). Then apply p_i-out-of-L rule to all neurons corresponding to each fictive task.

The equations to determine whether the tasks fully utilize the processor and to compute the execution time of fictive tasks are the following (where [x] means the integer part of x):

$$\mu = \frac{\sum_{j=1}^{T} p_j}{m \cdot L} \tag{10.2}$$

$$p_i = \begin{cases} \left[\dfrac{m \cdot L - \sum_{j=1}^{n} p_j}{m} \right] & if\, n < i < n + m - 1 \\ m \cdot L - \sum_{j=1}^{m+n-1} p_j & if\ i = n + m \end{cases} \tag{10.3}$$

Example

To test whether the ANN actually evolves to solutions of the scheduling problem, we propose to see its evolution from a random initial state for scheduling the following set of tasks (Cardeira and Mammeri, 1994).

Three real-time tasks must be scheduled using one processor. Their time constraints are as follows:

- Task 1: It must be active, for two units, every six time units.
- Task 2: It cannot start before the second time unit; it requires three time units and it must terminate no later than the sixth time unit.
- Task 3: Its ready time is equal to 0 and must terminate before the fourth time unit. It requires one time unit.

The net has 18 neurons. The neuron number is obtained by multiplying the number of tasks by the number of time units of the schedule length. To meet the timing constraints of each task, the k-out-of-N rules should be applied as follows:

- In the first line, apply 2-out-of-6 rule to all the neurons.
- In the second line, apply 0-out-of-2 rule to the first two neurons, and 3-out-of-4 rule to the last four.
- In the third line, apply 1-out-of-4 rule to the first four neurons and 0-out-of-2 rule to the last two neurons.
- Moreover, apply to each column 1-out-of-3 rule, because at most one task has the processor at a given time. In this case, we do not need to introduce slack neurons because this configuration fully utilizes the processor ($m = 1$).

External input vector

4	4	4	4	4	4	0	0	6	6	6	6	2	2	2	2	0	0

Connection matrix

0	–2	–2	–2	–2	–2	–2	0	0	0	0	0	–2	0	0	0	0	0
–2	0	–2	–2	–2	–2	0	–2	0	0	0	0	0	–2	0	0	0	0
–2	–2	0	–2	–2	–2	0	0	–2	0	0	0	0	0	–2	0	0	0
–2	–2	–2	0	–2	–2	0	0	0	–2	0	0	0	0	0	–2	0	0
–2	–2	–2	–2	0	–2	0	0	0	0	–2	0	0	0	0	0	–2	0
–2	–2	–2	–2	–2	0	0	0	0	0	0	–2	0	0	0	0	0	–2
–2	0	0	0	0	0	0	–2	0	0	0	0	–2	0	0	0	0	0
0	–2	0	0	0	0	–2	0	0	0	0	0	0	–2	0	0	0	0
0	0	–2	0	0	0	0	0	0	–2	–2	–2	0	0	–2	0	0	0
0	0	0	–2	0	0	0	0	–2	0	–2	–2	0	0	0	–2	0	0
0	0	0	0	–2	0	0	0	–2	–2	0	–2	0	0	0	0	–2	0
0	0	0	0	0	–2	0	0	–2	–2	–2	0	0	0	0	0	0	–2
–2	0	0	0	0	0	–2	0	0	0	0	0	0	–2	–2	–2	0	0
0	–2	0	0	0	0	0	–2	0	0	0	0	–2	0	–2	–2	0	0
0	0	–2	0	0	0	0	0	–2	0	0	0	–2	–2	0	–2	0	0
0	0	0	–2	0	0	0	0	0	–2	0	0	–2	–2	–2	0	0	0
0	0	0	0	–2	0	0	0	0	0	–2	0	0	0	0	0	0	–2
0	0	0	0	0	–2	0	0	0	0	0	–2	0	0	0	0	–2	0

In Fig. 10.5, we present two independent simulations of the ANN evolution (Cardeira, 1994). The simulator is a 216 lines C-coded program

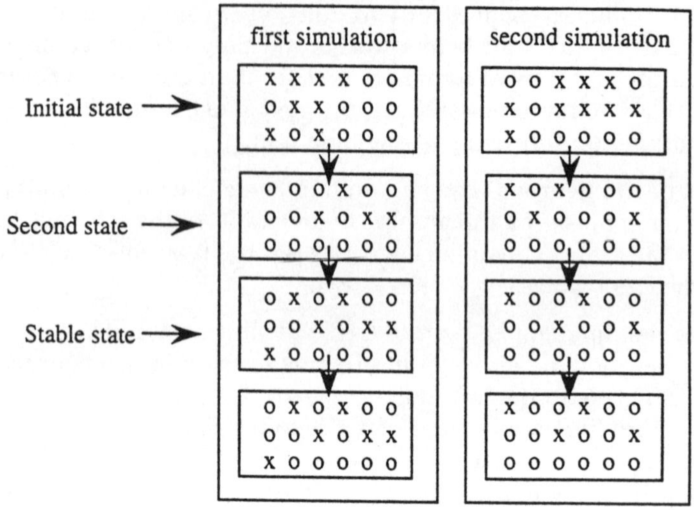

Figure 10.5 Simulation of the ANN.

which randomly fires the neurons to simulate an analog implementation. Each evolution starts from a random initial state of neuron outputs and, as we may see, the net evolves towards stable states meeting the timing constraints of the tasks. We draw attention to the fact that the convergence is obtained in only a few iterations, at most two, for this example.

When there are several possible solutions satisfying the constraints (as is the case for the previous example) it is impossible to determine which one will be chosen. The chosen solution depends on:

- the initial state of the neurons;
- the order in which the neurons are 'fired'. This order is not deterministic in an analog implementation;
- the amount of noise on the circuit implementing the ANN.

An initial state corresponding to a local optimum of the energy function prevents the ANN from finding the optimal solution.

Complexity analysis

An important feature of a scheduling algorithm is its complexity (i.e. the increase of its processing time relative to the size of the problem). The worst-case complexity of the existing algorithms for scheduling sets of independent tasks – i.e. Least Laxity (Dertouzos and Mok, 1989) and Earliest Deadline (Liu and Layland, 1973; Dertouzos, 1974) algorithms – is $O(n\log n)$, n being the number of tasks.

The most relevant parameter related to the time of an ANN to find a solution is the number of iterations needed to converge to a solution. This time is equal to the number of 'computations' of each neuron because all neurons 'compute' in parallel. In practice, the convergence time of the ANN is less or equal to the number of iterations, multiplied by the maximum propagation delay of the operational amplifiers of the analog computer.

An experimentation varying the number of tasks from 1 to 35, over a fixed schedule length (35 time units) yielded a total number of iterations never greater than 4, after 150 independent executions and the number of iterations remains almost independent relative to the number of tasks (Cardeira and Mammeri, 1994). This is a very interesting result because all existing algorithms have a complexity that is polynomial or, at most, linear.

Actually, most of the problem's complexity is translated into an increased number of neurons, but the execution time of the algorithm remains almost unaffected because the neurons 'compute' in parallel. Nevertheless, the number of neurons grows polynomially and this may lead to some implementation difficulties.

When a sequential digital implementation is chosen, one loses the advantage of finding fast solutions, and this provides only an algorithm with an $O(T^2L^2)$ worst-case complexity which is not very surprising for a problem that is polynomial.

10.3.4 Discussion

In conclusion, the lack of use of Hopfield ANNs for solving optimization or constraint satisfaction problems is mainly due to the following reasons:

- In spite of all the progress in finding energy functions for different problems like the TSP (Hopfield and Tank, 1985) and the graph partitioning problem (Ramanujam and Sadayappan, 1988), there is no standard way for translating a cost function into an energy function.
- Even if the mathematical basis of the Hopfield model is strong, there is no guarantee that the net will find the optimal solution. We only have the guarantee that the ANN will lead to the nearest local optimal and, unfortunately, it will stay there for ever.
- To take full advantage of the Hopfield model, a special hardware architecture is necessary, either an analog computer or a digital parallel architecture. As such architectures are not always available, ANNs are not efficient, especially when they are to be used on line.
- Even if a special hardware architecture like an analog computer with thousands of 'neurons' is available, the possibilities of reconfiguration are often complex, and in some cases only fixed connection weights are allowed.

However, when the previous drawbacks of the use of ANNs have been discarded, we have been able to show an approach to solve a task scheduling problem with ANNs. The main difficulty of ANN use for such problems is the elaboration of an energy function corresponding to the task constraints. We have restricted our study to independent tasks scheduled over a multiprocessor. These correspond to a very specific scheduling problem, not one of the difficult ones, as polynomial algorithms do already exist to solve it.

To really be a general-purpose technique for solving task scheduling problems, ANNs must be able to handle other constraints than the previously presented ones (i.e. non-pre-emptive scheduling, precedence constraints, resource constraints, etc.). This is still an open problem and some research is emerging in this domain. Hellstrom and Kanal (1992) deal with resource constraints, Chen (1992) and Cardeira (1994) deal with precedence constraints, and Cardeira (1994) deals with non-pre-emptive scheduling. The reader may find further details in the four previously mentioned references.

While waiting for the massive arrival on the market of parallel architectures enabling implementation of ANNs, the present use of ANNworks is reserved to specific critical applications where the following conditions are met:

- An approximate solution must be found in a few nanoseconds.
- The structure of the problem is more or less static.
- The host institution has enough means to support the costs of developing specific VLSI circuits implementing the ANN.

This is the main reason why the use of ANNs for scheduling purposes is currently limited to specific military applications (for instance, target allocation to weapons) or to physics for detecting particles in accelerators.

We have explained here how the general framework of parallel distributed processing could be used for optimization problems. This approach has lead to a lot of research but also to much criticism. A first criticism of this approach concerns the Hopfield model itself. It is clear that other connectionist models of content-addressable memory have larger capacities than this one (Tarassenko *et al.*, 1991).

A second criticism of this approach is related to the analogy that permits the application of the Hopfield model to optimization problems, as has been introduced above. Wilson and Pawley (1988) have deeply investigated the Hopfield and Tank work and have unsuccessfully tried to apply it to a larger system size. The same unsatisfactory conclusion was found when trying to improve this algorithm. These authors conclude: 'It would be unwise to dismiss the method as of no value as it contains many ideas which should be pursued with vigour, but the method as it stands lacks detailed justification.'

More generally speaking, Gee (1993) has investigated a large set of connectionist models applied to optimization problems and has rigorously analyzed their applicability, their convergence properties and the quality of the solution provided by these models on simple illustrative examples. He concludes, as we do here, that connectionist models cannot be considered today as a general tool for optimization problems. Indeed those three properties (applicability, convergence and quality of solution) can be successfully verified in the same manner as they can lead to failure, in the event of some special, difficult-to-handle characteristics of the problem.

10.4 CONCLUSION

The use of GAs and ANNs for solving scheduling problems is relatively recent. They have been developed either by GA or ANN specialists who have used scheduling as a new application field for their methods, but who were not generally scheduling specialists; or by scheduling specialists who found these new approximation schemes attractive and used them rather awkwardly, especially GAs which are very easy to design and implement but need careful choices concerning general GA structure, chromosome coding, genetic operators and various parameters. Unfortunately, about 40 years of scheduling results have not been used to improve the implementation of GAs or ANNs. We hope that specialists of these various research areas will meet and add their own knowledge so that more efficient specific GAs and ANNs will be created in the near future. These methods are known to be robust and rapid (in parallel version). But they have still to become more efficient by integrating knowledge about the problem and the solution space which must be explored.

REFERENCES

Ahalt, S., Chen, P., Chou, C., Kuttava, S. and Little, T. (1992) The neural shell: A neural network simulation tool. *Int. J. Eng. Applic.* (Special Issue on Neural Nets and Parallel Processing), **5**(3), 183–92.

Angéniol, B., De la Croix Vaubois, G. and Le Texier, J. (1988) Self-organizing feature maps and the travelling salesman problem. *Neural Networks*, **1**, 289–93.

Biegel, J.E. and Davern, J.J. (1990) Genetic algorithms and job shop scheduling. *Computers Ind. Eng.*, **19**(1–4), 81–91.

Cardeira, C. and Mammeri, Z. (1994) *Neural Networks for Multiprocessor Task Scheduling*. In Proc. 8th Euromicro Workshop on Real-Time Systems, IEEE CS, June, Vaesteraas, Sweden.

Cardeira, C. (1994) *Ordonnancement de tâches et messages dans les systèmes temps réel et répartis*. PhD thesis, to appear.

Chen, P. (1992) Design of a real-time AND/OR assembly scheduler on an optimization neural network. *J. Intell. Manuf.*, **3**, 251–61.

Chetto, H., Silly, M. and Bouchentouf, T. (1990) Dynamic scheduling of real-time tasks under precedence constraints. *J. Real-Time Sys.*, **2**, 181–94.

Chiu, C. and Yih, Y. (1995) The learning-based methodology for dynamic scheduling in distributed manufacturing system. *Int. J. of Prod. Res.*, to appear.

Cohen, M. and Grossberg, S. (1983) Absolute stability of goal pattern formation and parallel memory storage by competitive neural networks. *IEEE Transactions on Systems, Man, and Cybernetics*, **13**, 815–26.

Davis, L. (1985) *Job-Shop Scheduling With Genetic Algorithms*. Proc. 1st Int. Conf. on Genetic Algorithms and Their Applications, Lawrence Erlbaum, Hillsdale, NJ, pp. 136–40.

Davis, L. (1991) *Handbook of Genetic Algorithms*. Van Nostrand Reinhold, New York.

Dertouzos, M. (1974) *Control Robotics: the Procedural Control of Physical Processes*. In Proc. IFIP Congress, pp. 807–13.

Dertouzos, M. and Mok, A. (1989) Multiprocessor on-line scheduling of hard-real-time tasks. *IEEE Trans. Softw. Eng.*, **15**(12), 1497–506.

Djerid, L., Portmann, M.C. and Villon, P. (1995) *Performance Analysis of Previous and New Proposed Cros-Over Genetic Operators Designed for Permutation Scheduling Problems*. Int. Conf. on Industrial Engineering and Production Management, Marrakeck, April 4–7.

Djerid, L. and Portmann, M.C. (1995) *Comment entrecroiser des procédures par séparation et évaluation et des algorithmes génétiques: application à des problèmes d'ordonnancement à contraintes disjonctives*. Francoro, Mons, June 12–14.

Dorndorf, U. and Pesch, E. (1995) Evolution based learning in a job-shop scheduling environment. *Computer Oper. Res.*, **22**, 25–40.

Emmons, H. (1969) One-machine sequencing to minimize certain functions of job tardiness. *Oper. Res.*, **17**(4), 701–15.

Falkenauer, E. and Bouffouix, S. (1991) *A Genetic Algorithm for Job Shop*. Proc. IEEE Int. Conf. on Robotics and Automation, Sacramento, vol. I, pp. 824–9.

Ford, L. and Fulkerson, D. (1962) *Flows in Networks*, Princeton University Press.

Garey, M. and Johnson, D. (1975) Complexity results for multiprocessor scheduling under resource constraints. *SIAM J. Computing*, **4**, 397–411.

Gee, A.H. (1993) Problem solving with optimization networks. PhD thesis, University of Cambridge.

Goldberg, D.E. (1989) *Genetic Algorithms in Search, Optimization and Machine Learning*, Addison-Wesley, Reading, Mass.

Goldberg, D.E. and Lingle, R. (1985) *Alleles, Loci, and The Traveling Salesman Problem*. Proc. 1st Int. Conf. on Genetic Algorithms and Their Applications, pp. 154–9.

Gotha (1993) Les problèmes d'ordonnancement. *RAIRO Recherche Opérationnelle/Oper. Res.*, **27**(1), 77–150.

Hellstrom, B. and Kanal, L. (1992) Knapsack packing networks. *IEEE Trans. Neural Networks*, **3**(2), 302–7.

Hertz, J., Krogh, A. and Palmer, R. (1991) *Introduction to the Theory of Neural Computation*, Addison-Wesley, Cambridge, Mass.

Holland, J.H. (1975) *Adaptation in Natural and Artificial Systems*, MIT Press, Cambridge, Mass.

Hopfield, J. (1982) *Neural Networks and Physical Systems with Emergent Collective Computational Abilities*. Proc. National Academy of Science, vol. 79, pp. 2554–8.

Hopfield, J.J. and Tank, D.W. (1985) Neural computation of decisions in optimisation problems, *Biological Cybernetics*, **52**, 141–52.
Horn, W.A. (1974) Some simple scheduling algorithms. *Naval. Res. Logist. Quart.*, **21**, 177–85.
LeCun, Y. (1987) Modèles connexionnistes de l'apprentissage, Thèse de Doctorat, Université Paris VI.
Liu, C. and Layland, J. (1973) Scheduling algorithms for multiprogramming in a hard real-time environment. *J. ACM*, **20**(1), 46–61.
MacCulloch, W.S. and Pitts, W. (1943) A logical calculus of the ideas immanent in nervous activity. *Bulletin of Mathematical Biophysics*, **5**, 115–33.
Nakano, R. and Yamada, T. (1991) *Conventional Genetic Algorithm for Job Shop Problems*. Proc. Fourth Int. Conf. on Genetic Algorithms, Morgan Kaufmann, San Mateo, California, pp. 474–9.
Oliver, I., Smith, D. and Holland, J. (1987) *A Study of Permutation Crossover Operators on The Traveling Salesman Problem*. Proc. 2nd Int. Conf. on Genetic Algorithms and Their Applications.
Portmann, M.C. and Ghedjati, F. (1994) *Méthodes Approchées Pour Le Problème D'ondonnancement Avec Machines Non Identiques En Parallèles Et Contraintes De Précédence*. Proceeding of AGI'94, Poitiers, 2–3 June.
Ramanujam, J. and Sadayappan, P. (1988) *Optimization by Neural Networks*. Proc. IEEE Int. Conf. on Neural Networks, vol. 2, pp. 325–32, New York: IEEE.
Rosenblatt, F. (1962) A comparison of several perceptron models, in *Self Organizing Systems*, (eds M.C. Yovits, G.T. Jacobi, G.D. Goldstein) Spartan Books, pp. 463–84.
Rumelhart, D.E. and MacClelland, J.L. (1986) *Parallel Distributed Processing*, MIT Press, Cambridge, Mass.
Storer, R.H., Wu, S.Y.D., Vaccari, R. (1992) New search spaces for sequencing problems with application to Job Shop Scheduling. *Management Science*, **38**(10), 1495–509.
Tagliarini, G., Christ, J. and Page, E. (1991) Optimization using neural networks. *IEEE Trans. on Computers*, **40**, 1347–58.
Tarassenko, L., Tombs, J. and Reynolds, J. (1991) Neural networks architectures for content-addressable memory. *IEE Proc.*, Series F, **138**, 33–9.
Uckun, S., Bagchi, S., Kawamura, K. and Miyabe, Y. (1993) Managing genetic search in job shop scheduling. *IEEE Expert*, **8**(5), 15–24.
Widrow, B. and Hoff, M.E. (1960) *Adaptive Switching Circuits*. IRE WESCON Convention Record, New York, pp. 96–104.
Widrow, B. and Lehr, M. (1990) 30 years of adaptive neural networks: Perceptron, madeline and backpropagation. *Proc. IEEE*, Special issue on neural networks, **78** (Sept), 1415–42.
Wilson, G.V. and Pawley, G.S. (1988) On the stability of the travelling salesman problem of Hopfield and Tank. *Biological Cybernetics*, **58**, 63–70.
Whitley, Y., Starkweather, T. and Fuquay, D. (1989) *Scheduling Problems and Traveling Salesman: The Genetic Edge Recombination Operators*. Proc. Third Int. Conf. on Genetic Algorithms and Their Applications, Morgan Kaufmann, San Mateo, Calif., pp. 133–40.

FURTHER READING

Cerf, R. (1996) An asymptotic theory of genetic algorithms. *Artificial Evolution*, **1063**, LNCS, Springer-Verlag.

Cottrel, G., Munro, P. and Zipser, D. (1987) Image compression by back-propagation: An example of extensional programming. ICS Rep. 8702, University of California at San Diego.

Eiben, A.E., Aarts, E.H.L. and Van Lee, K.M. (1991) *Global Convergence of Genetics Algorithms: A Markov Chain Analysis*. Proc. of the 1st Workshop Parallel Problem Solving from Nature, Dortmund, October 1–3, (eds H.P. Schwefel and R. Männer), Springer-Verlag.

Gorman, R. and Sejnowski, T. (1988) Analysis of hidden units in a layered network trained to classify sonar targets. *Neural Networks*, **1**(1), 75–89.

Grossberg, S. (1982) *Studies of Mind and Brain: Neural Principles of Learning, Perception, Development, Cognition and Motor Control*, Reidel, Boston, MA.

Hebb, D.O. (1949) *The Organization of Behaviour*, Wiley, New York.

Page, E. and Tagliarini, G. (1988) *Algorithm Development for Neural Networks*. Proc. SPIE Symp. Innovative Sci. and Technol., vol. 880, pp. 11–19.

Starkweather, T., McDaniel, S., Mathias, K., Whitley, D. and Whitley, C. (1991) *A Comparison of Genetic Sequencing Operators*. Proc. Fourth Int. Conf. on Genetic Algorithms, Morgan Kaufmann, San Mateo, California, pp. 69–76.

CHAPTER 11

Production scheduling using FACTOR[1]

A.A.B. Pritsker and K. Snyder

11.1 INTRODUCTION

Manufacturing's business is to convert customer orders into products. To do this, manufacturing systems are required to process quickly individual orders with specific customer options. That is, they must be agile. Such responsiveness can only be achieved through a total coordination of the elements of the production process and with a clear understanding of the effect of each order on the manufacturing systems' performance. Essentially, it requires a common understanding of manufacturing processes and an ability to analyze quickly what happened in the past in order to forecast and control what will happen in the future. Planning and scheduling systems are at the heart of this understanding and have the unique ability to translate what has happened in the past and, through analysis, develop schedules to meet the large and changing demands on a manufacturing system. This chapter describes the use of FACTOR (Pritsker Corporation, 1992) for production planning and scheduling purposes.

FACTOR is a software program which uses simulation technology to analyze a manufacturing systems' ability and agility to satisfy diverse customer demands through the use of different planning strategies and scheduling technologies. Emphasis throughout this chapter is on practical production situations. Excellent textbooks and articles exist on the theoretical aspects of scheduling (Baudin, 1993; Baker, 1974; Askin and Standridge, 1993) and such material is not reviewed here.

11.2 GENERAL DESCRIPTION OF PRODUCTION SCHEDULING

Figure 11.1 presents a schematic of the manufacturing production-scheduling-operations environment (Pritsker, 1992). The functions

[1] FACTOR is a registered trademark of Pritsker Corporation.

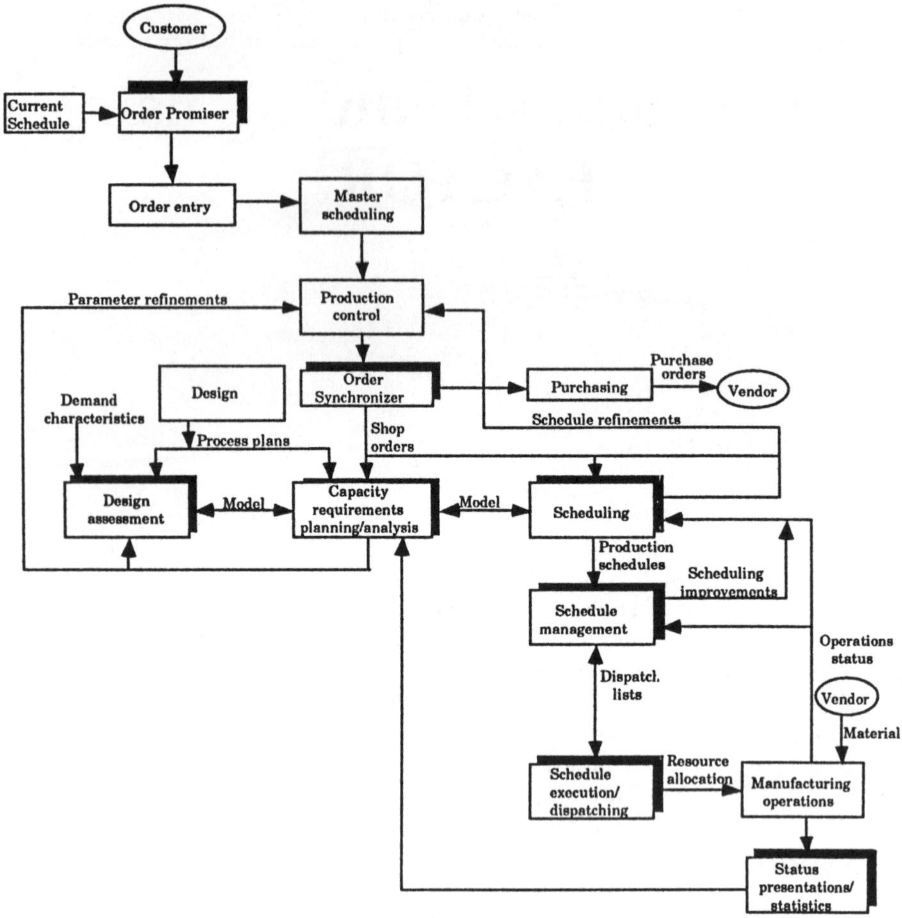

Figure 11.1 Manufacturing production–scheduling–operations environment.

indicated by the shaded blocks of Fig. 11.1 relate to FACTOR tasks: design assessment; capacity requirements planning and analysis; scheduling; schedule management; schedule execution and dispatching; status presentations and statistics; order promising, and order synchronization. The functions of master scheduling and manufacturing resource planning can also be performed using FACTOR. Before defining in detail the functions of planning and scheduling, brief descriptions of the tasks listed above are given.

Design assessment involves the use of a model of manufacturing operations to estimate the performance of the manufacturing system for different levels of demand. A key input to the model is the process plans that specify the jobsteps, including resource requirements, to make the product. **Capacity requirements planning and analysis** seeks to determine whether manufacturing operations can process the shop orders released in a timely manner. Before detailed scheduling can be done, a finite capacity analysis determines the level of resources required to meet current demand. When capacity levels are set, detailed **scheduling** can be accomplished by using the model to allocate available resources at specified start times to the actual jobs included in the shop orders.

If the model contains the detailed process plans or jobsteps, the start and completion times of each operation can be established. These schedules can then be distributed for **schedule management** which provides an ability; commonly through a graphical interface, for the user to easily view and modify the schedule. Modifications may be required in order to: (1) update the schedule to reflect the most current information, or (2) incorporate the expert knowledge of the user that is unique to the situation. Maintaining shop floor discipline when adjustments are made is important. The outputs of schedule management are dispatch lists for each resource detailing the scheduled time to perform each job. In addition, methods for improving the scheduling process through the collection of data and the parameterizing of rules to improve the scheduling procedures is part of the FACTOR continuous improvement process.

The dispatch lists are the basis for **schedule execution and dispatching**, that is, the actual resource allocations to jobs. Data and statistics on **operational status** are fed back to scheduling and schedule management to determine the frequency with which new schedules need to be prepared. The display of this status information provides a basis for ongoing decision making. The current status and up-to-date information on past performance is used in capacity requirements planning where additional simulations can be performed to assess future performance. Through this feedback link, continuous improvements in manufacturing operations can be made and information is gathered for future design assessments and new scheduling algorithms.

The feasibility of integrated planning, scheduling and execution relies heavily on the ability to build on existing data and models. The use of a common modeling language to build models across the functional tasks makes the integrated problem solving using the feedback control loops described above plausible. It allows for 'going to the gemba' (Weliever, 1992), where gemba is someone else's functional area in which models have been built, data collected, and analyses performed based on the needs of a different functional area.

An inherent advantage of the use of models is that they contain information about manufacturing processes, and by using and updating models continually, manufacturing processes will be better understood. It is worth stressing that understanding leads to improved manufacturing operations and information for improving design. FACTOR is built on a common modeling manufacturing language and thus, is a mechanism to achieve, using simulation, a new form of **kaizen** (Imai, 1986) by which the processes of manufacturing and decision making can be continually evaluated, changed, and improved. The need for such a mechanism is described in detail in Hayes *et al.* (1988).

Innovation also is enhanced, because a model developed in one functional area can be used to indicate the possibility of new constructs for another functional area. Thus, improvement cycles in a single functional area may be used to foster new models and concepts in other functional areas. The common model, **common data foundation**, when fully implemented, provides a basis for achieving world-class manufacturing.

11.3 PLANNING AND SCHEDULING FUNCTIONS

Planning is concerned with what to produce in what time period, whereas scheduling is concerned with the timing and sequence of how to produce products within the limitations of the production facility.

Planning is an analysis task. It is a process for assessing a manufacturing system's ability to meet demands measured in terms of an order's time of completion with respect to a targeted delivery date and the costs associated with producing products as listed on the order. Surrogate measures for costs typically relate to resource utilization, labor use, in-process and finished goods inventory, material stock and usage, materials handling mode and shipping requirements. When doing a planning analysis, an evaluation is made of the number of resources, ways of using the resource, process routings and different scheduling philosophies. A plan results from a selection of a specific alternative for each of these items based on both a performance evaluation and an understanding of management's objectives. The plan identifies the orders to be produced during a prescribed time period.

Scheduling produces a step-by-step procedure for the manufacturing system to produce the products identified by the orders in the plan. The step-by-step schedule could be in the form of a dispatch list or a specification of the times at which every operation should start and end.

From these definitions of planning and scheduling, a distinction is made between the two functions. In reality, many companies allow more options or alternatives to be included in the scheduling function and thus, there is a gray area between planning and scheduling. A convenient way to think about the difference is that scheduling begins with the plan and results in a detailed 'schedule' whereas a plan starts with information describing orders and the manufacturing system and provides a specification for what orders are to be produced in a given time period, and the resource requirements for that time period. If the planning function is performed using a detailed finite capacity model, the scheduling function should be able to successfully develop a schedule which conforms to the planned start and end times. If the plan is developed with a limited view of factory operations, there will be a high number of exceptions when attempting to meet the plan. Then the scheduling system will be called upon to perform some of the functionality of the planning system. For example, when the method used to approximate capacity in the planning function results in an overestimation of capacity, the scheduling system will be required to determine what orders will and will not be produced in a specific time period.

In a sense, scheduling is a special form of planning. However, most companies perceive they have scheduling problems, not planning problems. We believe there is a distinct reason for this perception: good scheduling provides operational directives which are verifiable and which produce benefits which are observable and directly tied to profits such as on-time performance without incurring premium shipments. The planning function is perceived as providing inputs to the scheduling function and the benefits are only realized in a nonobservable and indirect way. We do not believe this perception will change in the near term even though proper planning is prerequisite for good scheduling.

11.4 SIMULATION AS AN OPERATIONAL ANALYSIS TOOL

For the types of operational planning and scheduling described above, a manufacturing system can be considered as a producer of parts processed through jobsteps or operations which require one or more resources, material and associated auxiliary items. The auxiliary items could be tools, fixtures, pallets, CAD drawings, processing specifications and the like. The parts may have to be stored, assembled, batched and moved in loads for processing. All of these requirements can be described in terms of time delays, logical conditions to initiate or terminate a jobstep, and counting operations to keep track of inventories and resource use/availability.

In FACTOR, the manufacturing system is represented logically by a model, and an analysis of the flow of parts required to manufacture products is made. When a manufacturing system is represented in this form and a computer is used to run the manufacturing system, it is said that a simulation is being performed. During the simulation, data is collected on the time at which jobsteps are performed – just as if the manufacturing system was producing the parts. A simulation of a manufacturing system provides a realistic portrayal of its detailed operations from which its performance can be accurately estimated.

Because each aspect of the manufacturing system, including management's operating philosophies and procedures, are contained in the computer representation, a common understanding of the planning and scheduling functions is established. This allows for comparisons of alternatives and facilitates communication to managers and executives of any problem areas and improvement possibilities.

Because manufacturing systems are complex and large, FACTOR has been specifically developed to have the capabilities shown in Table 11.1. These capabilities when incorporated into a manufacturing system provide for an on-line planning and scheduling system.

11.5 FINITE CAPACITY PLANNING

There are many approaches to finite capacity planning with each having a specific goal in mind. Below are listed nine of the commonly used approaches/philosophies:

1. Minimize setup time to increase machine use/availability.
2. Just-in-time material delivery to decrease purchased inventory requirements.
3. Bottleneck analysis to increase the total production, that is, the flow rate of parts.
4. Theory of constraints to increase total production.
5. Kanban control strategies to regulate the flow rate of parts in meeting due dates with less in-process inventory.

Table 11.1 Planning and scheduling capabilities included in FACTOR

Allows diverse schedule-development philosophies
Demand-driven simulation using actual orders and build-to-plan forecasts
Multiple resource requirements for processing steps
Detailed process planning procedures
Menu-driven interactive query procedures to support reporting and problem analysis using manufacturing terminology
Future projections of order status and displays of operation contention in specific time periods
Concurrent and instantaneous Order Promise and Due Date Setting
Order Release that minimizes work-in-process and final inventory levels
Bill of Materials to any level of detail
Advanced built-in graphics with automatic chart generation
RDBMS with complete data structure specificity
Batch simulation operations
Manufacturing specific functions
Multi-Gantt chart outputs
Problem configuration and data collection specification
Capabilities for multiple performance measurement and estimation

6. Small batch sizes to decrease in-process inventory.
7. Continuous flow methods to balance the flow rate and reduce in-process inventories.
8. Local resource decision rules where rules are developed to achieve a goal, for example, minimize setup time.
9. Global sequencing rules where rules are developed to achieve a goal, for example, minimize late orders.

FACTOR can be used to produce a plan for one or a combination of these approaches. When this is done, detailed information regarding the operation of the manufacturing system in accordance with the approach is obtained. The data is stored in FACTOR's relational database system to provide information on why the manufacturing system model behaved the way it did. In a later section, an example will be used to illustrate the level of scheduling detail that FACTOR produces and the 'drill down' capabilities to make the information available interactively.

11.6 FINITE CAPACITY SCHEDULING AND ITS BENEFITS

Rather than elaborate on the details of producing a schedule from a plan, this section describes different uses of FACTOR relative to the application of finite capacity scheduling methods. FACTOR can perform one or more of the following three functions: schedule evaluation, schedule generation and automated scheduling. In schedule evaluation, the schedule is manually generated and then assessed by the computer to determine feasibility and to estimate performance measures. The forecasted results are manually reviewed to determine if problems exist with the schedule and to develop procedures to improve the schedule which are then input into the computer for further simulations. This cycle is continued until the schedule meets objectives, is approved and distributed. The advantage of schedule evaluation is that it keeps the manual scheduler in the loop but provides computer support for evaluation and for problem identification. When simulation is used as the evaluator for manually produced schedules, an excellent interaction between the manual scheduler and the software is possible; especially when the simulation evaluation employs a realistic model of the manufacturing system. Many non-simulation-based scheduling capabilities cannot perform this evaluation function.

A leitstand is a control center with an electronic scheduling board for manually generating and adjusting schedules. FACTOR works with the FI-2 Leitstand (Scheer, 1994) to provide the following capabilities:

1. Present Gantt charts similar to conventional manual planning boards familiar to many schedulers.

2. Support the schedule generation process by allowing the scheduler to graphically manipulate the schedule.
3. Provide 'real time' status display with interactive queries to obtain detailed operational status information.

As an example of schedule evaluation, a client used FACTOR to evaluate manually generated schedules. FACTOR accepted as input scheduled start times and determined if all the constraints of the manufacturing system were satisfied by the manually generated schedule. FACTOR presented all problem areas to the manual scheduler and also indicated the impact of the schedule on future shifts. The manual scheduler then assessed how to eliminate infeasibilities, potential problem areas and situations where future scheduling periods were negatively impacted by the schedule for the current time period. As a result of this application, the number of off-line occurrences at the next assembly area was reduced to zero.

A second level of scheduling has the initial schedule being computer generated and then being reviewed manually. Alternatives are then manually generated for review by the scheduling software. Advantages of this approach are the preparing of a good first schedule and a reduction in the time required to produce an initial feasible schedule. By employing a manual review, the knowledge of the manufacturing system and any special requirements for the scheduling horizon are taken into account. The human scheduler then uses judgment and management information to improve the schedule. In an application at Nucor Steel Company, FACTOR reduced the number of individuals performing scheduling from 12 to 1 by using this schedule generation approach (Largent, 1992).

When automated scheduling is used, the initial schedule is generated by the computer, problems are identified (sometimes with the aid of expert system techniques) and new and better schedules are generated. When using automated scheduling procedures, a realistic and accurate representation of the manufacturing system is a firm requirement. Only through an accurate representation of the manufacturing system can realistic schedules be generated. Without manual intervention for schedule review of alternative generation, care must be taken that infeasible schedules do not occur. In this automatic scheduling mode, FACTOR applications have been reported to produce the following gains: a 27% reduction in the number of setups (Credle, 1993); a 29% increase in throughput (Credle, 1993); an 80% reduction in the process of generating monthly production plans (Brill, 1994); and the incorporation of the finite scheduling system as the heart of a corporation's entire business planning system (Casella and Barnes, 1992).

With automated scheduling, graphic outputs for use by managers are also a necessity. Such outputs are required to assess the progress of the operating manufacturing system relative to the schedule. The use of simulation in conjunction with a leitstand provides this capability. Leitstands for different scheduling areas can display the effect of the schedule for all orders on the area leitstand. Coordination modules in the FI-2 Leitstand allow each scheduler to see the impact of all scheduling decisions on orders relative to his/her scheduling area. Through coordination, conflicts and potential lateness of orders can be resolved. A master leitstand capability in FI-2 provides a global view across the manufacturing plant of all order and material schedules.

11.7 FACTOR APPLICATIONS

11.7.1 Application 1: Fuel pump manufacturing

ACME Industries Incorporated (ACME) manufactures fuel pumps for oil heat furnaces. The control of operating expenses is extremely important to profitability in this mature market. ACME management wants to be proactive rather than reactive. To do this, they need to acquire insight into expected production performance. ACME realized a computerized finite capacity scheduling tool could provide this insight. FACTOR was selected to automate the scheduling process as well as assist in answering a series of key questions commonly asked but seldom answered under their current manual system. With confidence in estimates of performance, management could make decisions to meet customer service objectives while controlling manufacturing expenses. Through work schedules, i.e. dispatch lists, management communicated decisions to the people who must execute the plan.

The key questions that the ACME planners and schedulers were able to answer by using FACTOR were:

1. What orders are projected to be completed late?
2. What operations are the cause of such lateness?
3. What operator groups and machines are projected to be bottlenecks?
4. Will overtime help to get orders out on-time? How much overtime is needed? When?
5. How much time is spent on setup operations?
6. What is the impact on customer service of machine downtime?
7. What is the job schedule that meets desired performance?

Production environment overview

ACME's production facility consists of a number of manufacturing departments which make small pumps in high volume and large pumps in low volume. The large pumps provide a higher profit margin. Work orders flow within and between each department through a routing specific to that work order. Additionally, some operations such as chrome plating and heat treatment are subcontracted to outside suppliers. Upon completion, the component products are packaged and shipped to another facility for assembly.

Pre-FACTOR planning and scheduling

Work orders were generated from a specific customer order using an MRP II system. Work orders were typically released with from one to four weeks' lead time before scheduled shipment. Material was pulled from stock with the work order ticket when the order was to be started.

The departmental supervisor selected the work orders to run for the next day. In this decision, the supervisor considered due date, the work orders currently past their due date, the current setup of the resource, and operator skills and availability. All of the parts in an order were to be completed at an operation prior to moving the order to its next operation.

No system was available for accurately predicting when orders would be completed and what the impact of the capacity of the manufacturing system was on lateness for the mix of orders to be produced.

Planning and scheduling with FACTOR

Figure 11.2 displays an overview diagram of how FACTOR was integrated into ACME's manufacturing planning system.

FACTOR can run as a complement to MRP II through an interface to the MRP II database. Data which is updated and maintained frequently in the MRP II database is electronically transferred to FACTOR's database. This data is work order information, the status of released orders, and part routing (process plan) information. In a standard mode, FACTOR has utility programs to retrieve this information directly from a customer's database or file system.

FACTOR includes menu-driven editor screens to enable ACME production planning personnel to manually maintain information related to capacity. This capacity information includes shift schedules for equipment, operator group sizes, and holiday plant shutdowns. Production planners have the capability to adjust this information through the screen editors, and to analyze the impact of any changes on production throughput.

On a daily basis in this application, FACTOR simulates the factory including advanced scheduling strategies to process order, status and routing input data and to maintain capacity information to accurately predict how the current and future work order demand should flow through the shop. FACTOR then displays the predicted performance.

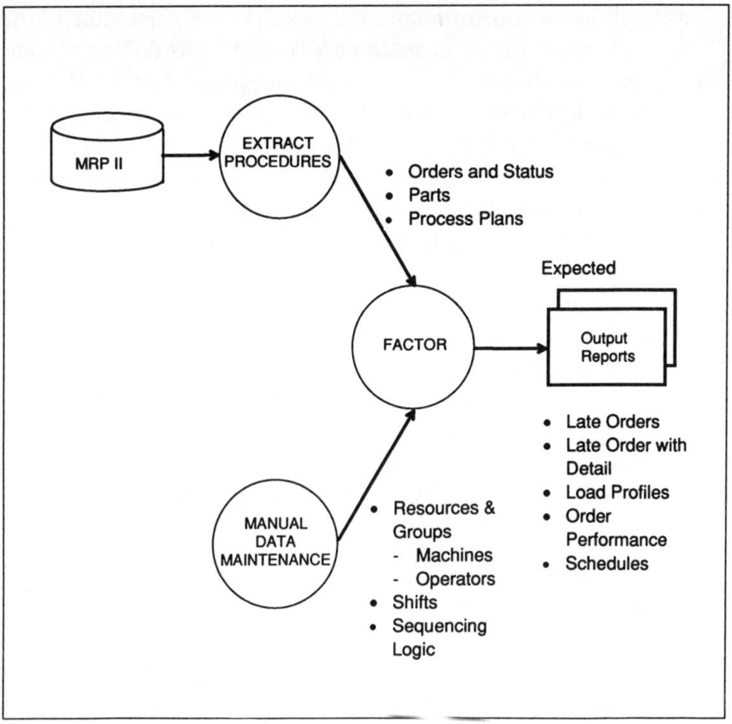

Figure 11.2 ACME's FACTOR Capacity Management System.

These FACTOR displays and corresponding reports provide ACME's production planners and manufacturing management with a high performance forecast into future manufacturing activities. If there are any problems indicated on the forecast, ACME can then deal proactively with any order completion dates and capacity problems by taking appropriate actions such as closely managing key orders and resources, approving selected overtime, or rerunning FACTOR with different order priorities or a change in scheduling philosophy. In this way, ACME can identify a potential problem and resolve the problem before it becomes a reality. FACTOR produces production schedules that are used on a daily basis by shop floor supervisors to control work assignments. These schedules guide production towards the performance predicted by the FACTOR-produced plan.

Usage procedures and output reports

The FACTOR system is used by production planning on a periodic basis (daily in this application) to support the planning and scheduling process as well as to communicate the results of a particular plan. The planner at any time has access to information stored in a relational database to answer questions and make decisions. FACTOR produces the latest status information for released orders that is available. The planning horizon for a run is set to approximately four weeks which, for ACME, is the longest production lead time of an order. In this way, FACTOR predicts completion dates for all orders.

To access the desired information, the scheduler interacts with a user interface menu which has seven main options (see Fig. 11.3). The scheduler can select an option to review expected results, try what-if scenarios, review resource load profiles, identify bottlenecks, and distribute schedules on lines. Shop floor users can log into the system to review work center schedules, preview planned completion dates of orders, and obtain simulated values of operational efficiencies. The reports answer key questions facing management. The five reports that were considered most important on this application are described below:

- **Late Order Report**: identifies work orders that are projected to be completed after their due date and indicates the number of days by which the order would be late.
- **Late Order Detail Report**: lists the production routing sequence of each late order and indicates the waiting time expected to accrue at each operation.
- **Resource Load Profiles Report**: identifies the expected load on each resource and operator unit for each daily period. A variation of this report, the Bottleneck Exception Report also identifies if and when a resource or operator unit becomes a bottleneck in future planning periods.
- **Order Summary Report** (Fig. 11.4): lists all orders and indicates the expected completion date for each order and indicates how many days early or late the order is expected to be.
- **Resource Schedule Report**: lists the orders to be worked by each workcenter and each operator unit and the day in which it is to be worked.

As a result of the analysis, the best alternative (scenario) is chosen and the schedule is put on-line for viewing. The schedule can be generated either in a textual or graphical format. A Gantt chart is used to display the operational status of jobs by changing the color or position of bars representing jobsteps over time. This type of display allows the user to modify the schedule graphically through point and click operations. The production planner reviews the simulated performance information and performs an analysis of the production plan.

An interactive query facility is available to the production planner as a tool to answer specific questions and to 'drill down' to get more

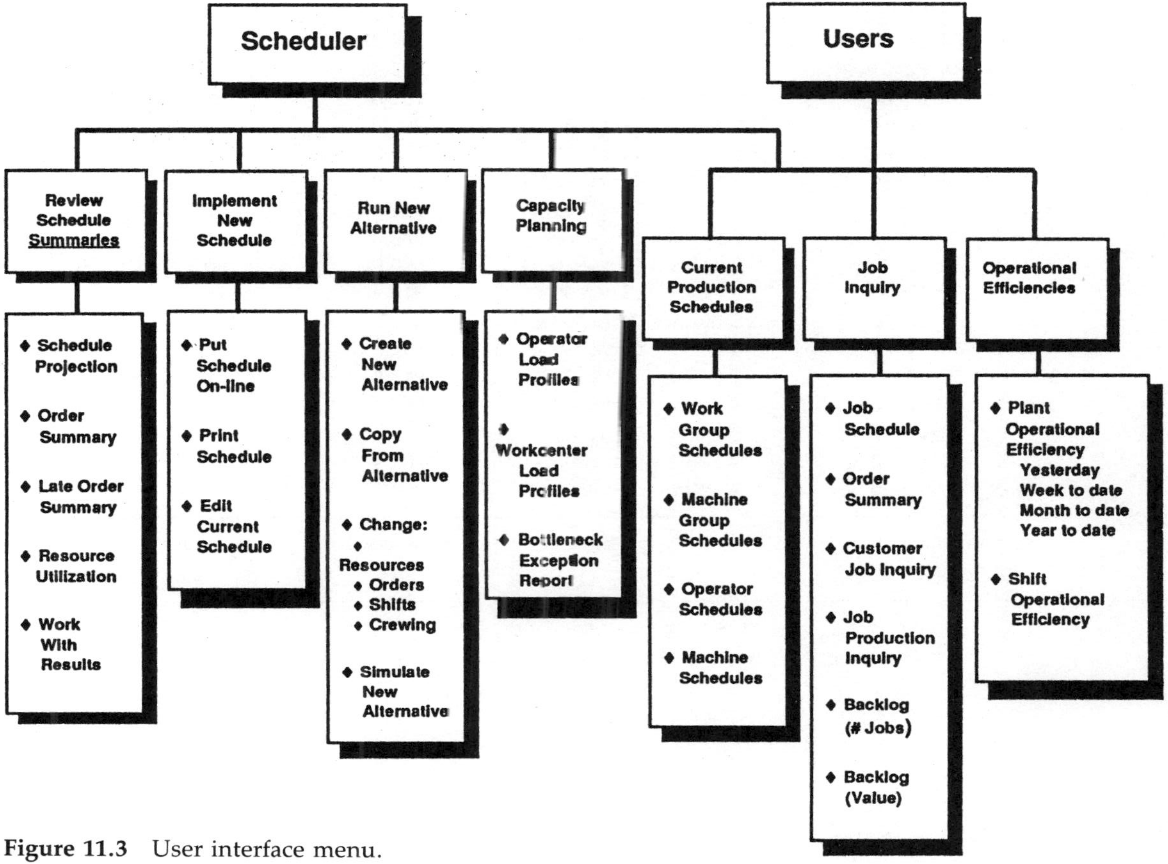

Figure 11.3 User interface menu.

FACTOR Order Summary
Date 04–17–05 10:02:40 Application BASE CASE
Alternative: 000 – BASE CASE

Order ID	Wait Time	Processing Time	Start Date & Code	Completion Date & Code	Hours Late
M210000	11.404	.694	09–14–92 11:24 S	09–14–92 12:06 C	–27.903
M210001	71.337	16.292	09–14–92 16:41 S	09–17–92 15:38 C	–264.371
M210002	170.193	9.574	09–15–92 15:16 S	09–21–92 11:46 C	–28.233
M210003	12.030	.613	09–14–92 12:02 S	09–14–92 12:39 C	–27.357
M210004	173.124	9.748	09–14–92 18:06 S	09–21–92 14:52 C	–25.128
M210005	11.329	.701	09–14–92 11:20 S	09–14–92 12:02 C	–27.970
M210006	103.924	8.128	09–14–92 18:06 S	09–18–92 16:03 C	–71.949
M210007	169.383	35.616	09–14–92 13:24 S	09–22–92 13:00 C	–531.001
M210008	235.051	13.007	09–14–92 18:17 S	09–24–92 08:03 C	–31.942
M210009	11.716	10.645	09–14–92 08:35 S	09–14–92 22:22 C	–209.639
M210010	172.941	16.879	09–15–92 16:46 S	09–21–92 21:49 C	–162.180
M210011	49.524	30.479	09–14–92 15:48 S	09–17–92 08:00 C	–607.997
M210012	905.590	53.410	09–14–92 07:00 S	10–23–92 23:00 C	–257.000
M210013	328.505	16.295	09–14–92 07:48 S	09–28–92 08:48 C	–7.201
M210014	108.895	8.554	09–14–92 14:32 S	09–18–92 21:27 C	–66.551
M210015	913.727	45.273	09–14–92 07:00 S	10–23–92 23:00 C	–65.000
M210016	239.020	36.155	09–16–92 08.11 S	09–25–92 11:11 C	–460.825
M210017	13.463	3.690	09–14–92 13:15 S	09–14–92 17:09 C	–70.847

M210018	402.612	21.147	09–15–92	22:05 S	10–01–92	15:45 C	–24.242
M210019	108.430	8.119	09–14–92	10:02 S	09–18–92	20:33 C	–67.450
M210020	190.429	10.116	09–21–92	14:26 S	09–22–92	03:33 C	–31.455
M210021	346.240	24.594	09–15–92	14:50 S	09–29–92	10:50 C	–149.165
M210022	46.590	17.141	09–14–92	07:00 S	09–16–92	15:44 C	–312.270
M210023	264.086	17.418	09–24–92	16:05 S	09–25–92	17:30 C	–94.496
M210024	15.586	4.835	09–14–92	07:00 S	09–14–92	20:25 C	–91.579
M210025	919.283	39.717	09–14–92	09:15 S	10–23–92	23:00 C	–113.000
M210026	53.617	6.170	09–15–92	21:37 S	09–16–92	11:47 C	–124.213
M210027	199.344	11.015	09–22–92	07:21 S	09–22–92	18:22 C	–45.641
M210028	536.635	28.688	09–14–92	17:33 S	10–07–92	13:19 C	–26.677
M210029	9.948	.754	09–14–92	10:42 S	09–14–92	09:57 C	–29.298
M210030	8.124	.920	09–14–92	08:07 S	09–14–92	09:03 C	–30.956
M210031	171.759	5.182	09–18–92	19:46 S	09–21–92	08:56 C	–7.060
M210032	29.138	2.842	09–14–92	21:08 S	09–15–92	07:59 C	–56.021
M210033	12.295	3.503	09–14–92	07:00 S	09–14–92	15:48 C	–72.201
M210034	216.788	11.024	09–14–92	07:00 S	09–23–92	11:49 C	–28.187
M210035	220.164	11.142	09–16–92	22:35 S	09–23–92	15:18 C	–24.694
M210036	99.637	14.532	09–14–92	14:08 S	09–18–92	18:10 C	–237.831
M210037	40,745	4.308	09–14–92	17:09 S	09–15–92	21:03 C	–66.946
M210038	7.00	1.124	09–14–92	07:00 S	09–14–92	08:07 C	–31.876
M210039	56.041	8.739	09–15–92	20:46 S	09–16–92	16:47 C	–143.220
M210040	12.0976	.530	09–14–92	12:06 S	09–14–92	12:38 C	–27.372

Figure 11.4 FACTOR order summary.

information. For example, when analyzing a plan, the ACME planners use the interactive query facility to investigate what orders are projected to be late, and to gain an understanding as to why they are scheduled to be completed beyond their due date. Fig. 11.5 is a result of a query showing all orders that are late. The query shows that there were six such orders. Of the six late orders the query shows that order M210008 was projected to complete on 09–27–95 13:56, which is 1.9 days after its due date. By interrogating the plan with respect to that order, a list of the projected waiting times for each of the job steps of the process plan for that order is generated as shown in Fig. 11.6.

From Fig. 11.6, it is seen that the process plan for this order consists of seven steps and the projected waiting time spent in each step is shown. For example, the order is expected to wait 7.7 days in the queue for the Grind workcenter. The Grind workcenter is made up of five individual grinding machines. By selecting that process step the user can choose to view a resource schedule which includes those orders that are planned to be processed from the time order M210008 arrived to the Grind workcenter until it is scheduled to be processed. Fig. 11.7 shows only the last seven orders of the 25 orders processed before M210008 waited at the Grind workcenter.

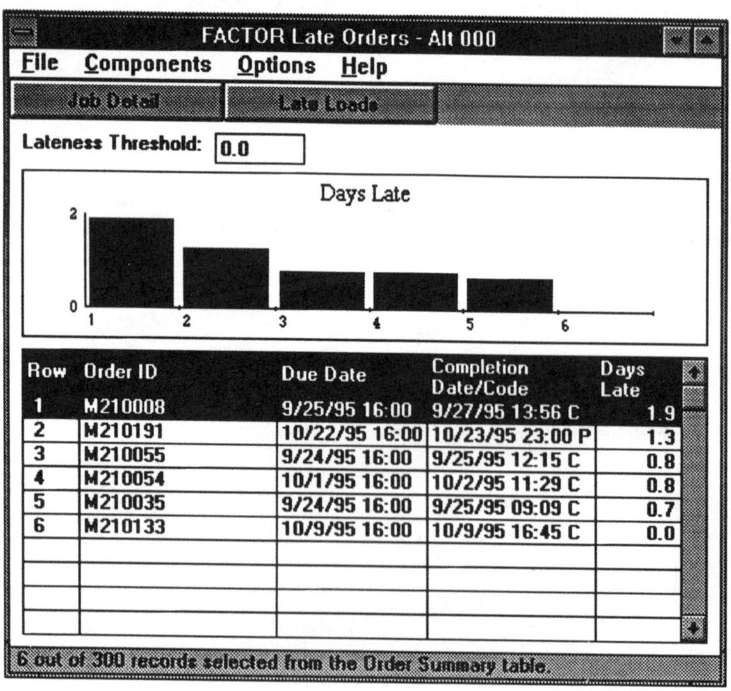

Figure 11.5 List of orders late by more than one day.

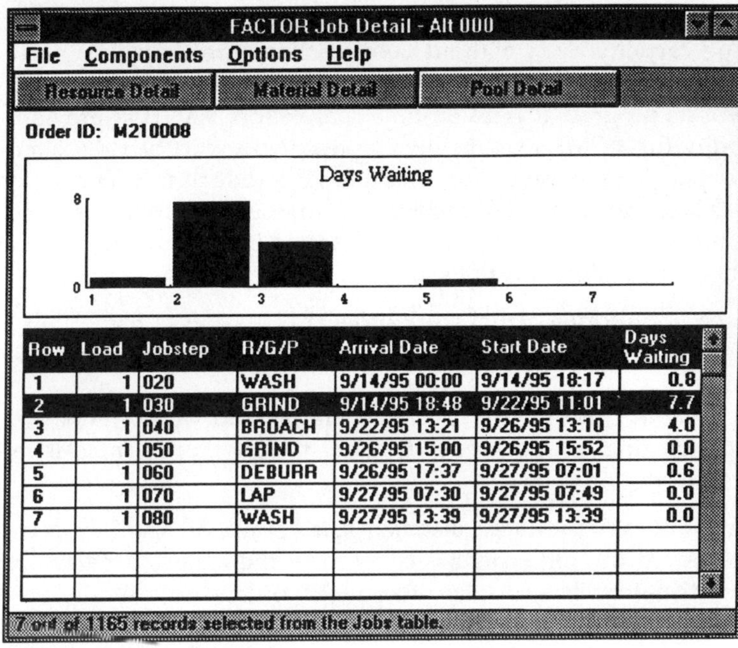

Row	Load	Jobstep	R/G/P	Arrival Date	Start Date	Days Waiting
1	1	020	WASH	9/14/95 00:00	9/14/95 18:17	0.8
2	1	030	GRIND	9/14/95 18:48	9/22/95 11:01	7.7
3	1	040	BROACH	9/22/95 13:21	9/26/95 13:10	4.0
4	1	050	GRIND	9/26/95 15:00	9/26/95 15:52	0.0
5	1	060	DEBURR	9/26/95 17:37	9/27/95 07:01	0.6
6	1	070	LAP	9/27/95 07:30	9/27/95 07:49	0.0
7	1	080	WASH	9/27/95 13:39	9/27/95 13:39	0.0

Figure 11.6 Projected waiting at each jobstep in the process plan for a late order.

Order ID: M210008 Arrival Date: 9–14–95 18:48
Workcenter ID: Grind End Date: 9–22–95 13:21

Order	State Date	End Date	Due Date	Dayc Late
M210022	9–19–95 17:26	9–20–95 07:07	9–23–95	–1.1
M210107	9–20–95 07:07	9–20–95 10:51	9–23–95	–0.4
M210050	9–20–95 10:51	9–20–95 16:35	9–23–95	–0.9
M210006	9–20–95 16:35	9–21–95 08:24	9–23–95	–1.4
M210064	9–21–95 08:24	9–21–95 13:08	9–23–95	–1.2
M210101	9–21–95 13:08	9–21–95 21:17	9–25–95	–0.9
M210072	9–21–95 21:17	9–21–95 11:01	9–25–95	–1.0
M210008	9–22–95 11:06	9–22–95 13:21	9–25–95	1.9

Figure 11.7 List of orders processed at the grind workcenter after M210008 arrival.

Upon analyzing Fig. 11.7, the cause of the delay is easily identified. Knowing that all jobs at the Grind workcenter are scheduled based on the earliest due date of the order, this report shows part of the queue for the Grind workcenter that have an earlier due date than M210008 at the time that it was scheduled to arrive.

The information available through FACTOR helps ACME to better manage capacity even without taking action directly in FACTOR. By knowing an order's expected status manufacturing management can monitor its progress toward on-time completion. FACTOR provides the capability for ACME management to test the impact of a decision in a future planning period. For example, if product mix fluctuation is expected to cause a high backlog and missed due dates in week 3, overtime can be added and the impact tested by resimulating. The most common management actions are:

- closely monitoring actual work order progression to plan,
- closely monitoring actual asset and operator group performance to plan,
- approval of overtime for specific operators and workcenters,
- re-prioritizing a work order to give it higher priority as it moves through the process.

A FACTOR run produces dispatch lists to answer the key question: What should the job schedule be to meet the expected performance? These dispatch lists display the work orders to be run at each workcenter and by each operator group for the next few days. The supervisors use the dispatch list to make final decisions about individual job assignments, material requirements, and timing within the daily period.

The implementation of the FACTOR finite capacity scheduler has allowed ACME to:

- improve customer services;
- reduce costs (overtime and setup);
- reduce manpower to reduce schedules;
- improve communication between departments;
- reduce risk by reducing the dependence on the current scheduler to produce a good schedule;
- anticipate problems in the future and provide sufficient time to react.

Today the finite capacity scheduling system in place at ACME gives the company an organized view of its operations. The system has automated a difficult, manual task of generating a production schedule and outputting it to the shop floor. The system eliminates tedious and unnecessary calculations of planners and production controllers and allows them to spend more time on continuous improvement activities related to customer order changes, inventory control and establishing customer delivery dates.

11.7.2 Application 2: Production control improvements

In an application of FACTOR at General Electric Aircraft Engine, FACTOR provided support to production control personnel to create a shop schedule. Each production control specialist supported production areas dedicated to part families with an average of 20 workcenters in each area and approximately 250 part numbers having an annual production volume ranging from 20 to 400 for each part. The system provided on-line access to status data, a daily schedule from FACTOR and performance feedback reporting. Fig. 11.8 portrays the enterprise level database of the company, the integration of FACTOR with these databases, and the general flow of shop floor data requirements and status.

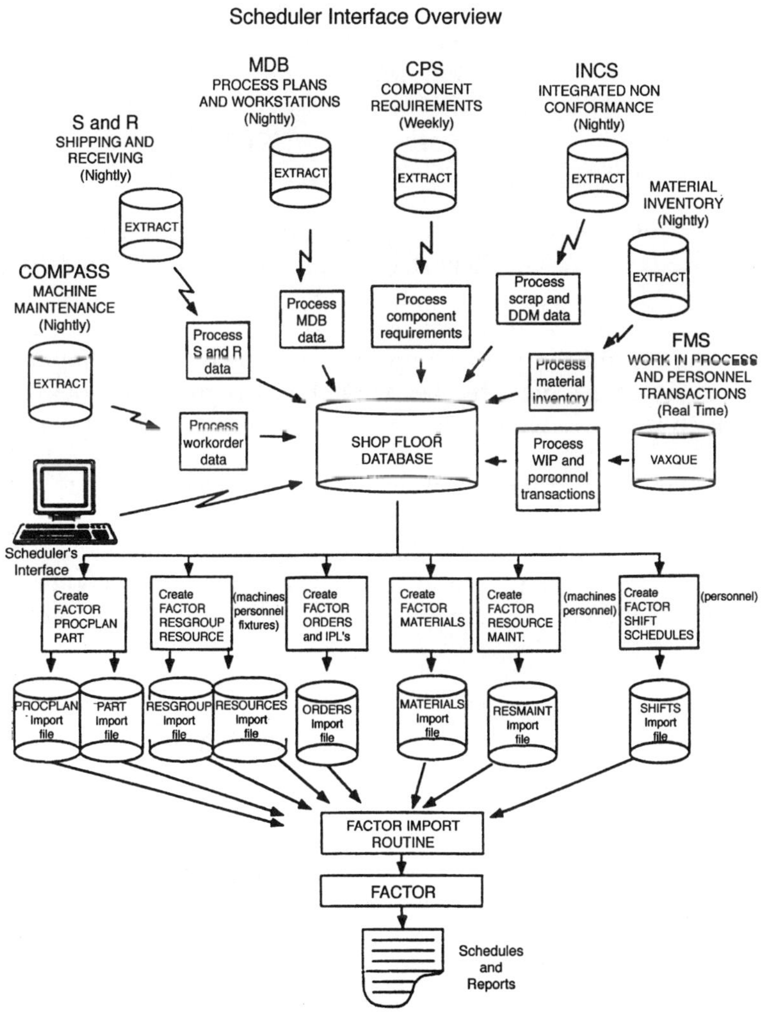

Figure 11.8 Data flows in FACTOR application at General Electric Aircraft Engine.

It has been reported (by GE Aircraft Engine at the Pritsker User Conference) that an overall improvement in production control productivity of approximately 33% was realized through head count reduction and redistribution of time to other value added tasks. A detail review of this improvement has been made (Schmahl and Amand, 1994). In this review, first line supervisors (the individuals most sensitive to changes in shop operations) were surveyed. Of the supervisors, 86% agreed with the statement 'the level of production control support from my unit matches well the needs of my unit' even though there was a significant reduction in the number of production control specialists after FACTOR was implemented. The remaining 14% of the first line supervisors neither agreed nor disagreed with the statement. In addition, the majority of the supervisors agreed that the quality of the schedules received from production control personnel had improved over the previous year and that production control had better data to support their scheduling decisions.

The following concluding remarks were made as part of the evaluation:

1. Productivity improvements in production control activity of approximately 26% could be tied directly back to system implementation.
2. The shift from a data collection/expediting function to a true scheduling and continuous improvement role is just starting to take effect. A positive trend has been noted relative to on-time delivery of parts to assembly with a first year reduction in delinquent deliveries of 14%.
3. A 37% increase in inventory turns has been realized.

These improvements were obtained at a time when shop supervisors indicated an enhanced perception of the value of the production control function and the number of production control specialists was reduced.

11.8 TRENDS IN PLANNING AND SCHEDULING

The trends in planning and scheduling are toward greater coordination across manufacturing operations; that is, scheduling areas and plants, and the integration of planning and scheduling systems as part of manufacturing execution systems. These trends will be supported by improved data collection and shop floor control systems, advanced relational database management systems, and improved software programs for developing graphical interfaces.

Through the use of detailed representations (models) of the manufacturing system as part of the planning and scheduling process, the barriers between engineering design, planning, scheduling and execution can be decreased significantly. Through the use of simulation and graphical tools for displaying the current and projected status of the manufacturing system, communication can be enhanced among the diverse groups performing these functions. Such communication inherently reduces the barriers that have long existed. By using manufacturing terminology to describe the computer representation of manufacturing, a common modeling language will result. This language can facilitate the use of data collected during operations in developing new planning and scheduling procedures.

Another trend will be that planning and scheduling can support the sales process by providing realistic shipment dates for sales representatives to promise to customers. This can be accomplished while the sales representative is on the phone talking to the customer. Planning and scheduling systems can and will support the purchasing of material by scheduling order release dates and optimizing the use of material with respect to the availability of resources while meeting promised ship dates. The operation of the production system can and will be coordinated with downstream logistics requirements to produce combined manufacturing and distribution cost savings. Lastly, the data from shop floor collection systems and from planning and scheduling systems will be organized to support the decisions of managers and executives. When this happens, a breakthrough will occur in classifying and categorizing the problems of managers and executives. This is the fundamental requirement to achieve continuous and significant improvement in the processes of planning and scheduling.

REFERENCES

Askin, R.G. and Standridge, C.R. (1993) *Modeling and Analysis of Manufacturing Systems*, Wiley, New York.

Baker, K.R. (1974) *Introduction to Sequencing and Scheduling*, Wiley, New York.

Baudin, M. (1993) *Manufacturing Systems Analysis with Application to Production Scheduling*, Yourdon Press, Englewood Cliffs, New Jersey.

Brill, S. (1993) Finite scheduling helps IBM satisfy its panel product. *APICS: The Performance Advantage*, October, 66–9.

Casella, B. and Barnes, J.W. (1992) Finite capacity scheduling is also a forecasting tool for BethForge. *APICS: The Performance Advantage*, January, 21–3.

Credle, R. (1993) Finite scheduling helps Eaton cut line setups and increase line throughput. *APICS: The Performance Advantage*, January, 30–2.

IDS Prof. Scheer CAM (1994) Intelligent Leitstand FI-2, Saarbrucken, Germany.

Imai, M. (1986) *Kaizen*, Random House, New York.

Hayes, R.H., Wheelwright, S.C. and Clark, K.B. (1988) *Dynamic Manufacturing*, The Free Press, New York.

Largent, T. (1992) Nucor steel runs lean on finite scheduling. *APICS: The Performance Advantage*, June, 28–30.

Pritsker Corporation (1992) *FACTOR Documentation Set*, Indianapolis, Indiana.

Pritsker, A.A.B. (1992) Manufacturing capacity management through modeling and simulation, in *Manufacturing Systems: Foundations for World-Class Practice* (eds J.A. Heim, and W.D. Compton) National Academy Press, pp. 204–14.

Scheer, A.W. (1994) *Business Process Engineering: Reference Models for Industrial Enterprises*, 2nd edn, Springer-Verlag.

Schmahl, K.E. and Amand, S. (1994) Impact of a decision support and finite scheduling system on a large machine shop. *Production and Inventory Management Journal*, 54–8.

Weliever, A. (1992) Going to the Gemba: in *Foundations for World-Class Practice* (eds J.A. Heim and W.D. Compton) National Academy Press, pp. 204–14.

Author index

Subject index